Y0-ASL-571

Flexible armoured revetments incorporating geotextiles

Flexible armoured revetments incorporating geotextiles

Proceedings of the international conference organized by The Institution of Civil Engineers and held in London on 29–30 March 1984

Thomas Telford Ltd, London

Organizing committee: I. M. Walker (Chairman), Dr M. Baldwin, A. J. M. Harrison, P. R. Rankilor and J. T. Williams

British Library Cataloguing in Publication Data

Flexible armoured revetments incorporating
geotextiles/proceedings of the international conference
organized by the Institution of Civil Engineers and held in London
on 29–30 March 1984.
1. Embankments — Design and construction
2. Shore protection
I. Institution of Civil Engineers
627'.42 TA760

ISBN 0-7277-0226-2

Published for the Institution of Civil Engineers by Thomas Telford Ltd, 26–34 Old Street, PO Box 101, London EC1P 1JH

First published 1985

Printed in Great Britain by Billing and Sons Ltd, Worcester.

Contents

1 Flexible revetments—theory and practice

C. T. BROWN, BSc(Eng), MICE, MIEAust, Tillotson Brown and Partners, and Seabee Developments, Australia

SYNOPSIS. Flexible revetments are defined as revetments that maintain an intimate contact with the underlying soil during any gradual settlement, and protect the slope from realignment by wave and current action. A brief summary is made of various failure modes before a comparison is made of the comtemporary theories of hydrodynamic failure due to waves and currents. Developments of the theory as regards armour geometry and porosity are made and some recent experimental results presented, which are particularly germane to the effect of currents on revetments. The practical applications of this are briefly touched upon.

INTRODUCTION

1. Since 1976, the Author has undertaken investigations into the behaviour of rip-rap, gabion and Seabee armoured revetments, mainly under orthogonal wave attack. Contemporaneously a review and reworking of the momentum theory was undertaken by considering the effect of a rotating jet of fluid instead of the usual fixed direction. The envelopes of the theoretical stability curves were found to bear strong similarities to the empirical relationships found by experiment for both uplift and sliding failure modes.

2. Further consideration of the similarities between initiation of erosion under currents demonstrates the strong kinship between the Shields parameter and the various stability numbers and coefficients to be found in coastal engineering.

3. This suggests that perhaps the onset of erosion in streams depends upon the generation of sufficiently large rolling eddies at the movable boundary.

4. In considering flexible revetments, the author considers this term to include all revetment systems whose elements can maintain intimate contact with settling under-

layers without rendering the revetment unstable or allowing erosion to occur.

5. Flexible revetments include rip-rap, dry laid blocks as well as gabions and tied block systems and asphaltic concrete layers. Impervious concrete slab systems do not fall within this category.

6. Failure Modes. Typically, revetments may fail by one or more of the following:

i. Vandalism, theft and faulty construction.
ii. Abrasion, weathering and chemical decomposition.
iii. Environmental hazard.
iv. Structural failure.
v. Scour at the edges or toe.
vi. Understreaming and loss of underlayer material.
vii. Extraction or uplift of the armour layer by currents or waves.
viii. Sliding of the revetment face.
ix. Slope failure.

7. Failure mode i. is hard to prevent and revetment systems should ideally be resistant to partial vandalism yet clearly demonstrate that such has occurred, so that the need for repairs is obvious. Faulty construction is this category.

8. Failure mode ii. is a result of inappropriate materials and/or expectations, and its avoidance requires the exercise of sound engineering judgement based on experience and experiment. In Australia we have found that resistance to salt crystal growth, thermal shock and transit of boats and trailers are the main agents of weathering and abrasion.

9. A revetment may also fail by being an environmental hazard, by harbouring noxious vermin or by being deceptively safe. Poorly finished gabions in a back beach revetment may be present a face of rusting wire-ends which can cause harmful cuts. Rip-rap and other large voided revetments may harbour rats. Other systems may become very slippery and dangerous to walk on.

10. Structural failure of individual elements is most frequently found where other criteria, particularly hydraulic performance, have been carried too far and is most usually associated with slender non-redundant elements. The ideal element is one that can fail structurally and either still function in the revetment or disappear completely, without damaging the adjacent elements or rendering them unstable.

11. Many revetments fail due to scour at the edges or at the toe, and detail design of these parts is of the utmost importance. The ability to accommodate peripheral scour is one of the chief advantages of tensile flexible revetments such as gabions. However, this tensile capacity needs to be an ultimate capacity, as otherwise the revetment may span over sublayer scour holes without much sign of distress, when early indication would allow an early remedy.

12. The direct result of loss of contact between the revetment system and its underlayers is understreaming which allows and causes the regrading of the underlayer material to a profile other than that designed. This will usually result in progressive readjustment of the slope and possibly a serious slope failure.

13. Similar results can occur with loss of underlayer material due to incorrect design or construction.

14. Although the modern use of filter fabrics has overcome the problems associated with multilayer gravel filters, other problems associated with the fabrics can ocur, namely unseen damage in construction; inadequate lapping; deterioration due to U.V. light; abrasion by sand; and fatigue due to working by wave action when installed too close to the surface of the revetment. Figure 1 shows a revetment under wave attack. The figure shows the differential pumping that can occur at the face of the filter cloth at the phreatic level.

15. Apart from the structural strength of the unit, the last two failure cases are the only two cases usually analysed mathematically, although all the derivations to date have required empirical calibration of the coefficients in the resulting equations.

THEORY

16. The strong similarities between the derivations for current and wave erosion (Shields[1], Irribarren[2] and Hudson[3]) are due to the common basic forces, the disturbing force being derived from the momentum of the water flow via the drag and lift forces, whilst the restoring forces are essentially those due to the weight of an armour element, although some have included the effect of inter-unit forces such as fricton, tension and shear.

17. The Irribarren and Hudson derivations for the effect of waves start with the water velocity in the breaking wave, transforming this to an equivalent wave height, the simpler and more usual design parameter. It should be remembered that it is this simplification that causes the effects of wave period to be lost.

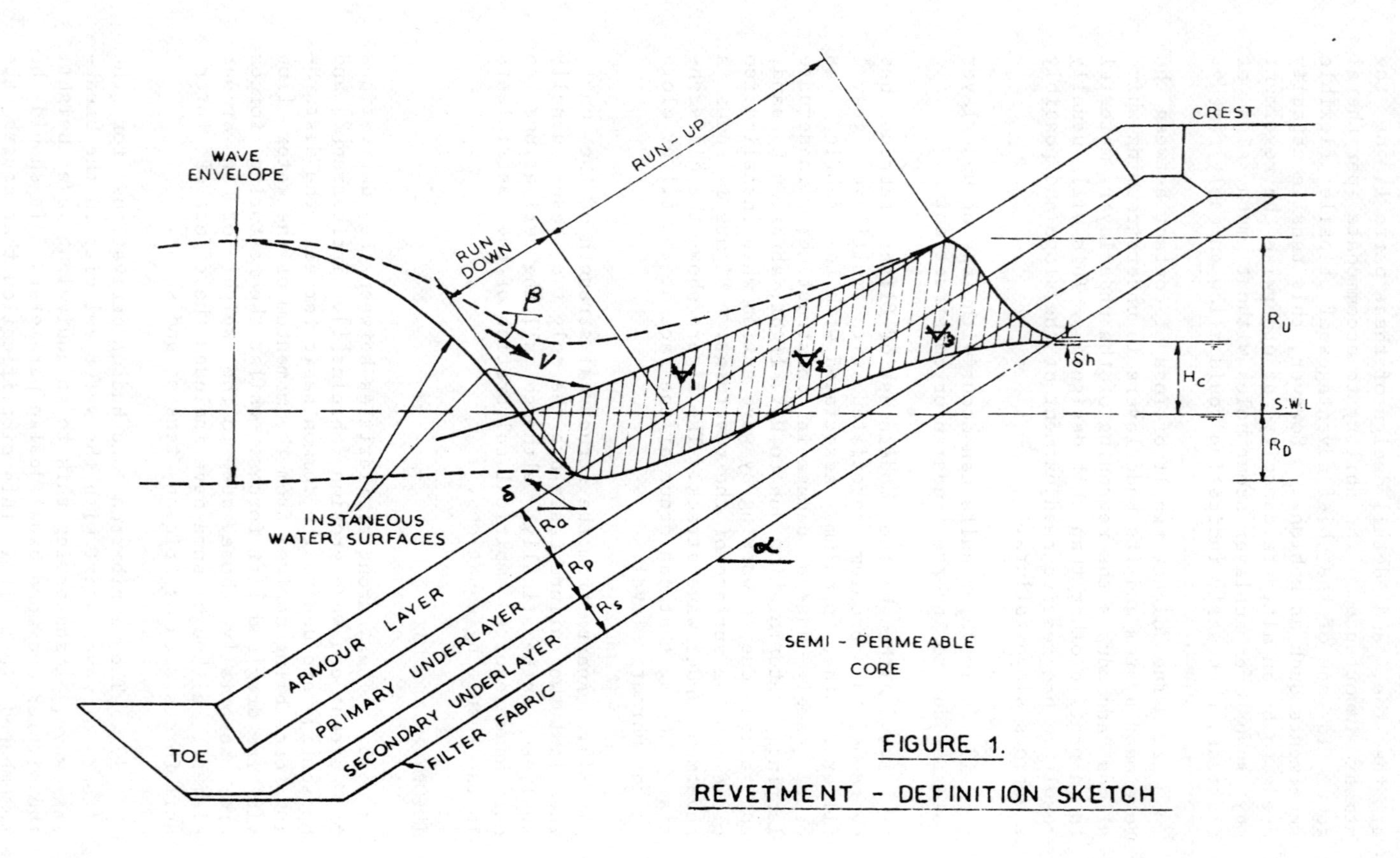

FIGURE 1.
REVETMENT - DEFINITION SKETCH

18. The relationships compare as follows:

Shields: $F_s = U_*^2/(S_r-1)gD = \tau_o/\gamma(S_r-1)D$

Irribarren: $W = K\gamma H^3/(S_r-1)^3(\mu\cos\alpha - \sin\alpha)^3 = C_v.\gamma D^3$

Hudson: $W = \gamma_r H^3/K_D(S_r-1)^3 \cot\alpha = C_v.\gamma D^3$

19. These may be reduced to a common form relating typical armour dimension with incident velocity. In both the Irribarren and Hudson cases, we take the reverse step from wave height to water velocity.

Shields: $D = U_*^2/F_s.(S_r-1)g$

Irribarren: $D = H/(C_v/K)^{1/3}(S_r-1)(\mu\cos\alpha - \sin\alpha)$

$= V^2/C_I.(S_r-1).g.(\mu\cos\alpha - \sin\alpha)$

Hudson: $D = H/(C_v/K_D)^{1/3}.(S_r-1)\cot\alpha^{1/3}$

$= V^2/C_H(S_r-1).g.(\cot\alpha^{1/3})$

All equations are of the form:

$$\frac{V^2}{gD} = C.(S_r-1).(\cot\alpha^{1/3})$$

20. In 1978,79 Brown (4,5,6) reworked the momentum flux derivation as a vector problem, allowing the jet to rotate on an element of a revetment. The basic forces are the disturbing force:

$$\vec{F}_D = p\vec{A}\vec{V}$$

and the restoring force:

$$\vec{F}_R = \vec{W} + \vec{S} + \vec{T}$$

21. The directionality of these forces was specifically considered and then resolved for two cases - uplift movement and sliding movement - to give the following results:

Uplift: $R(1-\rho) > \frac{V^2}{2g}\cos^2(\delta-\beta)/C(S_r-1)\sin\delta$

Sliding: $R(1-\rho) > \dfrac{\frac{V^2}{2g}\{K_1 \sin 2(\alpha+\beta)+K_2\sin^2(\alpha+\beta)-[\frac{S_n-T}{g\rho_w A}]\}}{(S_r-1)(\mu\cos\alpha \pm \sin\alpha)}$

By suitable assumptions, these derivations may be converted to either the Hudson or the Irribarren equation.

negative = downwash
positive = upwash

22. Ignoring the shear and tensile forces in the revetment, the envolopes of these theoretical equations are in close agreement with the empirical forms derived from laboratory tests.

Uplift:

$$R(1-p) > \frac{V^2}{2g} / C_{BU}(S_r-1)\text{Cot}^{1/3} \alpha$$

Downsliding:

$$R(1-p) > \frac{V^2}{2g} / C_{BS}(S_r-1)\text{Cot}\, \alpha$$

23. It can be seen that these formulae have isolated the plan shape of a revetment element from the stability equations. This enables suitable revetment elements to be designed for production and placement economy without affecting revetment stability. They have also introduced the porosity of the revetment as an independent variable.

24. The values of these coefficients have been determined by experiment for the cases of gabions and Seabees exposed to wave jets.

TABLE 1
STABILITY COEFFICIENTS FOR BLANKET REVETMENTS

For $H=V^2/2g$	Gabions	Seabees
Uplift	4	5 - 6.5
Sliding	7	Uplift dominates

25. The membrane and shear forces have been ignored on the basis that the area of concern is larger than an individual element and that the forces do not come into play in preventing instability of the revetment layers. However, they are mobilised during the failure process and may serve to limit the amount or control the rate of deformation of the revetment.

TABLE 2

TESTER		HANSEN & KEATS				PAGE	
GRADING		Fine Sand	Graded Sand	1mm Sand	2mm Sand	Poorly Graded	Well Graded
PARAMETER	d_{15}	0.24	0.29	0.59	1.22	0.28	0.26
	d_{30}	0.32	0.47	0.71	1.55	0.30	0.51
	d_{85}	0.41	1.55	0.98	1.82	0.41	1.10
$\frac{L}{D}$	Stage	A	B	C	D	E	F
0	1	<.21	<.22	<.27	<.37	<.21	
0.8	1	.35	.45	.51	.56	.32	.44
	2	.49	.55	.67	.75	.49	.52
	3	.60	.70	.82	1.19	.61	.67
1.2	1	.48	.63	.66	.81	.44	
	2	.56	.71	.82	1.09	.55	
	3	.68	.80	.95	1.40	.66	
1.6	1	.63	.78	.83	1.06	.62	.78
	2	.74	.88	1.01	1.34	.75	.88
	3	.80	.97	1.20	>1.4	.85	.90
2.0	1	.73	.89	.95	1.33	.67	
	2	.93	1.07	1.13	>1.4	.92	
	3	1.03	1.17	1.32		.99	
2.4	1					.74	.98
	2						
	3						
4.0	1	1.36	1.23	1.27		>1.0	
	2	>1.40					
	3						

RECENT WORK

26. Recent laboratory work has investigated the behaviour of underlayers under current action. Two series of flume tests by Page and Hansen & Keats (7, 8) have been undertaken to examine the effect upon the entrainment of a natural sand bed under a porous revetment without an intermediate filter layer.

27. For the most part, the porosity was normal to the bed but in one experiment a zig-zag horizontal porosity was provided by using two staggered layers of armour.

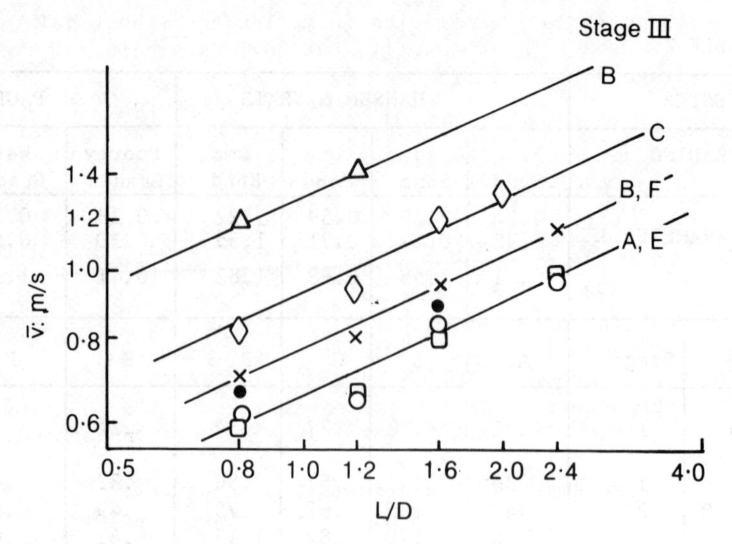

Fig. 2 Mean Velocity vs Cell Aspect Ratio

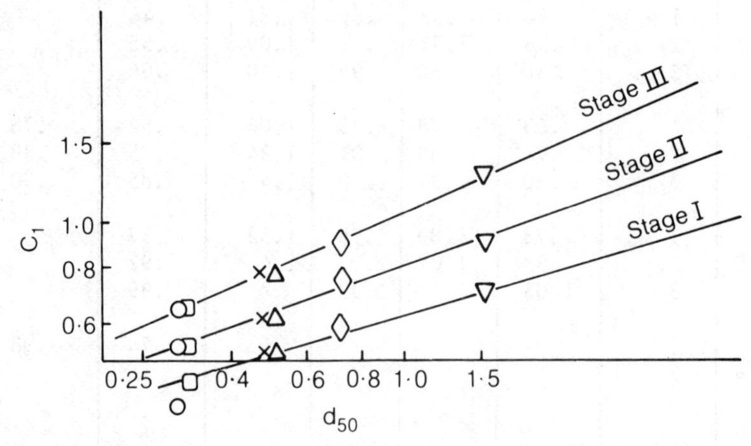

Fig. 3 Relationship between C_1 and Median Grain Size

28. The apparatus consisted of a tilting glass flume with a raised false floor containing the sediment bed protected by the armour layer and a sediment trap.

29. Three distinct phases of motion of the sediment were observed before significant entrainment of sediment occurred. These were:

Stage 0 - No movement.
Stage 1 - Hemispherical depression ocurs under void cell, but no sediment entrained.
Stage 2a - Lighter particles suspended in lowest vortex in void cell.

Stage 2b - Particles in motion throughout height of void cell, but no loss occurs.

Stage 3 - Vortices in cell rise above top surface of armour layer and entrained particles are lost from individual cells. Rate of loss increases with increased flow.

Stage 4 - Unidirectional flow occurs in void cells, uplifting sediment from beneath armour layer and removing. (Understreaming and/or rapid settlement occur).

30. The onset of each stage was found to be related to the non-dimensional aspect ratio of the void cell and the gradng of the sand. The actual height of the armour layer does not appear to affect the stability of the unfiltered bed material. The results obtained are shown in Table 2.

31. An empirical relationship of the form:

$$V = C_1 (L/D)^a$$

was found best to describe the relationship. C_1 was found to depend on the grading curve of the bed material with a relationship:

$$C_1 = C_2 \, d_{50}^{b}$$

Typical graphs are shown in Figures 2 & 3.

32. The final equation for the stages of motion were found to be:

Stage I: $V = 0.63 \, d_{50}^{0.30} \, (\frac{L}{D})^{0.8}$

Stage II: $V = 0.82 \, d_{50}^{0.35} \, (\frac{L}{D})^{0.75}$

Stage III: $V = 1.08 \, d_{50}^{0.43} \, (\frac{L}{D})^{0.45}$

At this time the second dimensionless parameter containing d has not been properly identified and the numerical coefficient has the dimensions:

$$L^{1/n} \, T^{-1}$$

33. The occurrence of macro turbulence or obstructions at the bed increases the local velocities so decreasing the

scour protection provided. The horizontal porosity also decreases the bed stability. Although velocities approaching 1.5 m/sec were achieved, 5 gm and 10 gm armour elements were not entrained in the flow, even adjacent to collapsing scour holes due to bed obstructions.

34. For practical purposes, the onset of Stage III is the important case and an adequate factor of safety must be provided against it. Stage II may be important when considering abrasion of an intermediate filter cloth.

35. It is hoped that work may proceed to consider cases where the armour layer can be rendered unstable before the underlayers, (i.e. by using very coarse material in the underlayer) but this will require substantially improved water flows compared to those presently available.

PRACTICAL APPLICATIONS

36. In Australia, approximately 1.3 km of revetments using gabions have been designed in accordance with the blanket theories outlined here and described in detail in ref. 5.

37. The Seabee armour unit, the progenitor of all this thought, has been taken through successive stages of development form model 10, 28 and 85 gram units to prototype ceramic units of 10 to 20 kg, and prototype concrete units for 0.5 to 4.0 tonnes mass. This armour unit allows the use of variable porosity both normal and parallel to the plane of the revetment as well as mass range in the order of 40:1 for any particular installation. It can also survive extremely poor quality control, providing a 'tough' design is utilised. The practical applications of this system are described in ref. 8 and some examples shown in the plates.

CONCLUSION

38. The great similarity between the elements of theories of incipient motion due to current and wave action are seen to derive from the common description of the active forces. What differences there are in nature are still locked up in our coefficients. Nevertheless, by exercising a degree of objectivity it is possible to gain more control of the multitude of variables involved in the design of coastal works.

39. It is hoped that, whereas it was usual to determine the size of rocks to be obtained (if possible) from the quarry, it is now quite feasible to manufacture revetment protection in any suitable size according to production and construction criteria, without sacrificing material economy. It is also possible to design various service criteria at

the same time, but I would recommend that we err on the side of structural integrity rather than hydraulic excellence, if err we must.

REFERENCES

1. STREETER V.L. Handbook of Fluid Mechanics, Section 18.7, McGraw Hill, London, 1961.

2. HUDSON R.Y. Stability of Rubble-mound Breakwaters, W.E.S. Vicksburg, T.M. 2-365, June 1953.

3. HUDSON R.Y. Rubble-mound Breakwaters, Laboratory Investigations of Proc. ASCE, Waterways & Harbours Division, Vol. 85, No. WW3, Sept. 1959.

4. BROWN C.T. Blanket Theory and Low Cost Revetments. Chap. 151, 16th ICCE, Hamburg 1978.

5. BROWN C.T. Armour-Units, Random-mass or disciplined Array? ASCE Specialty Conference, Coastal Structures 79, Alexandria Virginia, March, 1979.

6. BROWN C.T. Gabion Report. The Water Research Laboratory of the Univrsity of New South Wales, Research Report No. 156, Oct 1979

7. PAGE R. Erosion Control using Seabees. Undergraduate Thesis, N.S.W. Institute of Technology, Jan. 1983.

8. HANSEN S. & KEATS J. Investigation of Seabees as an Erosion Control Structure - Undergraduate theses, N.S.W. Institute of Technology, Dec. 1983.

9. BROWN C.T. Seabees in Service. ASCE Specialty Conference Coastal Structures '83, Washington D.C., March 1983.

NOTATIONS

A Area of element

C Coefficient a - exponent, b - exponent

C_v Volume Coefficient

C_I, C_H Volumetric coefficients for Irribarren and Hudson transformations

D	Characteristic dimension of armour elements, Diameter of porous cell
d_n	Sediment grain size, n% finer than this value (by weight)
F_s	Shields' parameter
g	Gravitational acceleration
H	Wave Height
K	Stability coefficient
L	Normal length of porous cell
P	Porosity of armour layer
S	Shear force
S_r	Relative density
T	Tensile force
U_*	Shear velocity
V	Velocity
W	Weight
α	angle of slope relative to horizontal
β	direction of streamtube relative to horizontal
γ	density
δ	angle of particle motion relative to horizontal
μ	coefficient of friction
τ	shear stress

2 Loads on beds and banks caused by ship propulsion systems

Dipl.Ing. H. U. OEBIUS, Versuchsanstalt für Wasserbau und Schiffbau, Berlin

SYNOPSIS. Using latest results from investigations concerning the velocity distribution in propeller jets with and without velocity head an attempt is made to deduct also loads to be expected from ship propulsion systems like water jets and propeller jets acting on beds and embankments of ports, channels and rivers.

INTRODUCTION

1. One of the most important presuppositions for the estimation of the degree of destruction of beds, embankments and revetments in harbours, channels and rivers caused by manoeuvring or cruising vessels, except that from collision, is the mathematical description of those currents which act as carriers of the essential and responsible kinetic energy. There are two sources for such energy carriers, i.e. the primary wave system of any vessel as well as the induced secondary wake field, and the propulsion jet as product of the necessary impulse system to push the vessel in the desired direction, including bow thrusters and similar propulsion systems.

2. It is obvious that wave and wake fields as result from the surmounting of the blockage resistance of the water body against its displacement by the moving ship are therefore causally connected with the displacement speed of the vessel relative to the surrounding water body (not relative to the bed!) and are the more significant the higher the speed or the blockage effect are. Under normal cruising conditions the influences from these currents exceed that from the propulsion system by far. Unfortunately universal solutions for the loads to be expected have not been found, yet, although these effects have been subject to diverse investigations (ref. 1).

3. Basically the degree of destruction due to the influence from propulsion systems is the more distinct the heavier this propulsion system is loaded (i.e. the greater the impulse of the system is), the easier the revetments can be removed by the impact of the jet and the longer the time of attack is, i. e. the lower the ship speed relative to the

attacked area is. Indeed the greatest destructions occur at those places, where vessels are standing, starting or manoeuvring with low speed, i.e. in front of piers, in basins, at turning basins, at outer harbours in front of locks and docks, etc. That they can reach enormous dimensions has been demonstrated by a test with an inland ship accomplished by the Bundesanstalt für Wasserbau, Karlsruhe, in a deserted channel (ref. 2) (see Fig. 1).

Fig. 1 Scour induced by inland ship

PHYSICAL BACKGROUND OF PROPULSION SYSTEMS

4. Physically seen the propulsion system's task is to accellerate air or water in such manner that an impulse of distinct force and direction is produced, which from modern propulsion systems generally will be a jet from either submerged nozzles or from propellers. Due to internal friction between this jet and the surrounding fluid with increasing distance from the orifice the diameter of the jet is increasing too, accellerating parts of the surrounding fluid, at the same time consuming kinetic energy from the core velocity of the jet, resulting in a decrease of the maximum velocity. Due to this spreading the jet eventually reaches the water surface and/or the bed and walls of the basin, where the jet dissolves, transferring its kinetic energy to the boundary. Loads from these energies are therefore directly connected with the actual axial and tangential velocities $u_{x,y,z}$ and $v_{x,y,z}$ at the location x,y,z in the interface. In rotational symmetrical jets $y^2 + z^2 = r^2$. The determination of the velocity distribution in a plane or circular water jet and a propeller jet with and without velocity head (which represents the current

of the fluid or the transition speed of the ship) will therefore be the first step towards the definition of the actual loads from propulsion systems.

5. Velocity distribution in jets without velocity head. From investigations by Kraatz (ref. 3) and Wiegel (ref. 4) we know that generally the velocity distribution in jets follows Gauß's law of the normal distribution of errors of observation (ref. 5). This law

$$f(\bar{x}) = \frac{1}{\sigma\sqrt{2\pi}} \cdot \exp\left[-\frac{1}{2}\left(\frac{\bar{x}}{\sigma}\right)^2\right] \qquad (1)$$

can be transformed with $f(\bar{x}) = u_{x,r}$, $\frac{1}{\sigma\sqrt{2\pi}} = u_{max}$ and $\bar{x} = r'$ (see Fig. 2) to

$$u_{x,r} = u_{max} \cdot \exp\left[-\frac{1}{2}\left(\frac{r'}{\sigma}\right)^2\right], \qquad (2)$$

which is valid for all kinds of jets and also for all sorts of velocity heads after definition and adjustment of the main parameters

$$u_{max},\ x_0,\ r',\ \sigma = f(u_0,\ u_\infty,\ w)$$

according to the individual flow conditions of the fluid.

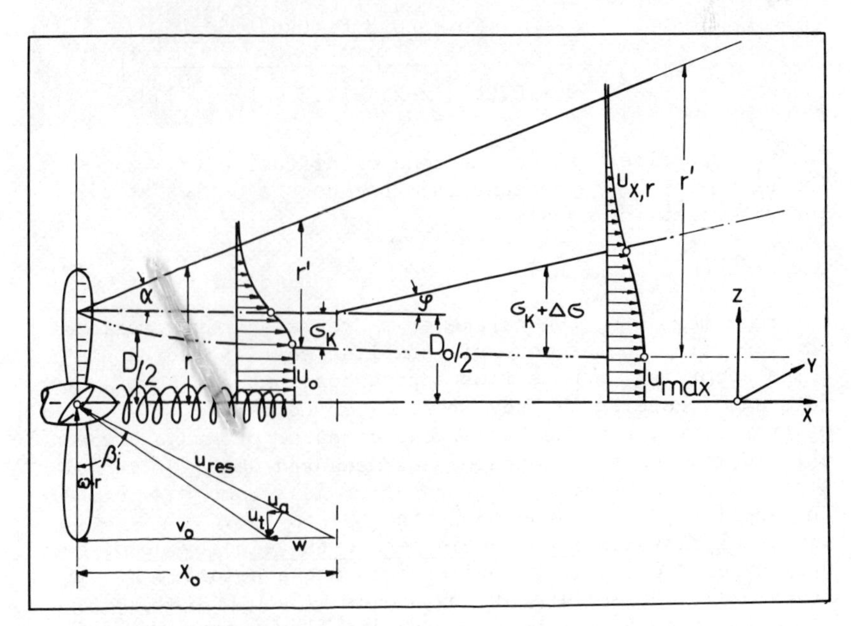

Fig. 2 Schematic outlay of the velocity distribution in a propeller jet

By extensive tests in the VWS, Berlin, these parameters have been defined empirically and lead to the equations

$$u_{x,r} = u_0 \cdot \exp\left[-\frac{1}{2} \left(\frac{r + \frac{D_0}{2} \left(\frac{x}{x_0} - 1\right)}{D_0 \frac{x}{2x_0}} \right)^2 \right] \quad (3)$$

for the circular water jet and

$$u_{x,r} = u_0 \cdot \exp\left[-\frac{1}{2} \left(\frac{r - \left[0.3\, D_0 \cdot \left(\frac{x}{D_0}\right)^{-0.3}\right]}{\frac{D_0}{2} - \left[0.3 \cdot D_0 \left(\frac{x}{D_0}\right)^{-0.3}\right]} \right)^2 \right] \quad (4)$$

for the propeller jet in the zone of establishment ($0 < x < x_0$; $x_0 = 2\,D_0$; see Fig. 2), as well as

$$u_{x,r} = \frac{D_0 \cdot u_0}{2\left[\frac{D_0}{2} + 0.0807(x - x_0)\right]} \exp\left[-\frac{1}{2} \left(\frac{r + \frac{D_0}{2}\left(\frac{x}{x_0} - 1\right)}{\frac{D_0}{2} + 0.0807(x - x_0)} \right)^2 \right] \quad (5)$$

for the circular water jet and

$$u_{x,r} = 1.5 \cdot u_0 \left(\frac{x}{D_0}\right)^{-0.6} \cdot$$

$$\cdot \exp\left[-\frac{1}{2} \left(\frac{r - \left[0.3\, D_0 \left(\frac{x}{D_0}\right)^{-0.3}\right]}{\frac{D_0}{2} + 0.0875(x - x_0) - \left[0.3\, D_0 \left(\frac{x}{D_0}\right)^{-0.3}\right]} \right)^2 \right] \quad (6)$$

for the propeller jet in the zone of diffusion ($x_0 < x < \infty$). The parameter D_0 represents the free nozzle outlet at circular water jets or equals

$$D_0 = 2\,(0.67\, R_F + R_N) \quad (7)$$

at propeller jets, the parameter u_0 represents the maximum core velocity in the zone of establishment. It can be computed according to pipe flow approaches in the case of water jets and according to Isay (ref. 6) or Lerbs (ref. 7) in case of propellers. For the latter one computer programs are available. Comparisons of computed and measured data for circular water jets (see Fig. 3) and propeller jets (see Fig. 4) show the applicability of the equations (3) to (6).

6. Velocity distribution in jets with velocity head. Here only experiences with propeller jets are available. But if the vessels are sailing at low speeds $v_0 = 0.5$ kts, the equations (3) and (5) may be used for circular water jets with velocity head, too, without causing too great errors. From the latest investigations the velocity distribution in a propeller jet with velocity head can be written according to equation (2)

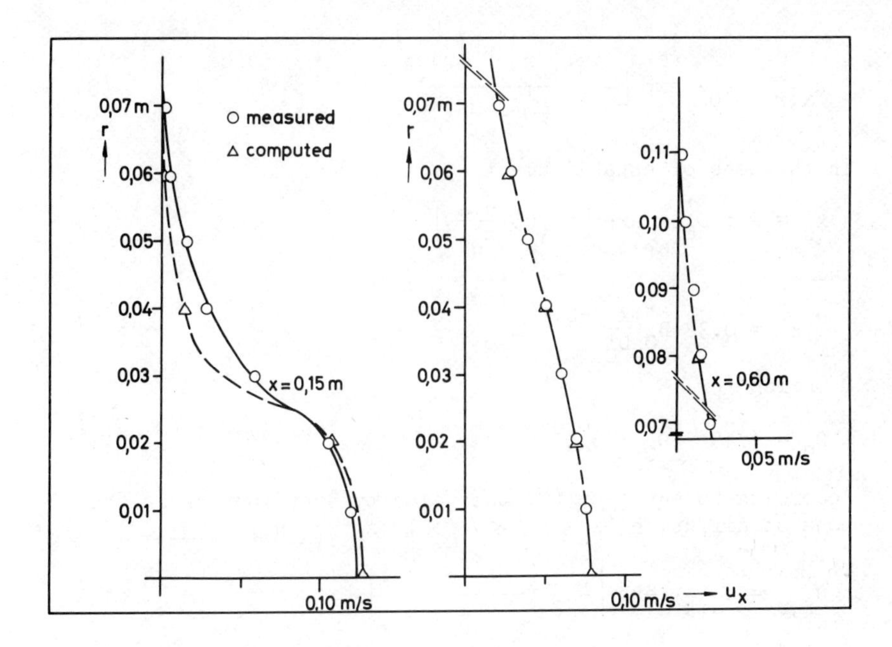

Fig. 3 Computed versus measured velocities in a circular water jet without velocity head

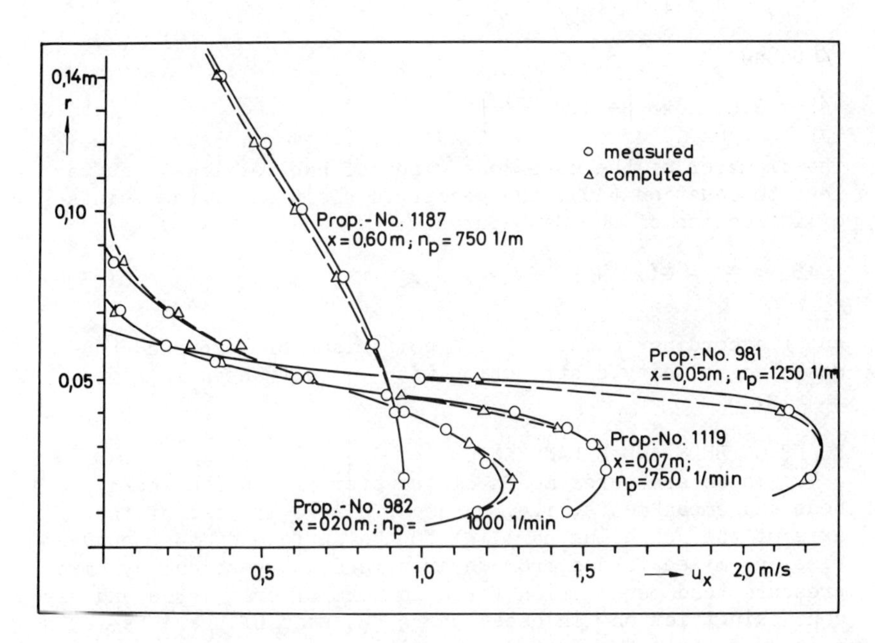

Fig. 4 Computed versus measured velocities in a propeller jet without velocity head

$$u_{x,r} = u_0 \cdot \exp\left[-\frac{1}{2}\left(\frac{r - \frac{D}{2}(x)}{\frac{D'}{2}0 - \frac{D}{2}(x)}\right)^2\right] \tag{8}$$

in the zone of establishment ($0 < x < x_0$) with

$$x_0 = 2 \cdot D_0' \cdot \exp\left[1.265 \frac{u_\infty}{u_0}\right] \tag{9}$$

and

$$\frac{D}{2}(x) = 0.32\, D_0'\left(\frac{x}{D_0'}\right)^{-0.1} \tag{10}$$

D_0' being

$$D_0' = 1.15 \cdot D_0 \cdot u_\infty^{0.055}. \tag{11}$$

According to equation (2) the velocity distribution in the zone of diffusion ($x_0 < x < \infty$) follows the basic law

$$u_{x,r} = u_{amx} \cdot \exp\left[-\frac{1}{2}\left(\frac{r - \frac{D}{2}(x)}{\sigma_D}\right)^2\right], \tag{12}$$

with

$$u_{max} = 1.5 \cdot u_0 \left(\frac{x}{D_0'}\right)^\beta, \tag{13}$$

β being

$$\beta = 0.6 \cdot \exp\left[-1.2 \frac{u_\infty}{u_0}\right] \tag{14}$$

the diameter of the core zone (zone of hub vortices) equivalent to equation (10), the geometric position of the points of inflection of the Gauß-curve

$$\sigma_D = \frac{D'}{2}0 - \frac{D}{2}(x) + 0.0875\,(x - x_0) \tag{15}$$

and x_0 according to equ. (9). A comparison of computed and measured velocity distribution with velocity head is given in Fig. 5.

LOADS ON BEDS AND EMBANKMENTS

7. There are three modes of transfer of kinetic energy to beds and embankments, i.e. by shear stress in case of the axis of the jet being parallel to the surface of the boundaries, by mixed forces from shear stress and vertical dynamic pressure (and percolation force in form of drag force and viscous skin friction) in cases where the axis of the jets is inclined to the embankments (per definition this does not or very seldom occur at the beds) and by mere hydrodynamic pressure (and percolation forces) in cases where the axis of the jet acts normally to the embankments (see Fig. 6).

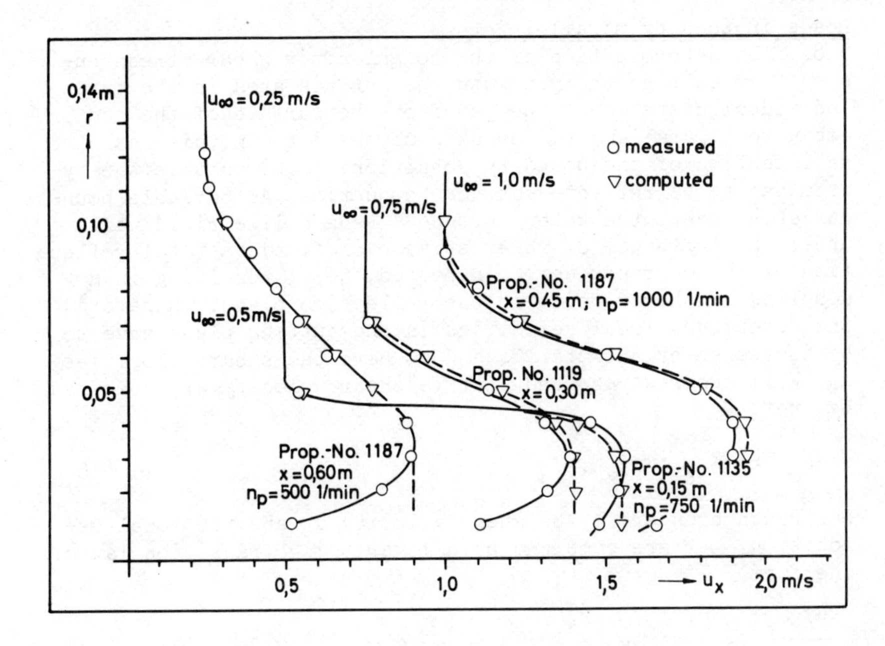

Fig. 5 Computed versus measured velocities in a propeller jet with velocity head

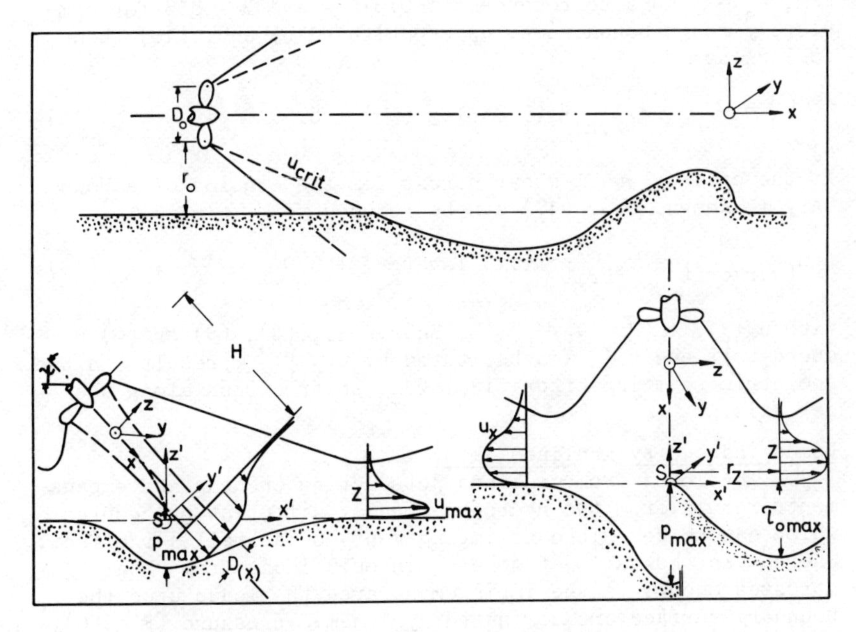

Fig. 6 Schematic presentation of the velocity fields in wall jets, inclined and impinging jets

Loads induced by parallel jets

8. The determination of the sought after shear stress ensues from cutting the jet with the surface area in the given individual distance of the jet from the surface of the bed or embankment parallelly to the axis of the jet. In this case the satisfaction of the boundary conditions requires that the system has to be reflected at the boundaries. As moveable boundaries - except the water surface - behave like rigid walls under the influence of shear stresses(ref. 5), a total reflection of the currents can be expected, here, resulting in a doubling of the velocity in the reflection plane. Therefore the determined local velocities in the cutting plane have to be increased by a factor 2, and so have the shear velocities u_*. Wall shear stress and fictive shear velocity are combined by (ref. 8)

$$\tau_0 = \rho_F \cdot u_*^2 \quad . \tag{16}$$

For rough boundaries the shear velocity u_* and the local velocity $u_{(x,r)}$ are combined by a logarithmic transition law of the form

$$\frac{u_{(x,r)}}{u_*} = \frac{2.3}{k} \log\left(\frac{r_*}{k_s}\right) + B \tag{17}$$

with k = Kármán constant = 0.4, $r_* \sim$ mean roughness diameter $\sim D_{50}$, k_s = Nikuradse roughness $\sim 0.5 \cdot D_{50}$ and B = 8.5 for completely rough boundaries. u_* substituted by equ. (16) then follows to

$$\tau_0 = \rho_F \cdot u_{(x,r)}^2 \cdot \left(5.75 \log\left(\frac{r_*}{k_s}\right) + 8.5\right)^{-2} \quad . \tag{18}$$

If the critical wall shear stress for the erosion of a boundary is known, equ. (17) can be reduced to

$$u_{(x,r)crit} = u_{*crit} \cdot 5.75 \log\left(\frac{r_*}{k_s}\right) + 8.5 \tag{19}$$

with $u_{*crit} = (\tau_{crit}/\rho_F)^{1/2}$. Equs. (3), (4), (5) or (6) reduced to r and $u_{(x,r)}$ substituted by $u_{(x,r)crit}$ result in the geometric position of critical wall shear stress along the jet axis.

Loads induced by inclined jets

9. The loads from inclined jets acting on beds and embankments result from two hydrodynamically different procedures which can be described as impingement and as wall jet effect. Whereas in the wall jet zone again only loads from shear stresses occur, in the impingement area the loads upon the boundary surface are dominated by dynamic pressure (Fig. 6). The effects in the impingement zone are the same as from jets acting normal to walls due to the fact, that the axis of inclined jets seems to bend to directions normal to the boundary

surface (Fig. 6). Beltaos (ref. 11) found a relationship between the maximum pressure and a resulting wall shear stress due to the deflection of the vertical current to be

$$\tau_{0max} = c_f \cdot p_s \cdot \sin \Psi \tag{20}$$

with $c_f = 0.166$, Ψ = angle of inclination and p_s = the momentum, due to the center velocity of the jet, acting on an area with the diameter D, which is defined by D_0 at free water jets and $D_{(x)}$ at propellers in the distance H from the orifice

$$p_s = \frac{\rho_F}{2} \cdot \frac{\pi \cdot D^2}{4} \cdot u^2_{xmax} \ . \tag{21}$$

The maximum actual shear stress in the impingement zone then follows to

$$\tau_{0max} = 0.166 \cdot \rho_F \cdot \frac{\pi \cdot D^2}{4} \cdot u^2_{xmax} \cdot \sin \Psi \ . \tag{22}$$

10. As from experiments it has been found that for angles greater than $\Psi = 45°$ the erosion in the impingement zone is significantly greater than in the wall jet zone, for estimations concerning loads from propellers only equ. (22) should be used. In all cases where $15° < \Psi < 45°$, the wall shear stress τ_{0max} should be computed according to Beltaos (ref.11):

$$\tau_0 = \rho_F \cdot u^2_{max} \cdot 0.098 \, R_{e0}^{-0.2} \ , \tag{23}$$

u_{max} being the maximum reflected velocity at the position x', equal to the center velocity of the jet at the distance x from the orifice, and R_{e0} being the Reynolds number at the orifice

$$R_{e0} = \frac{u_0 \cdot D_0}{\nu_F} \ . \tag{24}$$

Loads induced by impinging jets

11. Erosions by jets acting normally to the boundary surface result from two effects which have to be superponed (see Fig. 6), i.e. an impingement impact due to dynamic pressure only and a shear effect from the deflection of vertical to radial velocities. While the impingement effect is strongly dependent on the relative distance of the orifice from the boundary H/D_0, the shear effect is only dependent from the development of the velocities parallel to the boundary surface and reaches its maximum at the relative orifice Z (see figure 6) in the distance r_Z from the stagnation point S.

12. The dynamic pressure in the impingement area has been described by Kobus, Leister and Westrich (ref. 12) by

$$p_s = \rho_F \cdot \frac{u_0^2}{2} \cdot \left(\frac{H}{D_0}\right)^2 \cdot 57 \exp\left[-114 \left(\frac{r}{D_0}\right)^2\right] \tag{25}$$

and τ_0 resulting from

$$\tau_0 = c_f \cdot p_s = 0.07 \cdot p_s \quad . \tag{26}$$

The shear stress in the wall jet region, starting at the stagnation point, is described by Beltaos (ref. 11)

$$\tau_0 = \rho_F \cdot u_0^2 \left(\frac{0.0794}{\frac{r}{D_0} - 0.3}\right)^2 \quad , \tag{27}$$

with r being the diameter of the jet at the distance x from the orifice.

SCOURING

13. The erosion ε_m is a function of the reaction time T_ε, the velocity $u_{(x,r)}$, the critical shear velocity u_{*crit} and the distance r_0, H of the orifice from the boundary surface. Fig. 7 gives an example for the development of scours as function if these parameters. It can be seen that about 50 % of the final erosion depth is reached within half an hour, a relatively long time compared with the real reaction time. This means that the risk of damages in regions of low density of traffic is low, but extremely high in areas which are very near to the propulsion system or where the sequence of individual events is very short thus provoking long term effects.

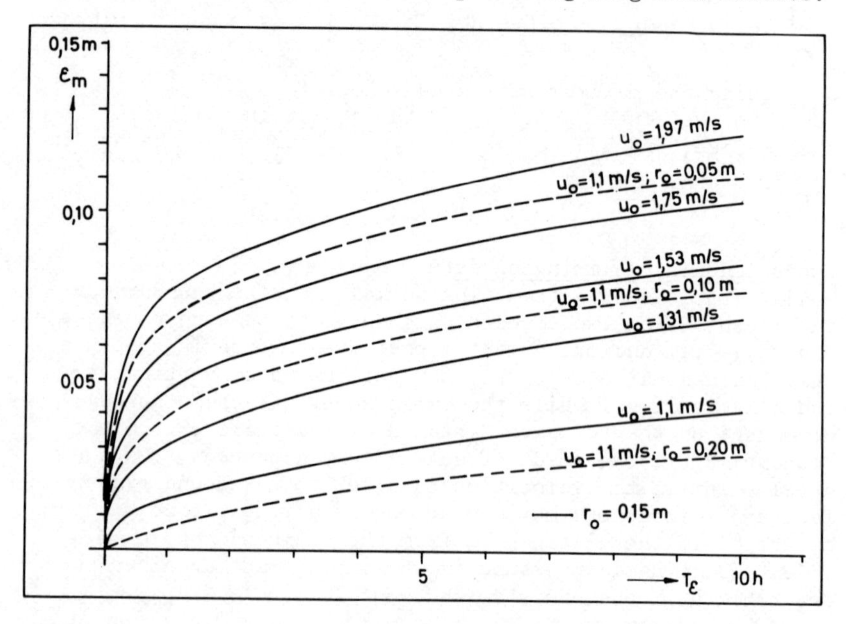

Fig. 7 Erosion depth ε_m as function of reaction time T_ε

CONCLUSION

14. The equations mentioned above are dimensionally correct and generally physically based but have to be regarded at as rough estimations of the loads to be expected. The formulae do not satisfy high standards, because they represent in nearly all cases simplifications for practical reasons. Those interested in more details are kindly referred to the original papers.

15. The developed equations enable the engineer to estimate damages at beds, embankments and revetments from propulsion systems although the definition of the critical shear stresses for the beginning of erosion will be difficult in many cases. They are known practically only for loose sedimments.

REFERENCES

1. SCHUSTER,S. Untersuchungen über Strömungs- und Widerstandsverhältnisse bei der Fahrt von Schiffen auf beschränktem Wasser. Jahrbuch der STG, 1952, 46, 244-280.
2. FELKEL, K.; H. STEINWELLER. Natur- und Modellversuche über die Wirkung der Schiffe auf Flußsohlen aus Grobkies (Breisacher Versuche). Die Wasserwirtschaft, 1972, 62, 8, 243 - 249.
3. KRAATZ, W. Strömungsverhalten und Verteilung horizontal an der Oberfläche eingeleiteter Beckenzuflüsse. Wasserwirtschaft - Wassertechnik, 1972, 22, 3, 102 - 104; 1972, 22, 5, 171 - 175.
4. WIEGEL, R.L.; J. MOBAREK; Y. YEN. Discharge of warm water jet over sloping bottom. Hydraulic Eng. Lab., University of Calif., Berkeley, 1960.
5. OEBIUS, H.U.; S. SCHUSTER. Analytische und experimentelle Untersuchungen über den Einfluß von Schraubenpropellern auf bewegliche Gewässersohlen. Versuchsanstalt für Wasserbau und Schiffbau, 1919, VWS-Report No. 848/79.
6. ISAY, A. Propellertheorie. Springer Verlag Berlin-Göttingen - Heidelberg, 1964.
7. LERBS, H. Ergebnisse der angewandten Theorie des Schiffspropellers. Jahrbuch der STG 1955, 49, 163 - 206.
8. SCHLICHTING, H. Grenzschichttheorie. G. Braun Verlag, Karlsruhe 1965.
9. OEBIUS, H.U. Shear stress and hydrodynamic pressure measurements at sea beds. Proc. of Oceans '83, San Francisco, 1983.
10. ZIPPE. H.; W. GRAF. Turbulent boundary-layer flow over permeable and non-permeable rough surfaces. Journal of Hydraulic Research 1983, 21, 1, 51 - 65.
11. BELTAOS, S. Oblique impingement of circular turbulent jets. Journal of Hydraulic Research 1976, 14, 1, 17 - 36.
12. KOBUS, H.; P. LEISTER; B. WESTRICH. Flow field and scouring effects of steady and pulsating jets impinging on a moveable bed. Journal of Hydraulic Research 1979, 17, 3, 175 - 192.

3 Proposals of flexible toe design of revetments

Dr Ing. G. HEERTEN, Naue Fasertechnik, Espelkamp and
Dipl.Ing. W. MÜHRING, Neubauamt Mittellandkanal, Osnabrück,
West Germany

SYNOPSIS. Revetments are frequently damaged by scour at revetment toe. If they are constructed by traditional methode, scour at the toe inevitable results in loss of stability. Scour is naturally more likely to occur at the transition between the protected and unprotected part of the canal bed.
Consequently, adequate toe protection is highly important for the entire revetment. This paper will provide the information on the fundamentals of revetment toe design, the findings of tests on different types of flexible toe construction on the Mittelland Canal in Germany, and new recommended toe constructions.

INTRODUCTION

1. One the construction of canals in Category IV, the Conference of European Transport Ministers recommends a waterway cross-sectional area at least 7 times the cross-sectional area of a fully loaded typical vessel. Depending on local conditions, construction is carried out in three typical cross-sections for technical and economical reasons (see Fig.1).

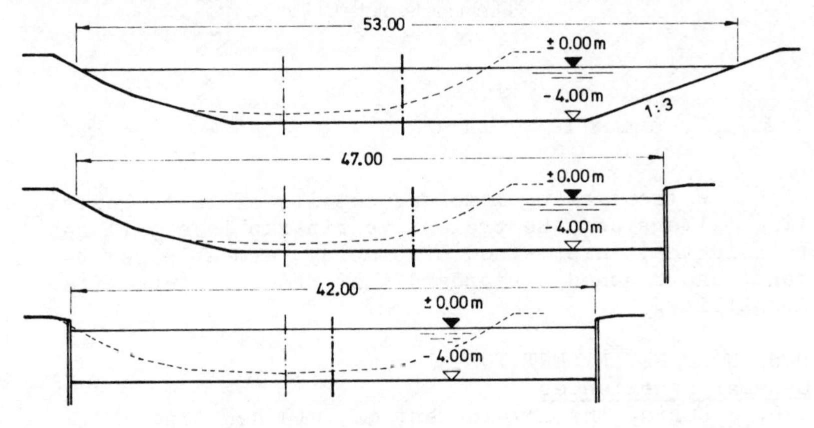

Fig. 1. Typical cross-sections

BANK PROTECTION

Revetment construction

2. The principles used in the design and construction of revetments in expansion work on canals are based on experiences accumulated in practice. In the course of expansion, difficulties arose because it was accompanied by a structural change in inland shipping brought about by the replacement of the old towed barges by self-propelled ships which increased the stress on revetments and had unforseeable effects on their durability.

3. Growing experience acquired during many years of work on the Mittelland Canal, for example, led to the development of standard revetment construction methods which guaranteed far longer life for aprons and similar bank protection constructions. This work had to be done without disrupting shipping and involved placement of a long-lasting filter layer covered by an apron that protected the revetment against the erosive effect of shipping and water.

4. Years of practical development work, for example, have shown that heavy multi-layer geotextile filters covered by bonded rip-rap or flat composite materials ensure adequate protection (see Fig.2).

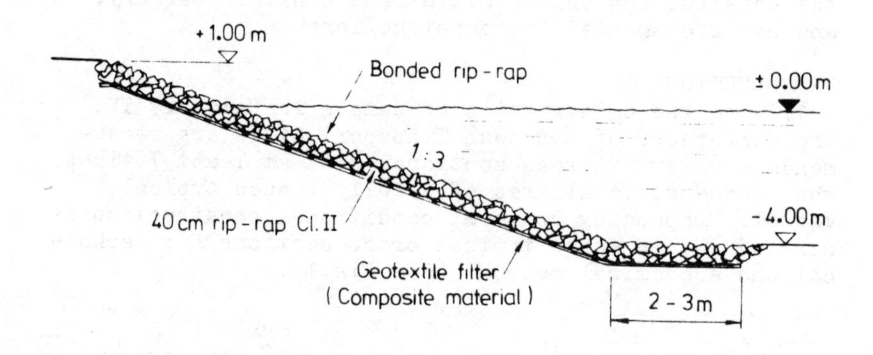

Fig. 2. Permeable revetment

5. Without going into the details of these geotextile filters and the protective rip-rap layer, it can be said that this method of placing permeable revetments has reached a standard that promises long-life durability.

DESIGN OF REVENTMENT TOE

General experiences

6. During the develpoment of this new type of revetment, attention was focussed for a long time on the construction of the elements on the actual slope.

Although it was generally realised that " a revetment is only as good as its toe", no specific consideration was given to its design.

7. Initially, this was unnecessary anyway, as the need for a new approach in revetment construction arose automatically as a result of negative experiences in bank protection.

8. The need to give closer attention to the design of the toes in revetment construction only became acute after the revetment on the slope had reached an acceptable standard. There were two reasons for this.

9. Firstly, scour had previously been delayed or left undiscovered because the toe was normally over-covered by stones sliding down the slope from unbonded rip-rap.

10. Underwater excavation ist done mainly with suction cutter-dredgers. Depending on the proportion of fines in the soil, part of the spoil is held in suspension and carried away from the workpit by the current from shipping into surrounding areas. It subsequently settles at the break point between the bank and canal bed, covering the stone pitching of the toe.

11. This sediment is transported further by canal currents and the wash from shipping until it reaches areas, like turning points or wider stretches of the waterway, where the lower velocity of the current allows the suspended matter to settle permanantly. As soon as this sediment reaches a critical height, it is excavated.

12. After several years the trough profile formed by the settlement of suspended matter again becomes a trapezoidal profile.

13. Under further stress from currents in the waterway, eroded channels and scour occur on the canal bed, and especially at the transition from the stone pitching of the toe to the unprotected section of the bed (see Fig.3 & 4).

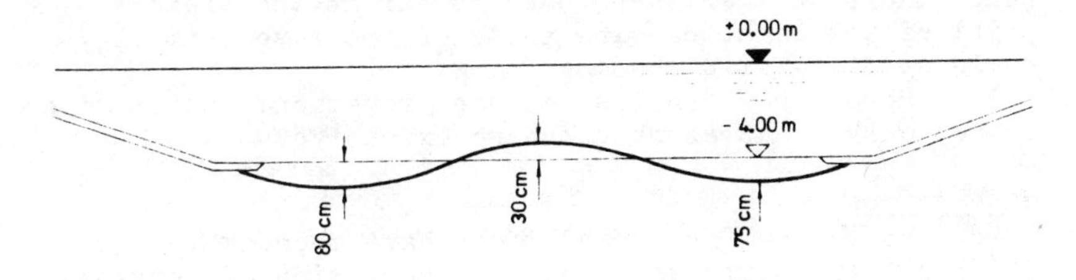

Fig. 3. Measured profile in canal bed - cross section

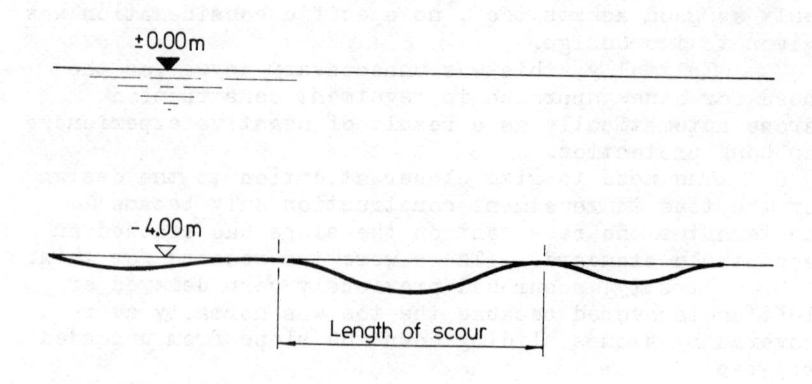

Fig. 4. Measured profile in canal bed - longitudinal section at the toe

14. Fig. 4 shows the typical trapezoidal profile of an eroded bed with irregular channels at the sides and sediment in the centre.

15. The situation becomes critical for the toe and, indeed, the entire revetment when scour spreads beneath the toe in the direction of its break point with the bank. When the scour reaches a length, such as that described in Fig. 4, the inadequate design of the toes causes its end to cave in. Subsequent damage is inevitable whenever the material from the collapsed toe-end fails to fill completely the depression created by scour in the canal bed. And this is normally the case, since the design of the toe (depending on thickness, weigth and type of bonding) generally causes collapsing material to break away in clumps, which often encourage further scour.

16. The higher the bending moment of the toe apron, the greater is the hazard to the whole revetment because the scour can then reach as far as the break point of the bank, causing parts of the revetment to collapse along with the toe.

17. Repair work on the damaged revetment and its toe is not only troublesome but also expensive.

Conclusions

18. Up to now field conditions have made it too difficult to obtain an accurate description of currents in the area of eroded beds, so there has been no way of calculating the expected depth and development of scour as a basis for design of toe approns.

19. Moreover, installation of a toe apron causes variations in the friction and stability coefficients

of the canal bed, which unfavourable influence bed deformation, especially in the area of the toe apron. Unlike 'conventional' scour, the oncoming flow directs its angle of attack not vertically but parallel to the protected end of the revetment, which acts as a baffle.

20. The only way of finding a solution to these problems, therefore, is to assess the degree of erosion unter a specific volume of shipping over a given time and utilise these parameters as empirical guidelines in design work.

21. One importent point that should be noted in this case is that on waterways with the crossectional area of the Mittelland Canal, for example, it has been established that bed deformation is caused more by current, i.e. the wash from shipping, than it is by the action of ship' propellers. These findlings have been confirmed by measurements carried out by the Bundesanstalt für Wasserbau in Karlsruhe (1). This finding applies only to the open waterway and not, of course, to mooring areas where the action of ships' propellers is an essential factor in the design of aprons on canal beds.

22. Another point that needs clarifying is whether the sediment in these areas is merely distributed or whether it is carried away by the current and causes permanent depressions in the waterway bed. In the latter case, either the entire bed should be paved or the waterway cross-sectional area should be increased considerably.

23. Consequently, it is essential to ensure that there is no longitudinal movement of suspended matter when the canal bed is not protected against erosion.

24. When construction work is done 'in the dry', the problem is easier to solve, either by extending the revetment along its axis into the canal bed to the extent of the assumed depression (see Fig.5).

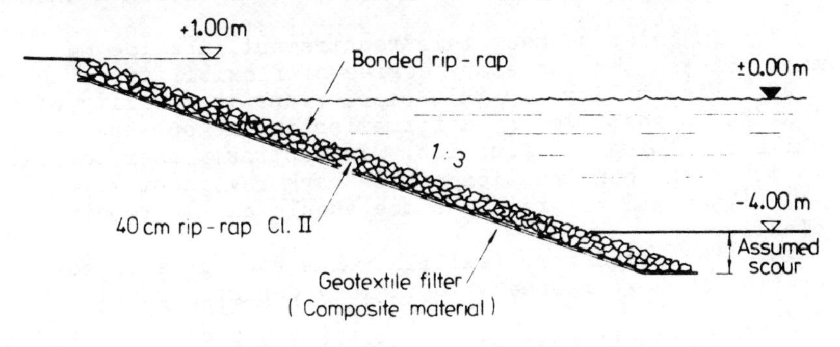

Fig. 5. Embedment in canal bed

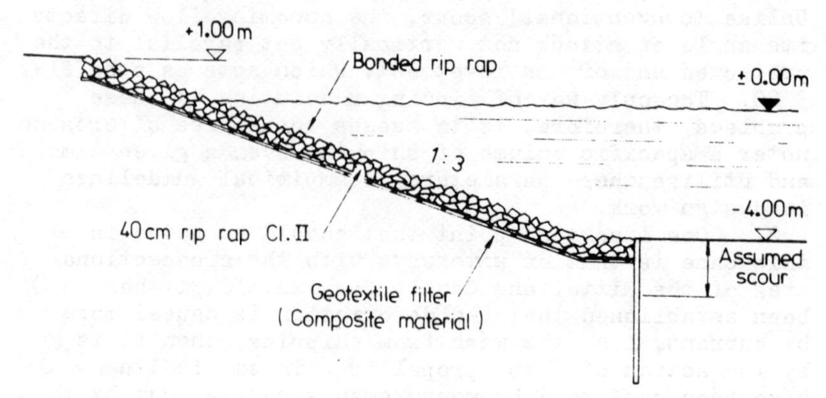

Fig. 6. Protection with a vertical wall

or by constructing a sheet pile wall, for example, at the break point of the slope or below it (see Fig.6).

25. If construction is carried out underwater, the revetment cannot be extended into the canal bed, because there is no way of adequately keeping an excavated hollow or depression open for placing the toe. Construction of a vertical retaining wall at the toe is prohibited on grounds of cost. But more about this solution later.

26. Consequently, a satisfactory solution on underwater construction of a revetment toe entails meeting the following requirements:
The toe mat must extend to a point in the canal bed that ensures that any erosion down to the deepest point remains covered, thus preventing any scour beneath the end.

27. In order to meet this requirement, the toe mat must possess the necessary degree of flexibility.

28. Flexibility, in this case, means the ability of the toe to adapt to any deformation in the sub-base while retaining its function. In contrast, therefore, to the rigid construction of the bank revetment with its filter and rip-rap, the toe should always remain elastic.

29. This need for flexibility has been underlined by previous experience with rigidly designed revetment toes.

30. Since all revetment toes on the Mittelland Canal have, up to now, all been of a rigid design (including rip-rap bonded with a bituminous mass), a test programme was formulated to establish ways of meeting these new requirements.

Tests

31. in 1981, 5 short test stretches were integrated in normal construction work on the Mittelland Canal and exposed to maximum stress from passing vessels.

32. On these test stretches, the toe mat was alternately 3 m or 5 m long and was given a multi-layer filter of geotextile composite material.

33. All used geotextile filter mats have to fulfill the strong requirements on geotextiles for hydraulic engineering of the Federal Institution of Waterways Engineering (BAW), Karlsruhe, Germany. That means for the given local soils that the geotextile has to proof e.g. its filtering efficiency in special tests and its penetration resistance in a test with a dynamic load of 600 Nm corresponding to a 300 N stone falling down from a heigt of 2,0 m.

34. Apart from its filter properties, this geotextile filter also absorbs any tensile forces and thus ensures that the toe mat is not severed from the revetment on the bank whenever settlement occurs at the toe end, i.e. it guarantees retention of the revetment function.

35. To meet the need for flexibility, three of the test stretches were given a top course of unbonded riprap stones of differing unit weight. In order to prevent any stones rolling off the mat after settlement of the toe end, the end was secured by a geotextile sack filled with "engineering" clay (see Fig.7).

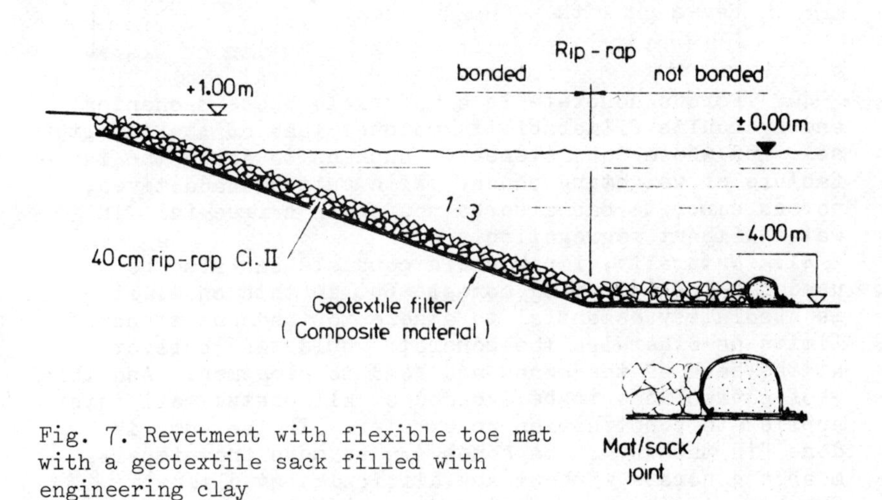

Fig. 7. Revetment with flexible toe mat with a geotextile sack filled with engineering clay

36. This engineering clay is made from chemophysical materials. If is permanently clastic and resistant to erosion.

37. In order to prevent the sack being displaced by even extensive settlement, it is tied at the rear to secure it in position.

38. On another of the test sections, the rip-rap was bonded by a small amount of pervious concrete whose quantity and strength was carefully selected to ensure the toe mat remained flexible.

39. On the fifth and final test stretch, the top of the filter mat was covered with a very permeable, rough fibre layer. This filter layer was weighted down with a ballast of porous concrete and helped interlock the stucture (see Fi.8)

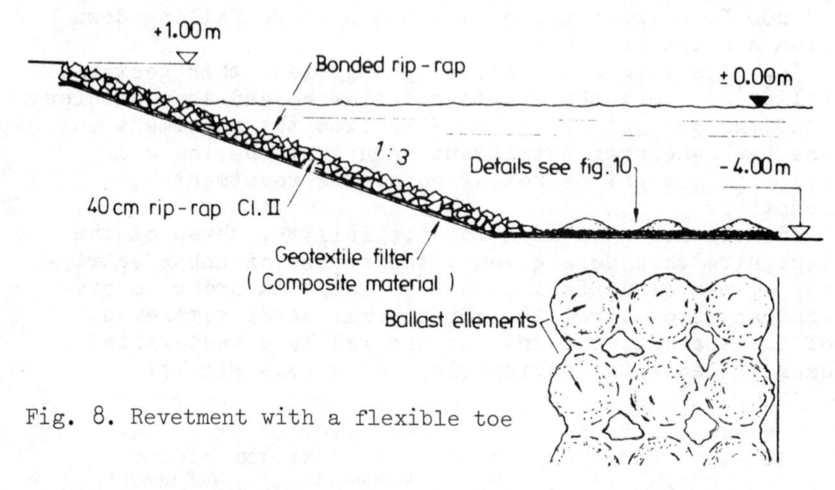

Fig. 8. Revetment with a flexible toe

40. Porous concrete is a material, whose mechanical and hydraulic filterability matches that of the in-situ soil and whose consistency is such as to allow the manufacture of weighting cones. With suitable additives, porous concrete can also be poured in a free-fall in water without segregation.

41. Naturally, impermeable concrete can also be used for the weighting cones. But in this case, it is absolutely essential to adhere to the consistency limits as otherwise the concrete would tend collect at the base of the cones and lead to clogging. And this, - of course, has to be avoided at all costs. All this applies to construction underwater. If the work is done "in the dry", the cones can be made from impermeable concrete without any difficulty at all. Checking and controls present no problem either.

42. These revetment toes are then exposed to stresses by radioing passing vessels, asking them to traverse the test stretch at full speed either in the centre of the canal or as near as possible to the bank.

The speeds of these passing vessels are recorded, and the stresses from currents and pressure changes are measured at three points near the canal bed with the help of current and pressure pick-off meters. Stresses from propeller action were also assessed in stationary tests.

43. The findings from these tests provided suitable evidence that the basis idea of constructing a flexible toe mat from a geotextile filter connected to the bank revetment and a geotextile sack at the toe end to secure sliding stones was correct. Under heavy stress from propeller action in the stationary tests, the rip-rap stones were considerably displaced, even in 35 cm-thick layers with unit weight of 3,5 - 3,7 kg/dm^3. The best resistance to propeller stresses was forthcoming on test stretch 4, where the rip-rap was bonded with porous concrete. In the stationary test, the ship was positioned with its longitudinal axis over the toe end (sack), so that the forces from the propeller were directed full-blast at the transition zone between toe end and canal bed. It turned out, however, that the vessel could not always be held stationary in this position because when the engine speed was increased, the propeller revs generated vibrations in the mooring cable and these, in turn, caused slight changes in the ship's position. Pressumable, it was this that also caused scour varying in depth from o,5 to 1,25 m in front of the sack. All in all, however, the sack was sufficiently flexible to adapt to any scour that occurres, irrespective of whether or not the rip-rap stayed in position on the geotextile filter.

44. The tests generally showed that the aforementioned toe mats did not adequately withstand stresses generated by ship propellers.

45. However, by stressing the toe mat sometimes to point of collapse, important findings were obtained on the behaviour of different solutions, especially as regards flexibility. In no instance, for example, even at the deepest scour of 1,25 m, was the sack filled with engineering clay destroyed. Thanks to its flexibility, the sack adapted to any hollowing-out caused in the canal bed by erosion and thus prevented scour beneath the toe end.

46. The multilayer, needle-punched geotextile filter dovetailed the movements of the sack, and adequately sealed-off and protected the sub-soil. But in areas, where propeller action caused considerable displacement of rip-rap, abrasion resistance was not entirely satisfactory. Damage also occurred on the top course in Test No. 5 (see Fig.8), because the rupture strength on this type of protective mat was too low and, consequently the ballast elements at the toe end were dislodged.

FURTHER APPLICATIONS

47. In 1983, repair work had to be carried out on an inflexible revetment toe of the type shown in Fig. 2, because scour beneath the toe apron had reached the break or buckling point of the bank revetment.

48. Utilising the test findings and the knowledge that crucial stresses arise only from the wash of passing vessels and not from propeller action, it was suggested that the toe be repaired by using a solution similar to that in Fig. 9a.

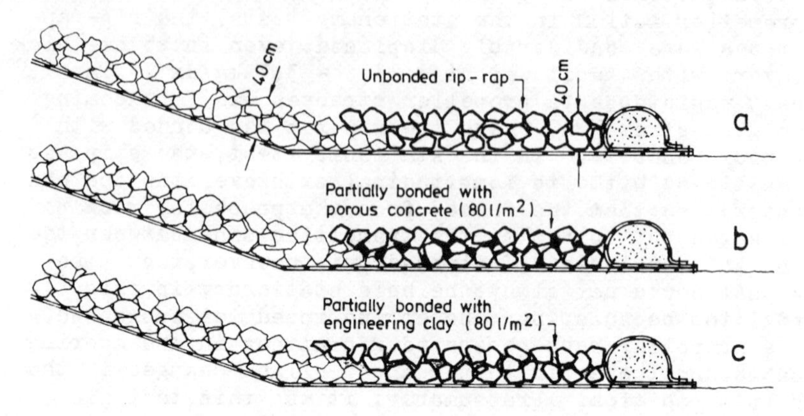

Fig. 9. Revetments with flexible toes

49 A Heavyweight, needle-punched non-woven fabric (1100 g/m^2) was used as a filter and this was covered by a 40 cm layer of rip-rap (arris length 15 - 25 cm). The toe end consisted of a sack (about 40 x 60 cm) filled with engineering clay.
In order to obtain further findings from alternative solutions for countering stresses from wash, additional test stretches were constructed on this stretch (No.4). The first of these involved a solution like that shown in Fig. 6. The sheet piling at the toe is 3.5 metres long. Special care was taken to ensure that the joint between sheet wall and bank revetment remained filterable. For this purpose, the rip-rap layer was fully grouted with porous concrete after installation of the filter mat.

50. On stretch No.2 (see Fig. 9b), this rip-rap was partially bonded with only 80 l/m^2 of porous concrete. The aim here was to increase the individual weight of the stones through pointwise bonding with neighbouring stones, but without reducing the flexibility of the entire structure. On stretch No.3 (see Fig.9c), the rip-rap layer was partially filled with engineering clay (80 l/m^2) to bond each stone as firmly as possible,

while retaining the permeability of the toe mat. In these applications, the engineering clay had to feature adequate flexibility combined with lasting resistance to erosion. As there is not yet any methodical way of determining these contradictory requirements for engineering clay, they were gauged empirically. On stretch No.4, the solution again was similar to that in Fig. 8.

51. In order to prevent the ballast elements being torn off, a geotextile material like that in Fig.10 was chosen.

52. In addition to these protective measures on revetment toes in the Mittelland Canal, similar work has been done an the River Weser.

53. With the increase in the size of ships many estuaries up to the important ports have been deepened. As well as the increased use of large scale dredging, river works to control flow and tidal range such as groins and training walls have been erected.

54. Several groins are under construction, to stabilize the shipping channel and to minimize the increase of the tidal range which was mainly expressed as a sinking of the tidal low water.

55. In a large river program 100 groins will be built, using a new construction method with geotextiles in the groin section down to tidal low water. In the underwatersection the conventional construction method using a willow fascine mattress for the groin foundation is used. Instability problems caused by scouring at the sides in the past lead to crest sinking and damage of the groins.

56. Fig. 10 is showing the groin cross-section with the foundation base above tidal low water using the new construction method:
On a heavy weight needlepunched nonwoven fabric ($1100\ g/m^2$) the riprap body of the groin is dumped. With an overlap of approx. 500 mm on both sides of the groin this special scour protection mat is installed.

57. This scour protection mat is composed of four parts

- woven or nonwoven filter layer
  soil-tightness, permeability and acting load have to be considered.
- sedimentation layer
  approx. 5 cm thick made of needlepunched and chemically bonded curled coarse fibres
  it reduces the drag forces in the boundary layer of the sea bed so that sedimentation takes place increasing the weight and stability of the structure
- reinforcement fabric
  by means of wide meshes (approx. 20 mm) and very

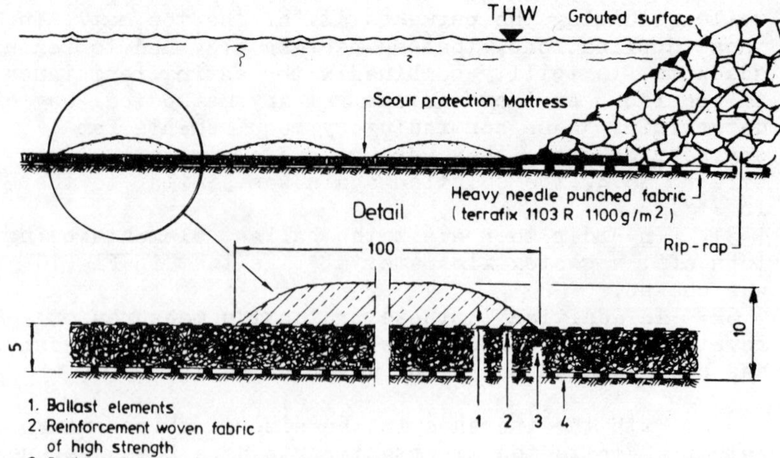

Fig. 10. Scour protection of a groin

high tensile strength the fabric combines and reinforces the ballast elements securely, providing a high degree of flexibility and adaptability with plenty of strength in reserve.

- ballast elements
  approx. 0.5 m in diameter and 0.1 m in hight with a minimum distance of approx. 0.2 m between the rims of the elements they stabilize the scour protection mattress during the sedimentation phase.

58. Fig. 11 is showing a groin under construction. The nonwoven geotextile and the scour protechtion are installed. The geotextile is fixed on the seabed by a first layer of stones. The special concrete vessel is still anchoring after ballasting the mat at the site.

SUMMARY

59. Toe mats are designed to protect revetments against underwash and scour. They form a transition zone between the unpaved canal bed and the baffle or energy dissipator in a flow system. Installation of a revetment toe presupposes that the non-erosionproof soil in the construction area is only re-distributed and not transported elsewhere by the current.
Revetment toes must be flexible, i.e. they must adapt to scour while retaining their function, and feature as steady a transition as possible into the hydraulic boundary areas (roughness, geometry).
Although design and development work on the construction of flexible revetment toes with geotextile filter has not yet been completed, there are already signs that this approach will produce a satisfactory solution to the problem.

Fig. 11. Groin under construction

REFERENCE:

Bundesanstalt für Wasserbau Karlsruhe :

Bericht über die Schiffahrtsversuche im Mittellandkanal 1981

4 Design of bank protection of inland navigation fairways

H. G. BLAAUW and M. T. de GROOT, Delft Soil Mechanics Laboratory,
F. C. M. van der KNAAP, Delft Hydraulics Laboratory, and K. W. PILARCZYK,
Hydraulics Division, Rijkswaterstaat

SYNOPSIS.
The collapse mechanism of bottom and bank constructions of fairways under attack of ship induced water motion has been studied from model and prototype tests. Transport relations and design criteria for subsoil, filter layers, and protection layers consisting of rip rap or blocks are formulated and verified, as far as they are at present.

INTRODUCTION

1. Ships sailing in inland navigation fairways produce a water motion which attacks the bottom and banks. During recent years power and ship size has increased considerably, resulting in more intensive attacks of fairway boundaries and, thus, to high maintenance and construction costs. Due to these reasons, the Dutch Public Works has charged the Delft Hydraulics Laboratory (DHL) with a long-term investigation to develop design rules for bank and bottom protection.

2. A protection layer should resist the hydraulic attack and at the same time prevent the movement of subsoil and/or filter material through the construction. Also, sliding of the subsoil or parts of the construction must be prevented. Both hydraulic and geotechnical aspects are thus of importance, and therefore, the investigations are being carried out in close cooperation with the Delft Soil Mechanics Laboratory (DSML).

3. Two basic design approaches can be identified: the deterministic and the probabilistic. In the deterministic approach a dominant design condition is selected. On the basis of this condition the dimensions of the protection layer and filter are determined for the criterion of 'initiation of motion' (1. no (or slight) displacement of individual stones of a rip rap top layer can be accepted; 2. lifting of individual blocks of a block revetment by pressure forces perpendicular to the slope cannot be accepted). The probabilistic approach aims at a calculation of the total damage of the construction on, for instance, a time (year)

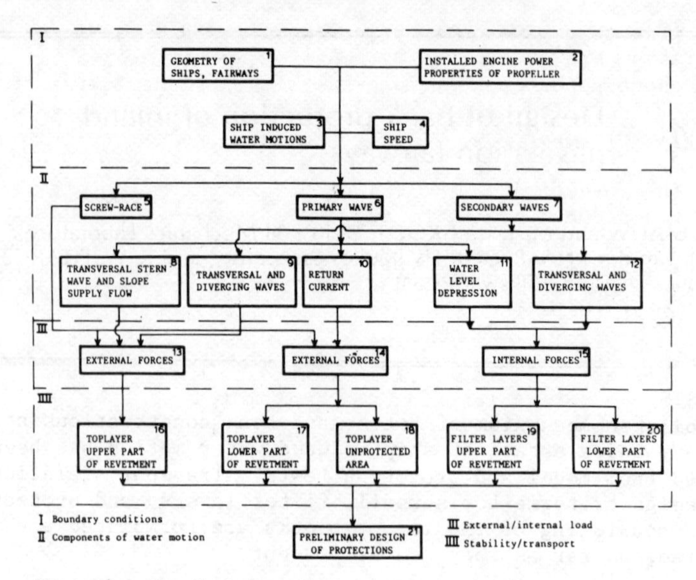

Fig. 1: Design process

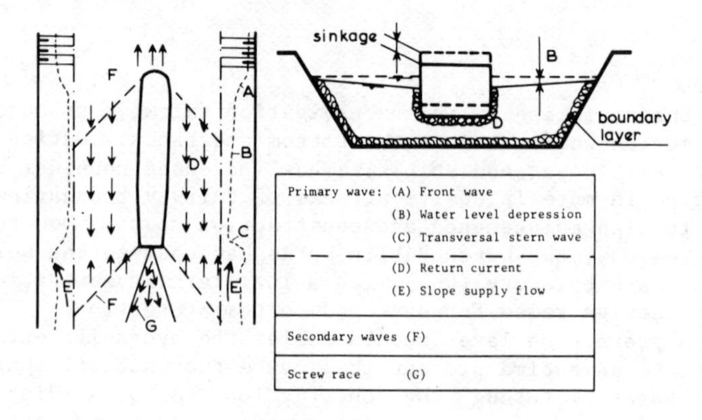

Fig. 2: Review of water motion components

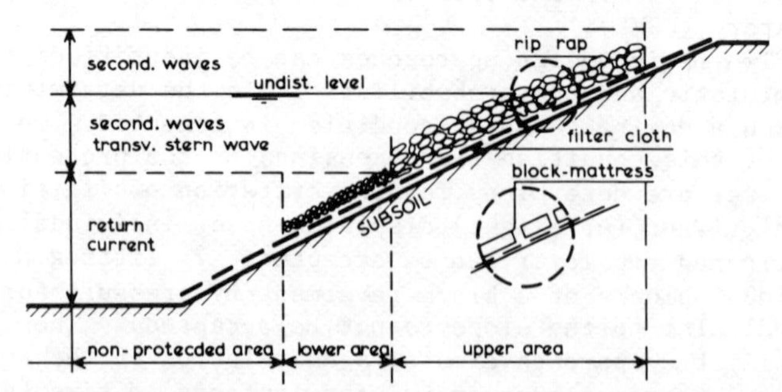

Fig. 3: Areas in the canal cross-section and dominant component of the water motion

base. The transmission functions required for this type of calculation are determined by means of deterministic tests. The present paper is therefore restricted to deterministic design rules. The transport relations, required for a probabilistic approach, will be presented as far as they are, at present, available.

4. The investigations carried out, in the prototype and also at a reduced scale, mainly concern push-tow canals, bank slope 1:4, a rip-rap protection and a geotextile or granular filter. Studies into the behaviour of block revetments have started recently, and some initial results are discussed.

DESIGN PROCESS

5. The design process, as presented in Fig. 1, forms the basis of the design technique adopted.
The ship-induced water motion and the corresponding ship speed are calculated (Blocks 1 through 7) using known ship dimensions, fairway cross-profile, and applied engine power.

6. The ship-induced water motion can be split up into screw race, primary wave and secondary waves (Blocks 5, 6, 7). The primary wave components are: return current, water-level depression, front wave, transversal stern wave (Blocks 8, 10, 11). The secondary waves are composed of diverging and translating waves which together form the well-known interference peaks. These waves are indicated in Blocks 9 and 12.

7. The various components of the ship-induced water motion are indicated schematically in Fig. 2.

8. The ship-induced water motion attacks the fairway boundaries. Basically the areas under attack, see Fig. 3, are: (a) unprotected bottom and part of the banks, (b) lower protected area, and (c) upper part of revetment. The return current (and screw race) are important, for Areas a. and b., whereas secondary waves and/or transversal stern wave dominate in Area c.

9. The external loads (transversal stern waves, secondary waves, return current) exert friction and pressure forces on the protective layer and are of prime importance for the determination of dimensions of this layer. In the subsoil the pore pressures respond to external variations of the water-level (front wave, water-level depression, secondary waves). The resulting forces determine the design requirements for the filter.

HYDRAULIC LOAD

General

10. Detailed calculations of the ship-induced water motion, based on the given geometry of a fairway cross-profile, the ship and the applied engine power are elaborated in this chapter.

11. In addition to the ship-induced water motion other hydraulic phenomena occur such as wind waves and tidal current. In the fairways considered in the present studies

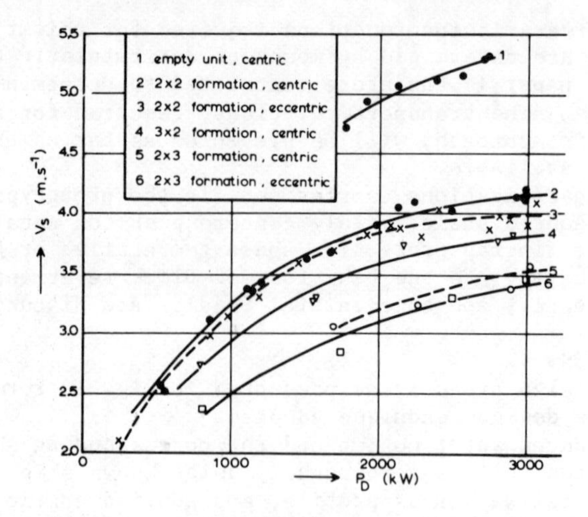

Fig. 4: Relation between shaft-horse power and ship speed

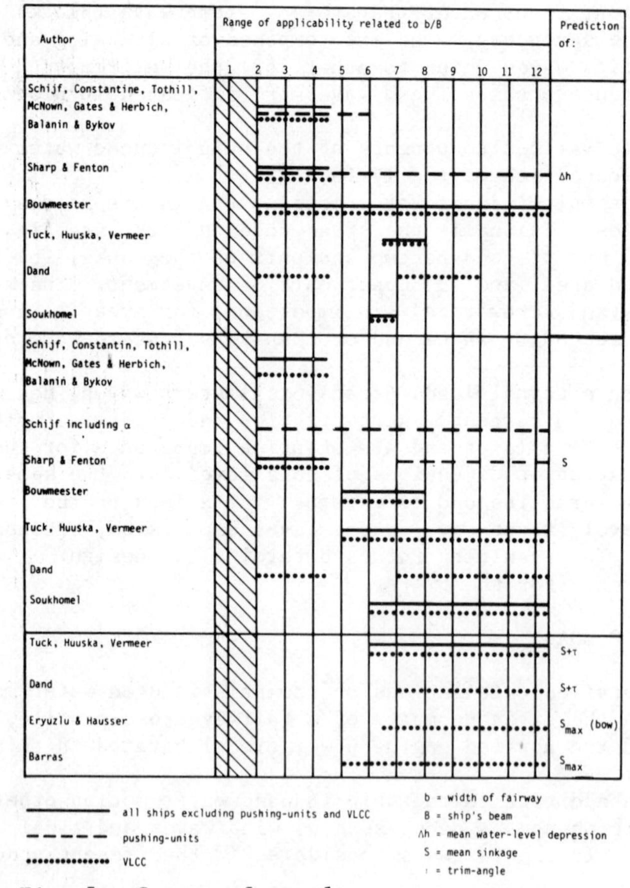

Fig. 5: Survey of results

ship-induced waves are more important than wind-induced waves. Therefore the wind-induced waves have not been taken into account particularly in case of a deterministic design. The effects of both ship-induced currents and natural currents are discussed.

Speed prediction

12. Frequently vessel speed prediction calculations are not required since the velocities of various types of ships are well known. For instance in case of the renewal of an existing protection (or large-scale maintenance) the design can be based on the existing conditions. However these calculations are indispensible, for new fairway design and/or introduction of new vessel shapes or more highly powered ships.

13. Speed prediction calculations are based on the equilibrium of the total required power, on the one hand, and part of the shaft horsepower, representing the thrust power, on the other, viz:

$$R_T(V_s + \overline{u}_r) = \eta_D \cdot P_S \qquad (1)$$

In ref. 1 this relation is extensively elaborated for pushing convoys.

14. Measurements of resistance and propulsion were carried out during the first series of tests on the Hartel-canal in 1981 for several configurations of pushing convoys. It follows that, for loaded convoys: $\eta_D = 0.85$. The relation between shaft horsepower and ship speed for investigated types of convoys, are given in Fig. 4.
It clearly follows that the speeds tend to limiting values for higher applied powers.

General ship-induced water motion

15. The calculation of the ship-induced water motion is very complicated and due to the full form of most of the ship types the presence of bottom and banks, and the free water level a three-dimensional calculation is necessary. This type of calculation has not been fully developed yet and for the present paper only a series of (most) one-dimensional calculation methods are discussed.

16. Generally, three main approaches can be distinguished, based on: conservation of energy, or momentum (one-dimensional; two-dimensional slender body theory; empirics. To get insight into the applicability of these methods, a thorough investigation was carried out, at the DHL, to determine which calculation method can best be used as function of width restriction of the fairway and ship type. These investigations are reported in ref. 2. The methods were verified with respect to their applicability to predict water-level depression, sinkage and, if possible, squat.

17. The results presented in ref. 2 were recently extended with results of prototype measurements (ref. 3). A survey of results is given in Fig. 5.

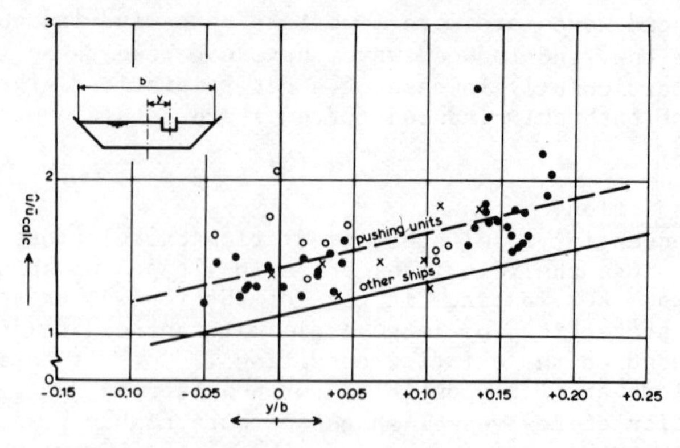

Fig. 6: Relation between extreme and average return current

For the calculation of areas of water-level depression it follows that, for pushing units, the method of Sharp and Fenton gives relatively good results, while for other ship types (including VLCC) the method of Bouwmeester proved to be satisfactory.

Detailed ship-induced water motion

18. Return current. The maximum return current ($\hat{u}_r$) and the simultaneously occurring maximum shear stress ($\hat{\tau}$) are decisive for the stability of a bank or bottom protection. During the conditions observed the return current had a more or less uniform distribution outside the boundary layers of ship and bottom, and banks. The current direction coincided approximately with the direction of the fairway axis (ref. 5).

19. The shear stress can be calculated according to:

$$\hat{\tau} = c_{fr}\tfrac{1}{2}\,\rho\,\hat{u}_r^2 \text{ in which } c_{fr} = 2.87 + 1.58 \log \frac{x}{k_s})^{-2.5} \qquad (2)$$

according to the Schlichting Formula for rough plates.
In this formula x is the distance over which a water particle near the embankment has moved due to the return current when a certain part of the ship-length (X) has passed. x is given by:

$$x = \frac{\bar{u}_r}{\bar{u}_r + V_s}\,X \qquad (3)$$

The maximum value of $\hat{u}_r$ occurs at a distance of 0.25 to 0.35 L_{OA} from the bow.
The value for the bottom roughness (k_s), including effects of unevenness, had quite a spread. A value of $k_s = 4.D_{50}$ is an acceptable average value.
In Fig. 6 the value of $\hat{u}_r$, related to $\bar{u}_r$ for a certain ship speed, can be taken as function of rate of "eccentricity"(y).

20. When a natural current prevails, in the fairway, the shear stress due to the return current and the natural current

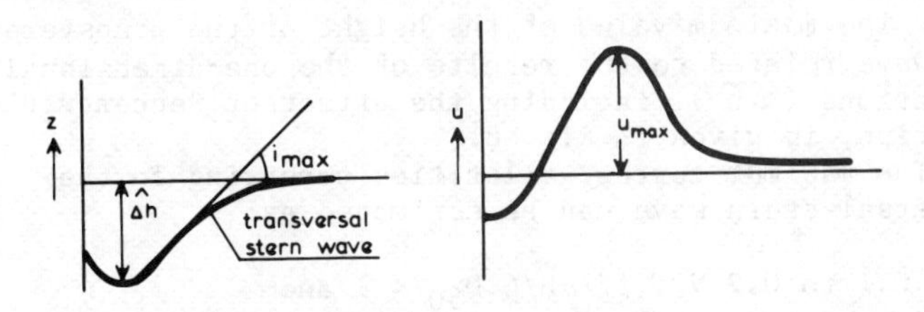

Fig. 7: Principal characters of transversal stern wave

can be calculated, according to ref. 4:

$$\tau = \tfrac{1}{2}\, c_{fc}\, \left(u_c + \frac{\sqrt{c_{fr}}}{c_{fc}}\, \hat{u}_r\right)^2 \text{ in which } c_{fc} = 0.06\, \left\{\log \frac{12h}{k_s}\right\}^{-2} \qquad (4)$$

21. Transveral stern wave. The principal characteristics of the transversal stern wave and related local current velocities at the side slopes are indicated in the sketch presented in Fig. 7.

22. The steepness of the transversal stern wave has been determined at both model and prototype scales. It follows from

$$i_{max} = \left(\frac{\hat{\Delta h}}{z_o}\right)^2 \qquad (5)$$

The steepness has a limiting value of i_{max} = 0.1 to 0.15. The factor, z_o, can be calculated according:

$$\frac{z_o}{b} = 0.04 - 0.158\, \frac{y}{b} \qquad (6)$$

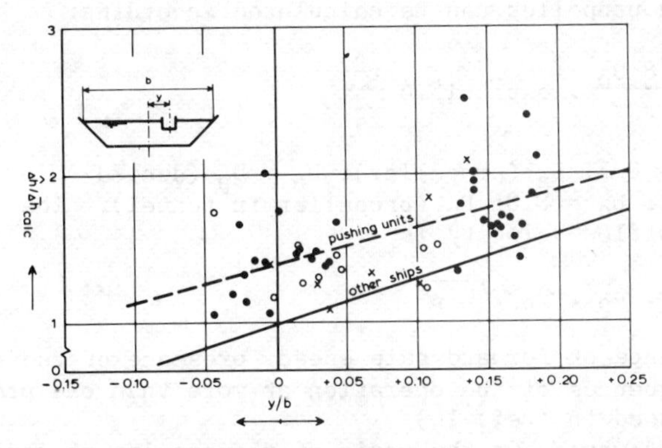

Fig. 8: Relation between transversal stern wave height and average water-level depression

Taking into account the effect of "eccentric" navigation (y/b). The maximum value of the height of the transversal stern wave related to the results of the one-dimensional calculations ($\overline{\Delta h}$), including the effect of "eccentric" navigation, is given in Fig. 8.

23. The maximum current velocities occurring in the transversal stern wave can be estimated as:

$$u_{max} = 0.1 \text{ to } 0.2\ V_s, \text{ if } \hat{\Delta h}/\Delta.D_{50} < 1 \text{ and}$$

$$u_{max} = (1 - \frac{\Delta.D_{50}}{\hat{\Delta h}})\ V_s, \text{ if } \hat{\Delta h}/\Delta.D_{50} > 1 \tag{7}$$

24. Secondary waves. Secondary waves are composed of transverse and diverging waves, which together form interference peaks. Interference peaks, and, to a less extent (behind the ship) transverse waves are of special interest in relation to bank attack. Secondary waves are elaborated thoroughly in ref. 7. A relation has been derived to determine the height of the interference peaks using the method of Gates and Herbich (ref. 8):

$$H_i = \alpha_i.h.(\frac{S}{h})^{-0.33}.(\frac{V_s}{gh})^{2.67} \tag{8}$$

From DHL prototype and model experiments it follows that : α_i = 0.80, pushing unit (loaded); α_i = 0.35, pushing unit (empty), tugboat, α_i = 0.25, conventional inland motorvessel. The wave length of the interference peaks can be described, ref. 7, as:

$$L_{wi} = 0.67 . \frac{2\pi}{g} . V_s^2 \tag{9}$$

25. Screw race. The velocities occurring in the screw race for ships manoeuvring and underway are extensively dealt with in ref. 9 and ref. 10. For a manoeuvring ship the velocities behind the propeller can be calculated according:

$$\frac{u_{x,\ r}}{u_o} = \frac{2.8\ D_o}{x_s} . \exp\ -15.4\ \frac{r^2}{x_s^2} \tag{10}$$

with: D_o = 0.71 D_p (propeller); D_o = D_p (ducted propeller); D_o = 0.85 D_p (propeller in tunnel). The limited outflow velocity is

$$u_o = 1.60 . n_p . D_p .\sqrt{K_{TP}} \tag{11}$$

The influence of forward ship speed, presence of rudders and the consequences of the operation of more than one propeller are discussed in (ref. 10).

26. Front wave. On the basis of the results of prototype and model experiments, the steepness of the front wave at the bank slope can be obtained from:

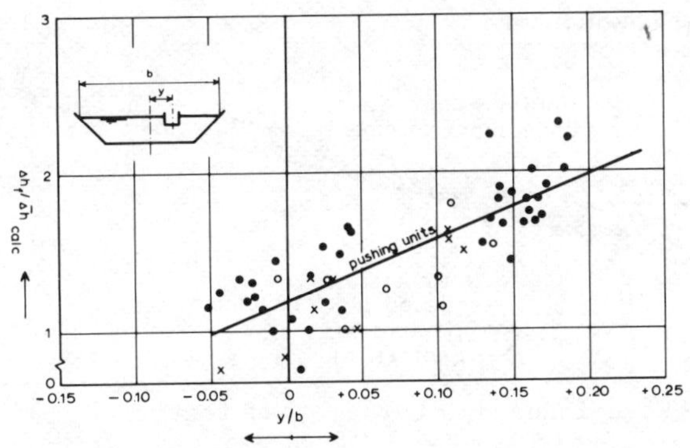

Fig. 9: Relation between front wave height and average water level depression

$$i_f = c(y) \cdot \Delta h_f \text{ in which}$$

$$c(y) = 4.06 \cdot 10^{-4} \cdot y + 1.79 \cdot 10^{-2} \quad (m^{-1}) \qquad (12)$$

The height of the front wave, related to the calculated water level depression, is presented as function of the rate of "eccentric" navigation (y/b) in Fig. 9. The front wave is important for the determination of the prevailing pressure gradients in the subsoil.

INTERNAL LOAD

27. Ship-induced waves in a channel constitute a direct external hydraulic load on the embankment but also bring about, indirectly, an internal load. Fluctuations in the water level due to passing ships affect water pressures under the top layer of the bank protection and in the subsoil. The influence of the fluctuations depends on wave frequency and amplitude and also on design characteristics, such as geometry of filter layers and on the permeability, density and stiffness parameters of the subsoil.

28. Wave-induced pore pressure under the revetment layer and in the subsoil constitute the internal load on the bank protection structure. The relation between external and internal loads plays an essential role when considering the strength of the bank protection, and the stability of bank protection can only be determined satisfactorily by considering the top layer and subsoil simultaneously. Optimizing a design only with respect to, say, maximum strength of the top layer against the external load, can give rise to loss of internal stability and thus lead to erosion. Meeting the requirements for external stability does not automatically imply internal stability or vice versa. In some cases a compromise has to be found see, for example, section 35.

29. Hydraulic boundary condition. Only fluctuations of the water-level will form the representative hydraulic boundary condition for the internal load. Both amplitude and speed of

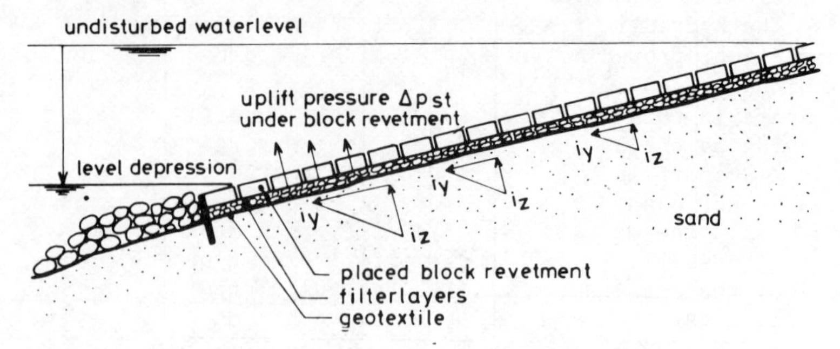

Fig. 10: Wave-induced internal load factors

these fluctuations have an impact on the induced groundwater flow.

30. Response subsoil. In general the response in the subsoil to channel water-level fluctuations can be described as follows. The phreatic level in the filter layer(s) cannot immediately follow a sudden lowering of the channel level. Under the top layer of the protection there remains an excess pore pressure, which results in hydraulic gradients in three main directions: (a) a hydraulic gradient, i_x, in the longitudinal direction of the waterway, (b) a hydraulic gradient, i_y, in the transverse direction of the waterway in the plane of the slope, and (c) a hydraulic gradient, i_z, in the direction perpendicular to the slope. Furthermore, uplift pressures, Δp_{st} against the slope revetment occur, when there are less permeable top layers, see Fig. 10.

31. Prototype measurements. An attempt has been made to establish a relation between the external hydraulic load and the induced internal load factors using a theoretical approach (analytical, numerical), see section 62, a scale model approach, see ref. 6, and prototype measurements. Prototype measurements have been carried out in the Hartel-canal in the harbour area of Rotterdam in 1981 and also very recently in 1983. External and internal loads induced by pushing units have been thoroughly investigated; several types of bank protection served as test sections for the experiments, see ref. 11.

32. Relation between external and internal load. Both the leading limits of the water-level depression (front wave) and its gradient with respect to time are representative for the internal load. Results of prototype measurements show, that the product of the front wave height Δh_f and the speed, with which the water level depression comes about is a satisfactory practical measure for the internal hydraulic gradients. Note that the ship's speed V_s, is implicitly represented by the quantity $\Delta h_f \cdot \frac{\partial h}{\partial t}$, viz.:

$$\Delta h_f \cdot \frac{\partial h}{\partial t} = \Delta h_f \cdot \frac{\partial h}{\partial x} \cdot \frac{\partial x}{\partial t} = \Delta h_f \cdot i_f \cdot V_s \qquad (13)$$

33. The hydraulic gradient, i_x, in the longitudinal x-direction, proved to be only of minor importance in the relation between internal load and strength. Any grain transport in this direction can be either positive or negative depending on the navigation direction. Generally the net grain transport in the x-direction can therefore be neglected. On the other hand, in the transverse direction, the internal response to the external load is of particular interest.

34. Figures 11, 12 and 13 show, for three types of protection, the measured transverse hydraulic gradient, i_y, directly under the geotextile as a function of the compound hydraulic boundary condition, $\Delta h_f \cdot \frac{\partial h}{\partial t}$. It follows, that application of a relatively permeable gravel layer under the top layer of concrete blocks reduces transverse hydraulic gradients considerably. Apparently the storage capacity of the gravel layer allows flow from the subsoil through the geotextile towards the toe of the slope. Figures 11, 12 and 13 show a more or less linear increase of the transverse hydraulic gradient, i_y, with $\Delta h_f \cdot \frac{\partial h}{\partial t}$ up to a value of $\Delta h_f \cdot \frac{\partial h}{\partial t}$ in the range 0.01 to 0.02 m^2/s; above this value almost no further increase in i_y was measured. Maximum values for i_y, in the order of 0.5, were recorded for concrete blocks without an underlying gravel layer. This maximum value was attained at the lowest outcrop point of the groundwater, following a water-level depression. Since, in general, the amount of pore water in a slope is large with respect to the flow discharge into the channel during a water-level depression, the transverse hydraulic gradient, i_y, will continue to be operative as long as the passage of the ship.

35. Hydraulic gradients, i_z, measured perpendicular to the slope, were very similar for the different test sections. The value of i_z, averaged over the upper first half a meter perpendicular to the slope was about 0.15. These uplift gradients occurred during the channel water depresson, that is, simultaneously with the transverse gradient, i_y.
Ref. 12 shows that this combination of "blowing" (flow in an upward direction) with flow parallel to the surface is more critical for grain stability than suction with flow parallel to the surface. Although blowing reduces the effect of the drag force exerted by the parallel flow, the reduction of the effective grain weight is the principal unfavourable factor contributing to loss of grain stability.

36. Uplift pressures under the top layer, as a consequence of a fall in the water-level, are to be expected in the case of revetments which are only slightly permeable, for example, concrete blocks laid with very narrow joints. Such a protection, with joint width smaller than 0.5 mm, has been examined in the Hartel-canal prototype measurements. Uplift pressures, Δp_{st} are shown in Fig. 14.
Here there is a conflict between the design for external stability and the design for internal stability. Application

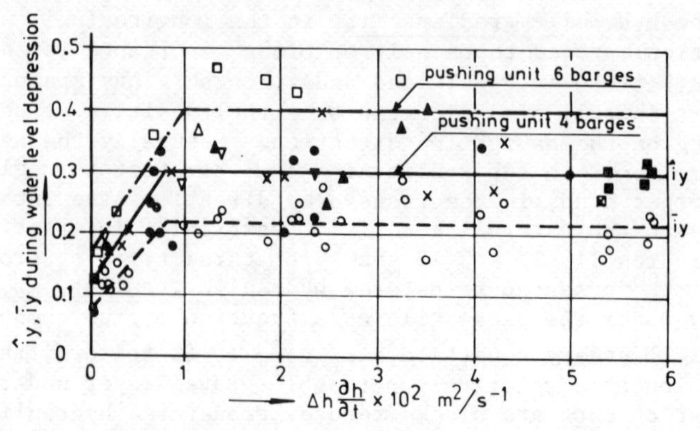

Fig. 11: Test section of placed block revetment upon gravel layer

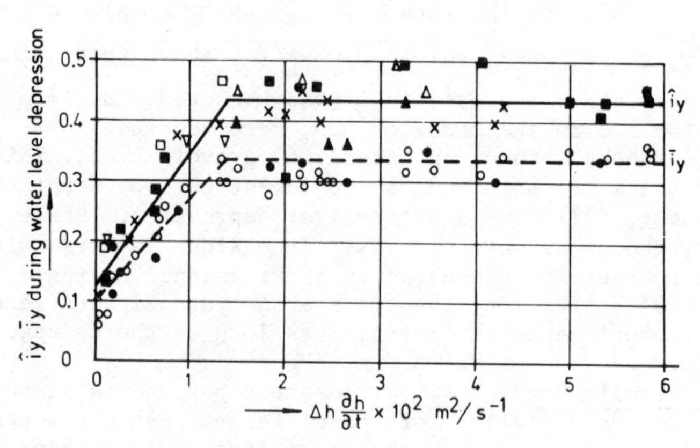

Fig. 12: Test section of placed block revetment

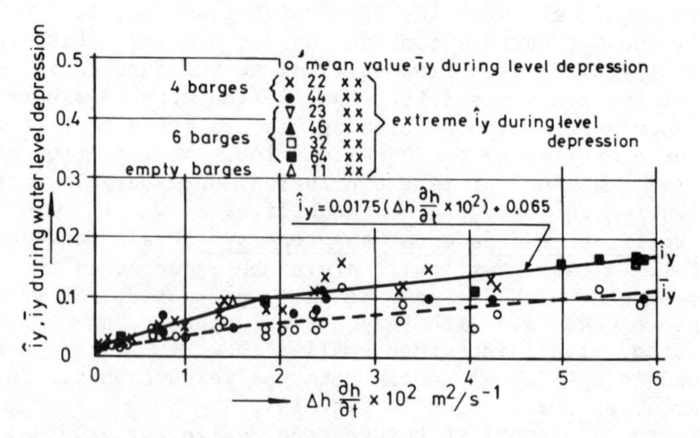

Fig. 13: Test section of rip-rap

of a gravel layer under the revetment increases uplift pressures but reduces transverse hydraulic gradients directly under the geotextile. Making the revetment more permeable, within limits, by enlarging the joints can possibly balance these conflicting design requirements.

37. The prevailing question about whether the magnitude of internal hydraulic gradients and uplift pressure will lead to loss of internal stability is connected with the relation between internal loads versus strength of the protection. Although complete answers cannot yet be given some aspects are discussed below in section 38 and following.

STRENGTH OF BANK CONSTRUCTION

Technical requirements

38. The technical requirements can be described briefly as follows: a good construction must be sufficiently stable, flexible and durable. Stability means that no part of the construction can be displaced (see Introduction); flexibility, on the other hand, means that the construction (or part of it) can deform to a limited extent without losing mutual connection. Durability concerns the resistance of the materials to weathering of any kind.

Internal failure mechanisms

39. The stability of bank protection can be endangered by external forces, and by the induced internal forces. It is desirable to determine critical values for typical internal load factors, which, in turn, depend on the external hydraulic load. The criterion for such critical values should be that internal loads exceeding this value lead to erosion of whatever kind. The kind of erosion depends to some extent on the type of bank protection. Questions that arise immediately are: what internal load factors can be taken as typical; what failure mechanism, brought about by the typical internal load, does one have in mind; how can one provide against possible internal erosion. Formulation of such critical values for internal load factors provides an essential tool for developing design criteria for bank protection.

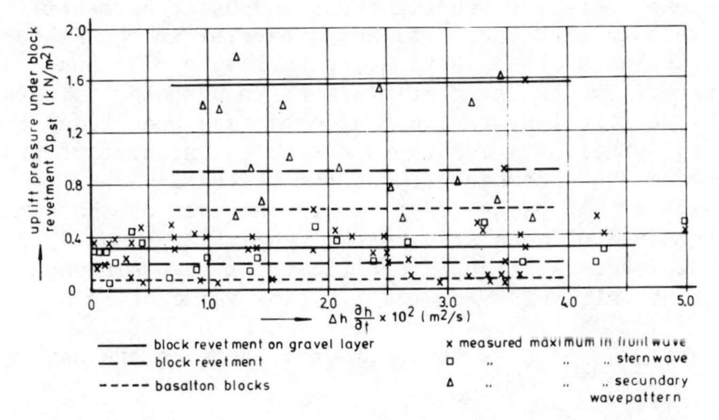

Fig. 14: Uplift pressures under block revetment

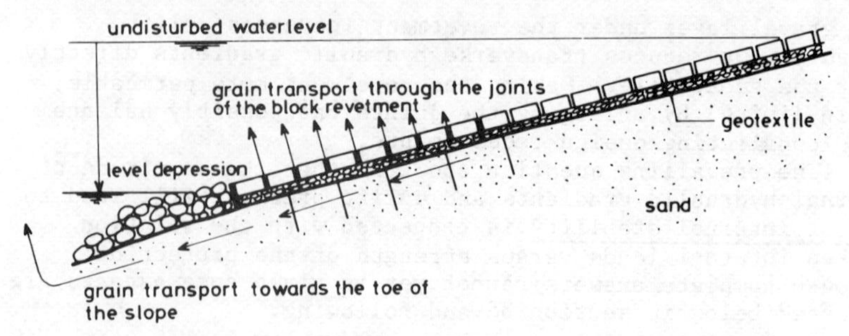

Fig. 15: Possible failure mechanisms by internal load

40. In the case or placed block revetments, initially there is a loss of stability when induced uplift pressure forces exceed the sum of the weight of single blocks and their mutual friction forces. One or more blocks can be lifted out from the revetment and the external hydraulic load can then act freely upon the protection, accelerating failure. To rely on uncertain friction forces is hazardous, since the uplift of one single block can lead to extensive damage of the protection. With blocks which do not interlock one has to take into consideration that some blocks in the revetment will be badly connected. In addition uplift pressures under neighbouring blocks will reduce any friction effects considerably. Both prototype measurements and analytical models, ref. 13, 14 and 15, show, that uplift pressures increase with the permeability of the subsoil filter layer and decrease with the permeability of the top layer.

41. The stability of block revetments can also be endangered when hydraulic gradients, i_y and i_z, on the interface between top layer and subsoil exceed the critical value. Depending on the storage capacity of the subsoil grain transport can take place through the geotextile or towards the toe of the slope, see Fig. 15. To provide against this long-term failure mechanism a sandtight geotextile should be placed on the subsoil. The determination of critical values for the internal load is being studied in the laboratory.

42. Design. For a filter to function properly it has to meet requirements for sandtightness and water permeability. Recent research for the storm surge barrier in the Eastern Scheldt indicates that vertical, parallel, cyclic and stationary flows can be distinguished in granular filters, see ref. 16. Critical hydraulic gradients have been found to be higher for stationary gradients due to the arching of grains. Furthermore, the permeability of the filter layer or geotextile should be at least as high as that of the subsoil. The percentage of open area of the geotextile, usually defined by the diameter O_{95}, is of specific interest. Although there is no uniformity for the limiting value of this parameter, $\frac{O_{95}}{d_{90}} \leqslant 2$ is considered to be on the safe side, see ref. 17.

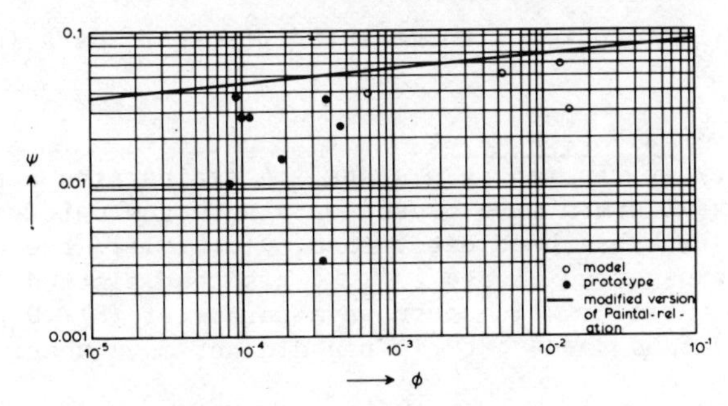

Fig. 16: Transport caused by return current

Stability and transport prediction for rip rap top layers

43. A number of computational models have been developed from the analysis of experiments in order to predict the stability and the transport of the top layer material under attack of the return current (including natural currents), the transversal stern wave, the secondary waves and the screw race of the top layer.

44. Return currents and/or natural currents. The stability against currents can be computed in different ways. If the current velocity, u_c, near the bank is known a quick estimate can be made with the stability criterion according to (ref. 18) using:

$$\frac{u_c}{\sqrt{g.\Delta.D_n}} = k_1 \quad (= \text{a constant}) \tag{14}$$

From analysis of experiments it follows that $k_1 = 1.2$ to 1.5.

45. A more accurate estimate of the stability against currents can be made with the criterion of Shields (ref. 19). In this case the maximum shear stress, $\hat{\tau}$, acting on the rip rap has to be known in order to compute the flow parameter, ψ , from the following:

$$\psi = \frac{\hat{\tau}}{p.g.\Delta.D_{50}} \cdot \frac{1}{k_D} \quad \text{in which } k_D = \cos\alpha \sqrt{1 - \frac{\tan^2\alpha}{\tan^2\varepsilon}} \tag{15}$$

Different values of ψ can be taken depending on the requirements, viz.: $\psi < 0.03$ - practically no transport of rip rap; $0.03 < \phi < 0.06$ - small transport of rip rap; $\psi > 0.06$ - rapidly increasing transport intensities.

46. In some situations some transport of material, caused by extreme loads, can be accepted. Such transports can be quantified with a modified version of the transport formula according to Paintal (ref. 5):

$$\phi = 1.64 \cdot 10^{10} \psi^{10.86} \tag{16}$$

in which ϕ = transport parameter, $q_s/\sqrt{g.\Delta.D_{50}}$, with q_s representing the transport of material per unit width.

Equation 16 is compared with measured transport data in Fig. 16.

47. Transversal stern wave.
The stability of rip rap on a slope 1:4 against the action of the transversal stern wave caused by a push tow unit sailing near to the bank has been determined. In Fig. 17 the measured number of transported stones (n_{meas}) has been plotted versus the characteristic stern wave parameter $\hat{\Delta h}/\Delta.D_{50}$. Clearly it can be seen that rip rap did not move when:

$$\frac{\hat{\Delta h}}{\Delta.D_{50}} < 2.3 \qquad (17)$$

48. Two methods have been developed to predict the transport caused by the transversal stern wave of a push tow unit. The first method is the most simple and has been based on measurements of shear stresses caused by a solitary wave. From the measurements of Naheer (ref. 20) it can be shown that:

$$c_{fw} = 0.62 \frac{g.D_{50}}{V_s^2} \qquad (18)$$

The flow parameter ψ_w can be determined from Equations (7) and (16), as follows:

$$\psi_w = \frac{c_{fw}.\ u_{max}^2}{2.g.\Delta.D_{50}} \qquad (19)$$

The model tests show, see Fig. 18, using the method of linear regression, that:

$$n = 7.2\ .\ 10^7\ \psi_w^{6.86} \qquad (20)$$

From Fig. 18 it can be seen that Equation (20) gives a good prediction of the measured transport in the prototype.

49. If a push tow unit is sailing some distance from the bank, the transport predicted with Equation (20) is overestimated. In these cases a second method is recommended which has been extensively described in ref. 6. This method is more complicated and has been based on the calculation of the shear stress distribution occurring under the stern wave. Using the transport relation of Paintal (ref. 21) and assuming a constant effective transport width, B_e, the total transport can be determined by integration over the stern wave length, L:

$$n = \frac{6}{\pi}\ .\ c_v^2\ .\ \frac{1}{V_s} \sqrt{\frac{g.\Delta}{D_{50}^3}}\ B_e \int_0^L \phi\, dx \qquad (21)$$

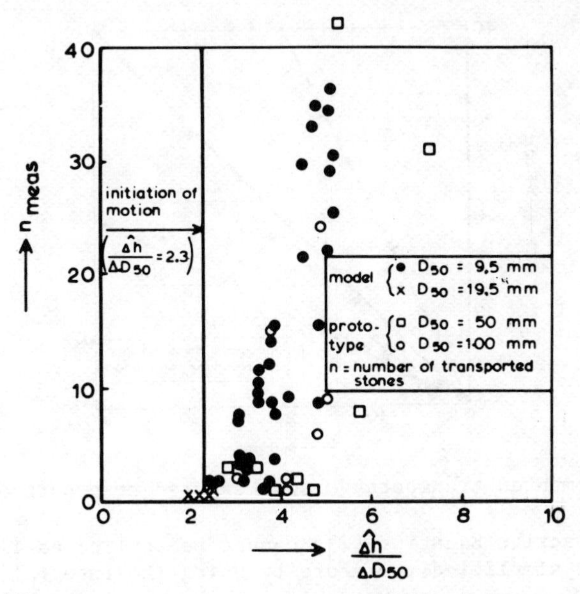

Fig. 17: Measured transport versus stern wave parameter

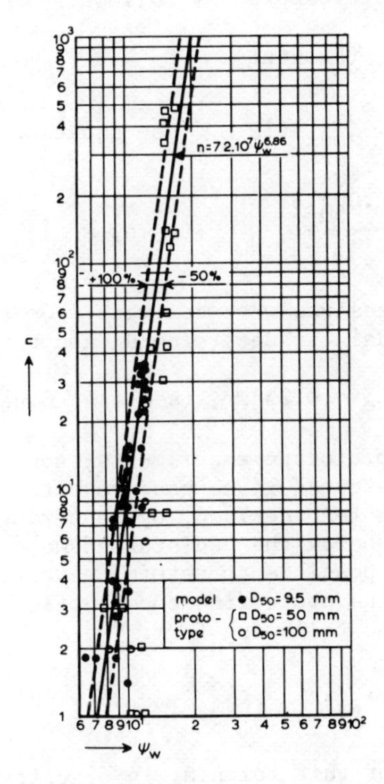

Fig. 18: Transport related to the flow parameter ψ_W

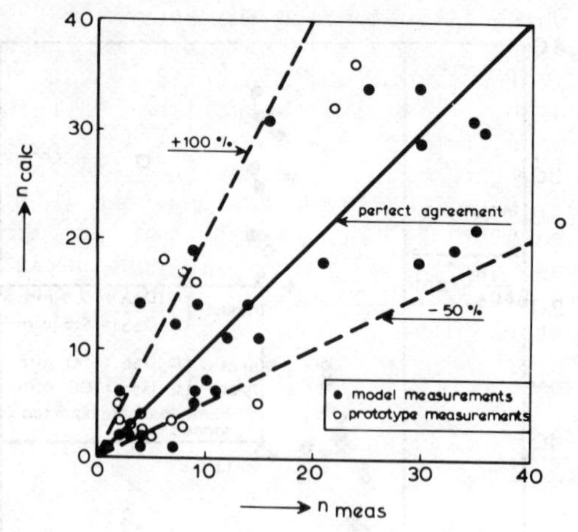

Fig. 19: Computed transport versus measured transport

50. In practice Equation (21) cannot be applied easily and it has been simplified therefore by using the integral value of ${}_0\int^L \phi\, dx = \phi_{max} \cdot L_e$ with ϕ_{max} representing the maximum transport parameter and L_e the effective transport length depending on the length L. The important parameters have been computed as follows:

$$\phi_{max} = 13\ \psi_{max}^{\ 2.5} \qquad \text{(Paintal)} \qquad (22)$$

$$\phi_{max} = \frac{\alpha_z}{2} \cdot \frac{\hat{\Delta h}}{\Delta . D_{50}} \cdot i_{max} \quad (i_{max} : \text{see Equation (5)}) \qquad (23)$$

$$\alpha_z = 1 - \frac{2V_s^2}{g . \hat{\Delta h}} \left(\frac{\Delta . D_{50}}{\hat{\Delta h}}\right)^2 \qquad (0 < \alpha_z < 1) \qquad (24)$$

The following expressions were derived, from the transports measured in the model, to determine B_e and L_e:

$$B_e = 1.8\ \hat{\Delta h} \text{ and } B_e L_e = 0.23\ z_o^2 \quad \text{(see also Equation (6))} \qquad (25)$$

The data from the prototype experiments, see Fig. 19, show that the method presented gives good results.

51. A formula has been derived, by measuring the lowest level of transport below the undisturbed water level. With this formula it is possible to determine the lowest level of the upper part of the protection construction, see Fig. 3. This formula has the following form:

$$\frac{y'}{D_{50}} = 4.4 \left(\frac{\hat{\Delta h}}{\Delta . D_{50}} - 1.2\right) \qquad (26)$$

The applicability of this formula, in practice, is determined by the requirement that the lower part of the protection construction which is subject to the attack of return currents, see Fig. 3, must be stable against the stern wave attack. So, if the D_{50} of the protection material on the lower part and $\hat{\Delta h}$, the maximum water level depression caused by the stern wave, are known, the lowest level, y', of the upper part below the undisturbed water level can be computed.

52. Secondary waves. A start has now been made to investigate the stability and transport related ship-induced secondary waves, see ref. 7. There is considerable information in literature about the stability of rip rap against the attack of waves perpendicular to the slope. In this context use has been made of the Hudson Formula (ref. 22):

$$\frac{H}{\Delta . D_{50}} \leqslant (K_{RR} \cdot \cot g\alpha)^{1/3} \cdot S_f^{1/3} \tag{27}$$

A value of 2.2 is given for K_{RR} in circumstances of breaking waves with heights less than 1.5 m. Assuming the value of K_{RR}, valid for the secondary waves, Equation (27) indicates that stability in the prototype and model ($\cot g\alpha = 4$, $S_f = 0.65$) is guaranteed if $H/\Delta . D_{50} \leqslant 1.8$. However in the tests, see Fig. 20, it was found that no material was transported for values of:

$$\frac{H}{\Delta D_{50}} < 3.0 \tag{28}$$

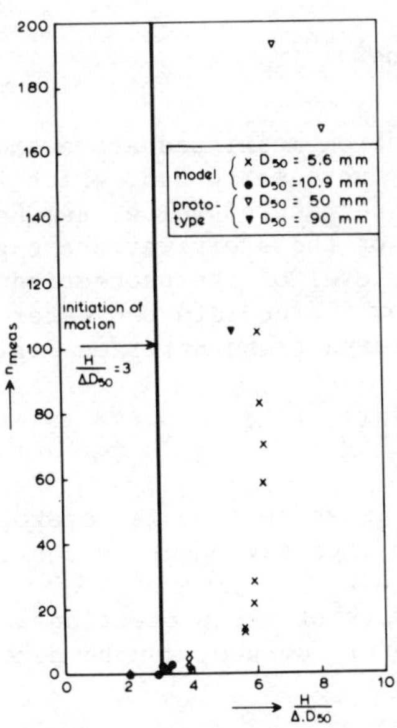

Fig. 20: Transport versus secondary wave parameter

53. The difference between prediction and experiment may be due to several factors including the influence of wave length, L_w, on the stability or the different wave propagation direction. In this respect use has been made of the work of Pilarczyk (ref. 23), in which the following stability criterion for perpendicular wave attack was presented:

$$\frac{H}{\Delta.D_{50}} < N_s \cdot S_f^{\ 1/3} \tag{29}$$

$$\text{with: } N_s = 0.54\ k_E \left(\frac{H}{L_w}\right)^{-0.25}, \ (\text{for } \frac{H}{L_w} \leqslant 0.05\ \text{tg}\alpha) \tag{30}$$

$$N_s = 2.25\ (\text{cotg}\alpha)^{0.5}\ k_E \left(\frac{H}{L_w}\right)^{0.25}, \ (\text{for } \frac{H}{L_w} \quad 0.05\ \text{tg}\alpha) \tag{31}$$

$$k_E = \text{tg}\varepsilon \cdot \cos\alpha + \sin\alpha \tag{32}$$

In Equation (32) ε is the natural angle of repose and can be taken at 45° for natural quarry stones, see ref. 23. From the tests it was observed that, in the critical situation - characterized by $H/\Delta D._{50} = 3$ in Fig. 20 and Equation (28) - H/L_w had a value of 0.08. This means, with cotgα = 4, that $N_s.S_f^{\ 1/3} = 2.5$, which is less than the value of 3.0 given in Equation (28).

54. It can be assumed that this small difference is due to the direction of wave propagation, $\Theta = 54°$. It is recommended that in such cases the wave height should be reduced to $H.(\cos\Theta)^{0.5}$. Substitution of this reduced value of H in Equation (29) gives:

$$\frac{H}{\Delta.D_{50}} < N_s \cdot S_f \cdot (\cos\Theta)^{-0.5} \tag{33}$$

In the critical situation mentioned above the stability value for $H/\Delta.D_{50}$ will now increase to 3.3, which is in good agreement with the experimental value, see Equation (28).

55. As in the case of the stern wave an expression to determine the lowest level of the protection zone against the secondary waves, below the undisturbed water surface, has been derived from the measured transport, see Fig. 21:

$$\frac{y'}{D_{50}} = 3.0\ \left(\frac{H}{\Delta.D_{50}} - 1.5\right) \tag{34}$$

The applicability of Equation (34) is determined by the requirement that no damage may occur in the lower protection zone, see Equation (26).

56. The upper boundary of the protection zone with respect to the undisturbed water surface, can be determined with the wave run-up formula

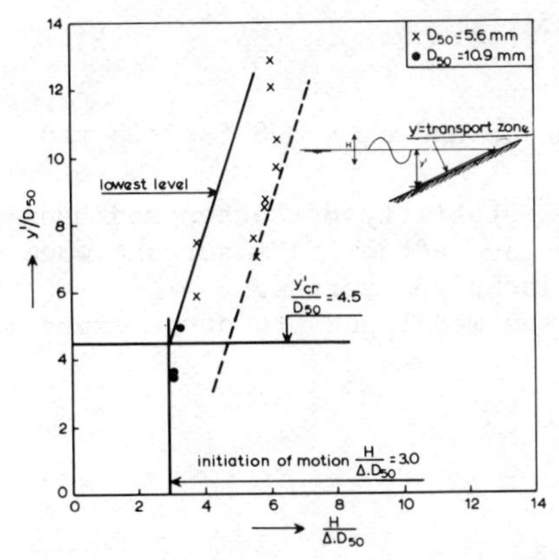

Fig. 21: Lowest level of transport by secondary waves

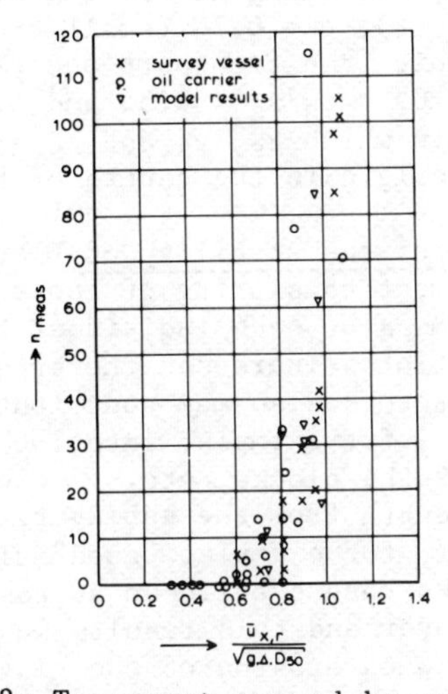

Fig. 22: Transport caused by screw race

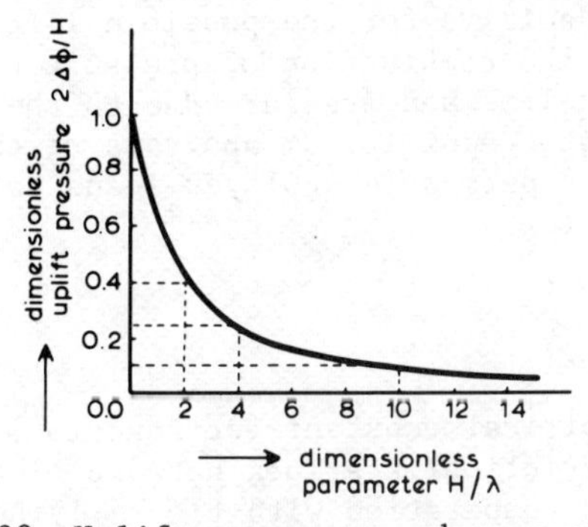

Fig. 23: Uplift pressure under revetment

$$\frac{R_u}{H} = 2\ c_s\ \left(\frac{H}{L_w}\right)^{-0.5}\ tg\alpha\ = 2\ c_s\ \xi\ (\text{for}\ \xi < 3) \qquad .(35)$$

with R_u = wave run-up and c_s = 0.6 for rip rap

57. Screw race. Stability of bottom and bank protection against the screw race attack is important when ships are manoeuvring near locks and berths.
For these situations model and prototype experiments indicate that:

$$\frac{u_{x,r}}{(g.\Delta.D_{50})^{0.5}} < c \qquad (36)$$

in which $u_{x,r}$ represents screw-induced current velocities computed with Equation (10) and c is a constant. From Fig. 22 it can be seen that, according to the experiments (ref. 10): c = 0.55 (no transport); c = 0.70 (small transport).

58. The area in which the transport occurs is characterized by, see ref. 10,: $0.05 < z_s/x_s < 0.35$ and $-0.2 < y_s/x_s < 02$, in which x_s, y_s and z_s are ordinates with the origin in the centre of the screw.

Experimental model for the stability of block revetments

59. A slope revetment consisting of loose blocks derives its strengths from the mass of each individual block. Friction between individual blocks increased the strength of the slope revetment. Other factors also may contribute to the strength of a slope revetment, for example, interlocking between blocks, clenching of the blocks, etc. A slope revetment may also derive its strength from the sublayer. In the case of an impermeable sublayer, for example, "good" clay, the pressure underneath the blocks cannot build-up as easily as in the case of a permeable sublayer and this results in a higher revetment strength. However, when erosion of the clay occurs, for example, "poor" clay, the strength of the slope revetment is reduced. In this case, therefore, the strength of the clay, that is, resistance against erosion, is the weakest link.

60. For wave attack (wind waves or ship waves) the downsurge stage is mostly decisive for the possible lifting-up of blocks, that is, the combination of pressure due to the high level of phreatic line and pressure due to the oncoming wave front. A simplistic equilibrium analysis of the stability of blocks placed on a permeable sublayer leads to the following strength equation:

$$\frac{H}{\Delta.D_b} = \frac{\cos\alpha}{K} \qquad (37)$$

where K is an empirical constant (or function) depending on revetment type (friction/interlock between blocks and porosity of revetment) and cooperation with blocks lying above.

61. In the case of ship-induced loads, for example, transversal stern wave and/or secondary waves, the value of K can be roughly taken equal to 0.20 for free blocks and 0.15 for grouted revetments. However, in the latter case, the stability of the filter and/or sublayer may be more critical. The absolute height of a block must not be less than about 0.10 m for it to retain its stability. More exact relationships on the aspects will probably be available when the recent prototype data have been compiled and evaluated.

62. The upper boundary of a block revetment with respect to the undisturbed water surface is directly related to the wave run-up, which can be computed with Equation (35) taking into account a value of 1.0 for the constant c_s.

Mathematical model for the stability of block revetments

63. A mathematical model has been developed by the DSML for the calculation of pore pressures in the layer underneath a block revetment, ref. 15 and 24. The model is based on the solution of the equation for groundwater flow in the layer underneath the blocks, with leach terms to include the seepage through the revetment. The variation of the phreatic line within the filter layer is included by a simultaneous solution of the mass balance equation for the flow to and from the phreatic surface. By using a finite difference code a realistic representation of the revetment as an alternation of blocks and joints is possible. The permeability may be a function of the local hydraulic gradient, thus allowing for turbulent or semi-turbulent flow. Formulae for flow in narrow joints have been derived from special permeability tests. The geometry of the protection may be rather arbitrary in the model; a succession of different slopes is possible. The hydraulic boundary conditions may also be arbitrary, for example, it is possible to use a tape with measured wave pressures as input for the programme. The programme calculates both pore pressures and the phreatic level as a function of time and place. The following conclusions can be derived from the calculation study: a. The risk of damage to the surface layer decreases with: more permeable revetments and with less permeable or thinner (or even completely absent) underlying (filter) layers; b. The elevation of the mean level of the phreatic surface above its original level increases the more permeable the revetment; however, the pore pressures are then smaller; c. An important parameter for the determination of the quasi-static pressures underneath the revetment is the leach length, defined as $\lambda = \sin\alpha \sqrt{bdk/k'}$, where α is the slope of the dam, b is the thickness of the (filter) layer underneath the revetment, d is the thickness of the revetment and k' and k are the permeability of the revetment and of the filter layer respectively.
A pressure-difference curve is given in Fig. 23 based on a horizontal free water surface, which varies sinusoidally in time, with amplitude H.

RECOMMENDATIONS

64. The design rules, presented in this paper, concern the subsoil, the filter layers, the bottom and bank protection constructions of fairways. As input the predicted values of the water motion near the banks and bottom are needed. Until now both the prediction of the water motion and the developed design rules are mainly based on results of measurements with pushing units. To give wider applicability to the design rules presented it is recommended to: (a) make a verification for more varied circumstances as different ship types, channel cross-sections, subsoils; (b) take the influence of more ships at a cross profile into account as well as a verification of the effect of natural currents; (c) elaborate and adapt the formulas given for speed prediction; (d) study the long term effects on the behaviour of the protections and subsoil; (e) check and possibly improve the relations to determine the dimensions of block revetment; (f) develop critical values for internal hydraulic gradients; (g) develop a 3-dimensional model to calculate the ship-induced water motion, especially near to the bottom and banks.

ACKNOWLEDGEMENTS

65. The authors like to thank Rijkswaterstaat (Dutch Public Works Department) for their permission to publish results of research carried out at the Delft Hydraulics Laboratory and the Delft Soil Mechanics Laboratory. Also the authors wish to thank mr. H.J. Verhey and mr. M van der Wal for their support.

NOTATION

B_e	effective transport width	m
b	waterline width of channel	m
c_{fc}, c_{fr}, c_{fw}	shear stress coefficients	-
c_s	porosity coefficient	-
C(y)	front wave coefficient	m^{-1}
c_v	shape parameter	-
D_b	block height	m
D_n	nominal diameter	m
D_o	effective outflow diameter	m
D_p	propeller diameter	m
D_{50}, d_{90}	characteristic diameters of graded material	m
g	gravitation acceleration	ms^{-2}
H	wave height	m
H_i	height of interference peak	m
$\underline{h}$	water depth	m
$\overline{\Delta h}$	average waterlevel depression	m
$\hat{\Delta h}$	transversal stern wave height	m
Δh_f	front wave height	m
i_f	front wave steepness	i
i_{max}	maximum stern wave steepness	-

i_x, i_y, i_z	hydraulic gradients	-
K_{RR}	stability factor	-
K_{TP}	thrust coefficient of propeller	-
k_D, k_E	coefficients	-
K_s	roughness	m
L	stern wave length	m
L_e	effective transport length	m
L_{OA}	length overall	m
L_{wi}	length of interference peak	m
N_s	stability coefficient	-
n	number of transported stones	-
n_p	number of revolutions	-
O_{95}	characteristic open area in geotextile	m
P_S	installed engine power	W
R_T	total resistance	N
R_u	wave run-up	m
r	radial distance to screw centre	m
S	distance to ship's side	m
S_f	shape factor	-
u_c	current velocity	ms^{-1}
u_{max}	maximum current velocity in transversal stern wave	ms^{-1}
u_r	average return current velocity	ms^{-1}
$\hat{u}_r$	maximum return current velocity	ms^{-1}
$u_{x, r}$	screw-induced current velocities	ms^{-1}
u_o	outflow velocity	ms^{-1}
V_S	ship's speed	ms^{-1}
X	distance from ship's bow	m
x_s	distance from screw centre	m
y	eccentric distance from canal axis	m
y'	lowest level of transport	m
z_o	stern wave coefficient	m
α	slope angle	degrees
α_i	coefficient for interference peaks	-
α_z	coefficient for transversal stern wave	-
Δ	relative density	-
η_D	efficiency	-
ε	natural angle of repose	degrees
Θ	angle of wave propagation	degrees
$\tau, \hat{\tau}$	shear stresses	Nm^{-2}
ϕ, ϕ	transport parameters	-
ϕ, ϕ_w^{max}	shear stress parameters	-

REFERENCES

1. KAA E.J. VAN DE Power and speed of push tows in canals. DHL Publication, No. 216, August 1979

2. BLAAUW H.G. and KNAAP F.C.M. VAN DER Prediction of squat of ships sailing in restricted water. DHL Publication, No. 302, April 1983

3. BLAAUW H.G. and KNAAP F.C.M. VAN DER Prediction of water level depression and squat of ships sailing in restricted water (in Dutch). Report on model investigations, M1115-Part VA, September 1983

4. KAA E.J. VAN DE Hydraulic attack of bank protections (in Dutch). Kust en Oeverwerken, in praktijk en theorie, March 1979

5. WAL M. VAN DER Bottom and bank erosion by the ship-induced return current (in Dutch). Report on model investigations, M1115-Part XB, to be published in 1984

6. KNAAP F.C.M. VAN DER Attack of transversal stern wave on rip rap bank revetments (in Dutch). Report on model investigations, M1115-Part XC, April 1982

7. VERHEY H.J. Attack of secondary ship waves on rip rap bank revetments (in Dutch). Report on model investigations, M1115-Part VI, to be published in 1984

8. GATES E.T. and HERBICH J.B. Mathematical model to predict the behaviour of deep-draft vessels in restricted waterways. Texas, A and M University, Sea Grant College, Report TAMU-SG-77-206, 1977

9. BLAAUW H.G. and KAA E.J. VAN DE Erosion of bottom and sloping banks caused by the screw-race of manoeuvring ships. DHL Publication, No. 202, July 1978

10. VERHEY H.J. The stability of bottom and banks subjected to the velocities in the propeller jet behind ships. DHL Publication, No. 303, April 1983

11. PILARCZYK K.W. Prototype tests of slope protection systems. Conference on flexible armoured revetments incorporating geotextiles, London 29-30 March 1984

12. OLDENZIEL D.M. and BRINK W.E. Influence of suction and blowing on entrainment of sand particles. J. Hydraulics Division, July 1974, 935-948

13. GROOT M.T. DE THABET R. and KENTER, C.J. Soil mechanical design aspects of bank protections (in Dutch). KIVI-Symposium, Delft, 25 May 1983

14. GROOT M.T. DE and SELLMEYER J.B. Wave-induced pore pressures in a two layer system. DSML-mededelingen, Delft, Part 2-4, 1979, 67-78

15. SELLMEYER J.B. Uplift pressures under a block revetment (in Dutch), Internal report DSML, Delft, CO-255780, August 1981

16. GRAAUW A. DE MEULEN T. VAN DER and DOES DE BYE M. VAN DER Design criteria for granular filters, DHL-publication 287, January 1983

17. VELDHUYZEN VAN ZANTEN R. Geotextiles in coast and bank protections (in Dutch) Publication Dutch Association of Coast and Bank protections, Rotterdam, 1982

18. IZBASH S.V. and KHALDRE K.Y. Hydraulics of river channel closure, Butterworths, London, 1970
19. SHIELDS A. Anwendung der Aehnlichkeitsmechanik und der Turbulenzforschung auf die Geschiebebewegung. Preussische Versuchsanstalt für Wasserbau und Schiffbau, Berlin, 1936
20. NAHEER E. The damping of solitary waves. Journal of Hydraulics Research, Vol. 16, no. 3, 1973
21. PAINTAL A.S. Concept of critical shear stress in loose boundary open channels. Journal of Hydraulics Research, Vol. 9, no. 1, 1971
22. HUDSON R.Y. Design of quarry stone cover layers for rubble mound breakwaters. Waterways Experimental Station (WES), Research report No. 2-2, July 1958
23. PILARCZYK K.W. and BOER K. DEN Stability and profile development of coarse materials and their application in coastal engineering DHL Publication, No. 293, January 1983.
24. BEZUYEN A. STEENZET, a model for calculating uplift pressures under block revetment (in Dutch). DSML, Delft, Internal report CO-258901/91, 1983.

Discussion on Session 1: Hydraulic aspects

Mr K. Pilarczyk, Delta Department, Netherlands

Mr Brown's paper is restricted only to the highly permeable toplayers. One may get the impression that this is an ideal solution, replacing the other types of revetments. This, however, is not true. The highly permeable toplayer can only be used on permeable sublayer or rockfill-body (i.e. breakwaters, gravel islands, thick stone layer on sandy subsoil, etc.). Such a highly permeable toplayer system is not advisable for placing directly on a clay or sand.

Of course, it is always possible to place a thick stone layer on clay or sandy subsoil and to apply a permeable sublayer. However, the level of a hydraulic load and the cost analysis have to give the best solution for a particular case.

The idea of a permeable toplayer as a measure to improve toplayer stability is not a new one. There are a number of block-types on the market with high permeability (e.g. Dycel-system, 50% open area; Armorflex, 20%, etc.). However, I don't think that Mr Brown's method can be applied for stability calculations of these systems.

Mr Brown uses Hudson's formula on rock stability as a basis for his derivations; thus Hudson's formula and Brown's criteria give one stability number ($H/\Delta D$) for a particular slope. For the past few years it has been generally known that because of the influence of wave period, the stability of an armour unit is described by a function (and not by a number) for a given slope (see Fig. 1). This is also supported by model and prototype tests on high and less permeable revetments carried out in the Netherlands.

Moreover, for rockfill structures the stability increases with decreasing slope, because the drag forces are dominant. In the case of block revetments the stability can even be less for milder slopes than for steeper slopes, because not the drag forces but the uplift forces are dominant, and support by the blocks lying above increases for steep slopes.

The correction coefficients (C_{BU}, C_{BS}) should be also a function of sublayer characteristics. Filter velocity (and thus also build-up of pressure) can be, for example, expressed by the relationship given by Cohen de Lara (Coefficient de

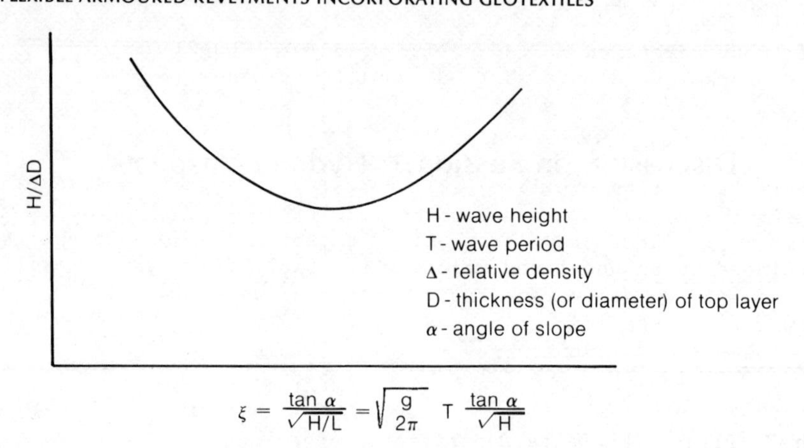

$$\xi = \frac{\tan\alpha}{\sqrt{H/L}} = \sqrt{\frac{g}{2\pi}}\ T\ \frac{\tan\alpha}{\sqrt{H}}$$

Fig. 1.

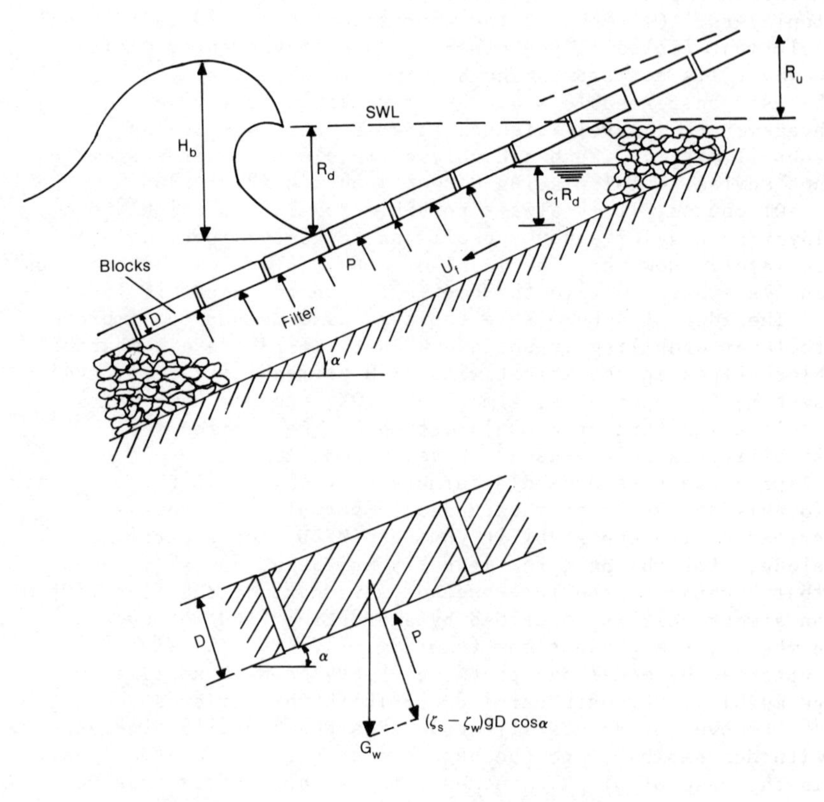

$P = \zeta_w g\, C_1\, R_d$ or $= \zeta_w g\, C_2\, (R_d + R_u + H_b)$

where R_d, R_u, H_b are functions of $\xi = \tan\alpha/\sqrt{H/L_0}$ (breaker index)

and $H/\Delta D = \cos\alpha/C_2\,\xi$

(in this derivation the friction forces between the blocks are neglected)

Fig. 2. Equilibrium conditions for a single block

perte de charge en milieu poreux basé sur l'equilibre hydrodynamique d'un massif. La Houille Blanche, 1955, No. 2.)

$$U_f^2 = 2g \frac{D}{C} \varepsilon^5 i$$

where U_f = filter-velocity, ε = porosity, D = grain diameter, i = gradient and C = resistance coefficient (= f (Reynolds number)). In Mr Brown's stability criteria the wave height is replaced by an undefined velocity (probably for test purposes: to be able to apply flow tests instead of wave tests). However, if one wishes to use these criteria one must know how to translate this undefined velocity into a real wave height. In my opinion, the general application of these criteria is rather doubtful.

A more simple black-box approach to this problem can be illustrated by equilibrium conditions for an individual block as shown in Fig. 2. However, the best approach will be one based on an analysis of all physical factors involved in the phenomena under consideration.

Mr C. T. Brown

Mr Pilarczyk makes a number of assertions about the content of the paper which, if true, are only so because of severe pruning to fit within the confines of the paper and presentation.

In the original consideration of permeability of armour layers, the effect of variations of surface porosity and underlayer porosity were considered. A literature review in 1978 showed a variety of block type armour systems showing porosity of between 10% and 60%, and it was apparent that low porosity units were less efficient on a material usage basis, as well as having higher run-up and reflection characteristics.

Contrary to Mr Pilarczyk's assertion, the theoretical equations (refs. 4, 5 and 6 of the Paper - known as the Blanket Theory) used to examine the relationship between element size, layer thickness and water motion were developed from first principles using the momentum theory as applied to a water jet of variable direction. Previous analyses had, for some reason, always ascribed a horizontal direction of motion to the jet, or not considered its direction at all. Evaluation of the terms of the equation demonstrate three modes of failure:

(a) upslope sliding
(b) downslope sliding (and slope instability)
(c) outwash.

Evaluation of the angular function of the outwash case, shown here in Figs 3 and 4 (taken from ref. 6) shows for each particular slope and block system a parabolic stability

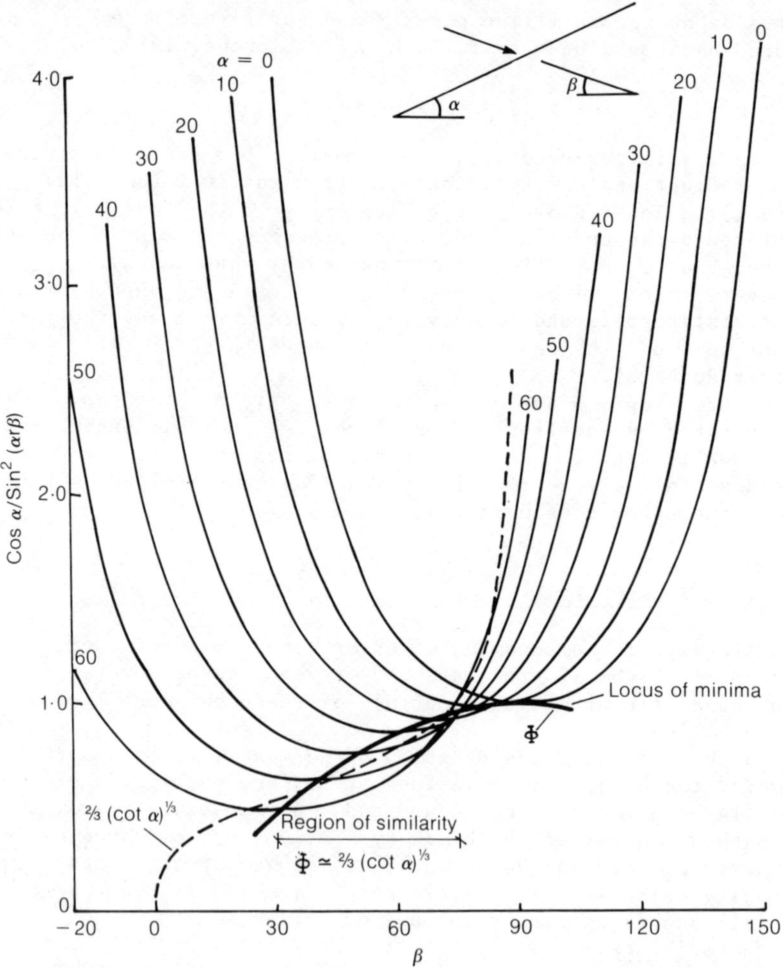

Fig. 3. Variation of uplift angle function with flow angle

relationship between wave height (jet velocity) and the joint angle function derived from jet angle and the direction of incipient motion of particle. For a given block system, there is a direct relationship between this angle and the slope angle. Plotting the minimum values of these curves gives a curve of the locus of minima which agrees very closely with the experimentally derived relationship reported by Hudson.

A review of these references will show that the effects of wave period and bed slope, etc., as they affect the direction of motion and velocity of the jet, are essentially described in the derivation, and then, through lack of data, simplified to a single coefficient.

However, for preliminary design purposes, it is wise to assume that the critical wave conditions will occur and that the minimum stability conditions of the revetment or armour

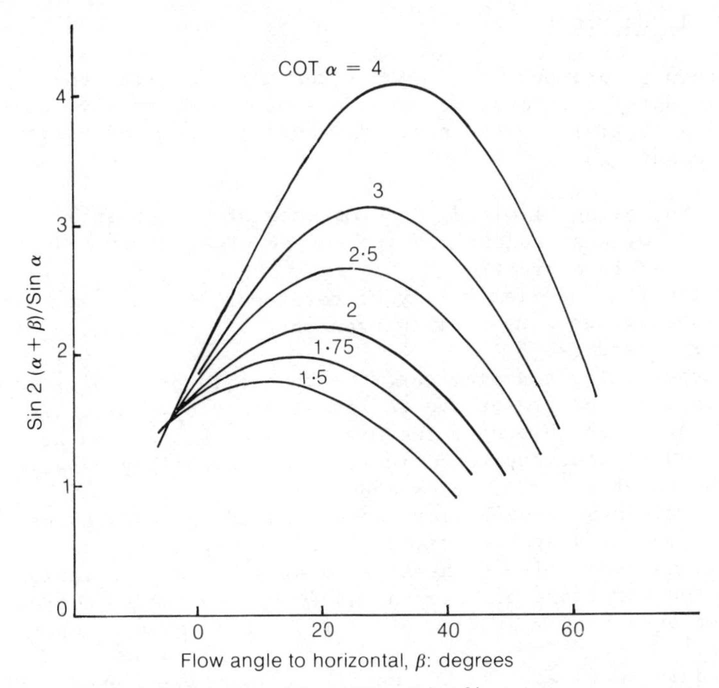

Fig. 4. Variation of function $\frac{\sin 2(\alpha+\beta)}{\sin \alpha}$ versus flow angle

system will be reached. In this condition, as described above, the curve of interest is the locus of minima of the parabolic stability curves, and it is found that the minimum surcharge for the slope can be expressed in terms of the incident wave height, the relative density and the joint angle function:

$$R(1-P)\Phi(\alpha) \propto H/(Sr-1)$$

Due to the close similarities between this curve and the well known empirical Hudson Equation, it seems quite natural to present this more general expression in terms of Hudson's Equation. Outside the area of plunging waves, the formula suggests that a reduction of over 50% of armour surcharge may be made for a suitable block, and this has been confirmed by model studies of several projects (see ref. 9 of the Paper).

In the expression of paragraphs 21 and 22 of the Paper, the velocities are the velocity of the water in the jet. A transformation route from a particular wave height is given in ref. 4 of the Paper. In paragraphs 31 onwards the velocity is the mean velocity of flow. Kolkman gives a good examination of typical velocity profiles in apparatus similar to that used in the experiments. (Kolkman P. A. An artificial ceiling for free surface flow reproduction in scale modelling of local scour, Int. Conf. on hydraulic modelling of civil engineering structures, Coventry, England, September 1982.)

Ir M. T. de Groot

Mr Brown points out in his paper that "The ability to accommodate peripheral scour is one of the chief advantages of tensile flexible revetments...". Indeed, the flexibility has the advantages of

(a) following (limited) deformations of the subsoil
(b) showing at an early stage any deformation or change in the slope profile.

To rely upon the flexibility of revetment layers only, without considering the causes of changes in the slope profile, is rather hazardous.

Suppose, for example, that the change in profile is caused by migration of grains due to wave-induced hydraulic gradients. In this case the migration of grains can be expected to continue, and finally the flexibility capacities of the revetment will be exceeded.

So, although flexibility can be considered as a necessary condition for a stable design, it certainly is not a sufficient one. It is therefore desirable to gain insight into the magnitude of internal wave-induced loads in order to design such that migration of grains can be expected not to occur.

Requirements will be put on filter layers and geotextiles concerning tensile strength, permeability, thickness, and how far towards the toe of the slope the protection is to be applied. It is preferable to provide against scour in combination with the advantages of flexible revetment layers properly observed by the author.

Mr C. T. Brown

I concur with Mr de Groot in describing flexibility as an insufficient condition for stable design. It does, however, render the onset of substrate failure more noticeable and is hence a desirable quality, as is a thick aquifer between the armour surface and the filter means.

Mr J. Nomes, N.V. Uco, Belgium

Would Mr Mühring please explain how geotextile sacks at the bottom of the protection are joined together linearly along the embankment. How was the filling of the mats achieved?

Dipl. Ing. W. Mühring

The geotextile sacks were sewn to the filter mattresses. The filter mattresses with the fixed bags were laid to the toe of the canal using a pontoon with a chute (steel sheet) at an inclination of 1 : 3. The mattress was lying on the chute of the pontoon, and the pontoon was torn in the direction of the canal axis so that the mattress was slipping over the chute to the toe of the canal. While the mattress was on the pontoon it was filled with the permanent flexible and erosion resistant artificial clay. The sack was cut so that the pump hose could be inserted and the material be pumped in. After the sack was completely filled, the hole was sewn.

Mr N. Ordman, Noel Ordman & Associates Ltd., London

What is the meaning of the term 'chemophysical' as applied to the material used to fill the sacks used for toe protection (paragraph 36 of Mr Muhring's paper)?

Dipl. Ing. W. Mühring

The material used to fill the bags is prepared by using chemically and physically treated raw materials. The properties of the filling material are permanent elasticity, erosion resistance against screw jets, pumpability, and self compaction without mechanical energy.

Erosion resistance is required because anchors may damage the sacks. If the material is not erosion resistant against screw jets, material would be sucked out. The erosion resistance is tested according to the testing regulations for filter mattresses. The material is exposed to a high turbular current which simulates the screw jet. After six cycles of 30 min no material should be lost.

Mr H. Blaauw

The prediction of scour holes due to the propeller action of manoeuvring ships presented in Mr Oebius' paper seems quite satisfactory. Generally speaking, these scour holes will only arise when ships use their propellers repeatedly at the same location in the same direction (e.g. ferries). In these cases protection might be required (see Paper 4, paragraph 57) especially if the stability of e.g. quay walls is endangered. The time factor is important when considering scour holes. Final results of model tests may properly represent corresponding situations in prototype, but the time in which a final condition is reached differs considerably due to scaling effects. This time aspect is important: if the time in reality is larger than the scaled up model time, problems in reality might be less than expected because the scour develops more slowly.

Dipl. Ing. H. U. Oebius

There is no question that the time factor plays an important role in scouring processes, as mentioned in paragraph 13 of my paper and shown schematically in Fig. 7. Unfortunately, there is no way of predicting the historical development of scours, which severely impedes the estimations of local damage at given times. One therefore has to rely on boundary-value considerations.

It is also known that there may occur scale effects between the results from model tests compared with those gained under natural conditions, among others concerning the time factor. Mr Blaauw apparently assumes that the results given in the paper are based on scaled experiments, and under these circumstances his remarks are absolutely correct. But if one assumes that the experiments were not scaled, which in fact was my intention, and that the results are universally valid as long as they are based on hydrodynamically similar processes, which in connexion with ship propulsion systems seems to be acceptable, scale effects lose their importance. This applies also to the time factor, although the considerations mentioned before even then have to be taken into account.

Mr H. Blaauw

In Mr Mühring's paper, the protection of the banks of inland navigation fairways is extended such that it includes the toe of the protection. The dominant attack is generated by ships, especially the return current occuring in the lower areas of the wetted cross-profile. Transport of granular materials due to the return of the current (and, to a lesser extent, the wake) will occur. However, a certain transport of granular material (giving rise to a deformation of lower parts of the banks and bottom) is acceptable. In any case, care should be taken that the lower part of the construction is able to follow local deformations of the subsoil due to migration before an equilibrium condition is established. A protection layer applied only at the upper part of the bank provides a far less expensive solution.

Dipl. Ing. W. Mühring

The extension of the bank protection of a waterway is a question of stress from currents in the wetted cross-profile. A waterway cross-sectional area about 7 times the cross-sectional area of a fully loaded typical vessel requires the extension of the revetment into the canal bed. How far the revetment should be extended into the canal bed depends on the transport of the granular material. If the transport causes permanent depression, the entire bed must be paved when it is not possible to increase the waterway cross-sectional area considerably.

When the ratio of the cross-sectional area of the waterway to the section of the ships is much larger than 7, it may be sufficient to protect only the upper part of the bank.

Mr H. Blaauw

The Pianc working group on design of bank protections will mainly deal with and comment on existing design rules in various countries, to be able to obtain some standardization. A great deal of these can be found in the proceedings of this conference. It is worthwhile to mention that a working group also dealing with the design aspects of inland navigation fairways is already operational and recently had its first meeting. This working group is dealing with probabilistic design. The main aims of this working group are to formulate probabilistic design approaches including risk-analysis also of events which may not be expressed in terms of load and strength. It should be studied whether the application of such methods would lead to more cost-optimal solutions, with a defined reliability, or to the possibility of optimalization of probability of failure (Fig. 5).

When designing a protection the total expected cost during the lifetime of a construction has to be taken into account to generate cost-optimal solutions. To be able to apply a probabilistic approach, several transfer functions are required. Can Mr Van der Knaap say how much progress has been made to determine these transfer functions which give the relation between loads on the one hand and transport on the other?

Mr N. Ordman

In the papers and discussion there has been no reference to costs. An important consideration in the selection of revetment systems is cost effectiveness. It is necessary to have regard not only to initial cost and maintenance expenditure but also to the real cost of interrupting or interfering with the normal operations of the waterway.

A related consideration is the definition of failure. It is desirable to have a clear understanding of what constitutes failure. Deformation of bank profile or even the displacement of revetment constituent materials do not necessarily constitute failure if they do not cause a deterioration of function. Functional failure is an essential concept, particularly in the context of cost effectiveness.

An appreciation of these concepts has a place not only in design and construction, but also in research.

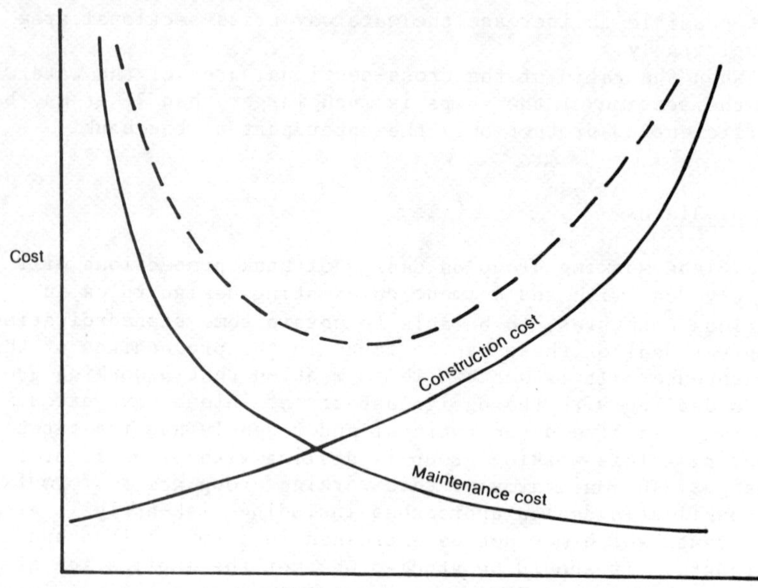

Fig. 5.

Mr K. Pilarczyk

Definition of failure and the need for proper stability criteria regarding revetments are important aspects. Clear distinction has to be made between failure and stability criteria for riprap and for revetments. In the case of thick layers of riprap, there is nearly 50% stability reserve between the stage of beginning of movement and the stage of total failure (between these two stages the profile development will take place).

In the case of block revetments the margin between the beginning of block instability and lifting out is rather small. Moreover, if the block is lifted out, the sublayer will become immediately exposed to the wave and current attack and, very often, the failure of the total slope will follow.

Mr F. Van der Knaap

A programme has been started at the Delft Hydraulics Laboratory (DHL) to study the probabilistic approach. The basis for this approach is formed by the deterministic method presented in our paper. For example, the transport relations derived from model and prototype tests will function as the input for probabilistic design. It seems possible to compute the damage allowed as well as the future maintenance costs with these kinds of relationships. The DHL probabilistic design programme started last year, and will continue over the next couple of years.

Mr B. J. Hook, C. H. Dobbie & Partners, London

The protection of the toe of a flexible revetment against scour should also incorporate a robust construction to withstand damage caused by dredging operations. Clients are becoming more conscious of the need to reduce maintenance costs by providing a better job in the first instance. Vital elements of construction, such as the toe of a revetment, should only be subjected to severe cost effectiveness studies if the realistic costs of carrying out repairs, disruption and future maintenance are known.

Mr D. P. Brady, Salford University

Are there any detrimental streamflow changes caused by flow over revetments? Does flow over the hollow revetment units induce any vortex action within the hollow which would erode the bedding layers?

Mr F. Van der Knaap

In our paper the return current velocities induced by ships are treated. These current velocities form the determining attack on the bottom and the lower parts of the revetments. Test have been carried out to determine design rules. During these tests both return current velocities and the amount of transported material were measured. Transport relations have been derived from these measurements. The secondary (turbulence) effects are included in the coefficients of these relations.

Prof. D. M. McDowell, Peter Fraenkel and Partners, London

The papers in this session are the only ones to be concerned with scour caused by ship propellers. They present theories and results of experiments which demonstrate the damage that propeller jets can do. However, there is little comment on engineering solutions for this very important problem. The units presented by Mr Brown and used for protection of coastal revetments appear to offer the possibility of protection. Do the authors have any comments on this or other engineering solutions?

Mr C. T. Brown

The entrainment relationships for currents described in my Paper are for unconcentrated flow in a rectangular flume. Placing constrictions or weirs in the flow, giving the effect of jets directed at the bed, considerably lowered the

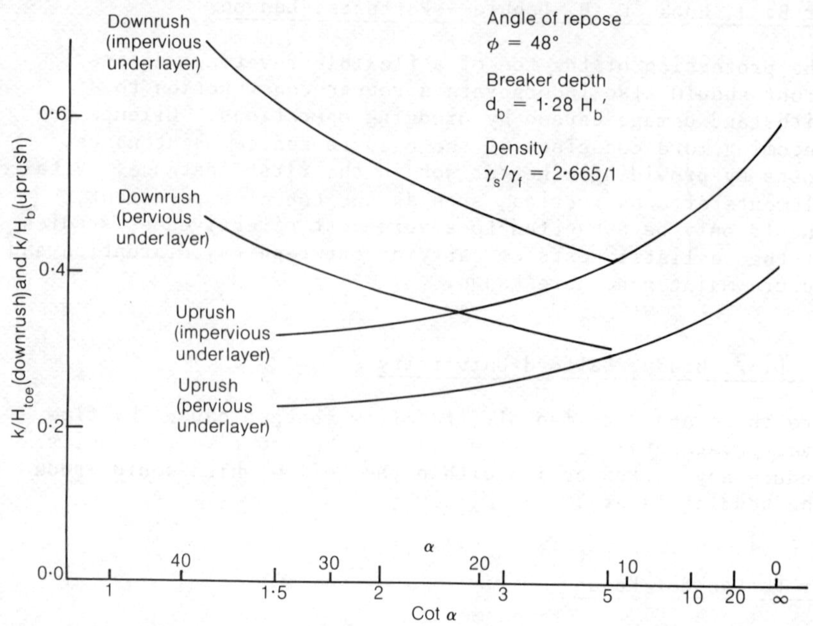

Fig. 6. Block size

threshold of entrainment. Reference to Hansen and Keats (ref. 8 of the Paper) should be made for experimental results.

Mr P. A. Hedar, Goteborg, Sweden

I would like to present a diagram relevant to the discussion between Mr Brown and Mr Pilarczyk concerning the stability of the armour layer (Fig. 6). The diagram has been published before, but is presented here in an updated form.

Notation:

k = mean diameter of the block
Q = weight of an individual block
H = design wave height
γ_s = density of rock
γ_f = density of fluid
ϕ = angle of repose (degree of interlocking)
α = angle of armour slope
e = Napier's number

$Q = \frac{\pi}{6} \gamma_s k^3$ (the blocks equal spheres, k is a fictitious diameter of the block)

UPRUSH

Pervious underlayer:

$$k = \frac{0.33\ (d_b + 0.7H_b)(\tan\phi + 2)}{\left(\frac{\gamma_s}{\gamma_f} - 1\right)\left(3.6 - \frac{1}{e^{4\tan\beta}}\right)\cos\alpha(\tan\phi + \tan\alpha)}$$

Impervious underlayer:

$$k = \frac{0.41\ (d_b + 0.7H_b)(\tan\phi + 2)}{\left(\frac{\gamma_s}{\gamma_f} - 1\right)\left(3.3 - \frac{1}{e^{4\tan\beta}}\right)\cos\alpha(\tan\phi + \tan\alpha)}$$

DOWNRUSH

Pervious underlayer:

$$k = \frac{2.3\ H_{toe}\ (\tan\phi + 2)}{\left(\frac{\gamma_s}{\gamma_f} - 1\right)\left(e^{4\tan\beta} + 13.7\right)\cos\alpha\ (\tan\phi - \tan\alpha)}$$

Impervious underlayer:

$$k = \frac{3.7\ H_{toe}\ (\tan\phi + 2)}{\left(\frac{\gamma_s}{\gamma_f} - 1\right)\left(e^{4\tan\beta} + 16.5\right)\cos\alpha\ (\tan\phi - \tan\alpha)}$$

where $\beta = \alpha + 48^\circ - \phi$

Mr H.U. Oebius

Satisfactory approaches have been developed concerning the erosion depth, the scour length and width, and the location of the scour relative to a propulsion system acting in parallel to channel beds or embankments. These approaches are empirically based and only require the computation of the maximum velocity u_o at the nozzle or the propeller plane, as well as the determining diameter D_o. The velocity u_o can easily be obtained from equation 18 of Paper 4 of this conference. D_o is available from equation 7 of my paper. The historic development of the scour depth has been found to depend on the maximum velocity u_o, the time of influence T_k, the distance of the nozzle (propeller plane) from the bed

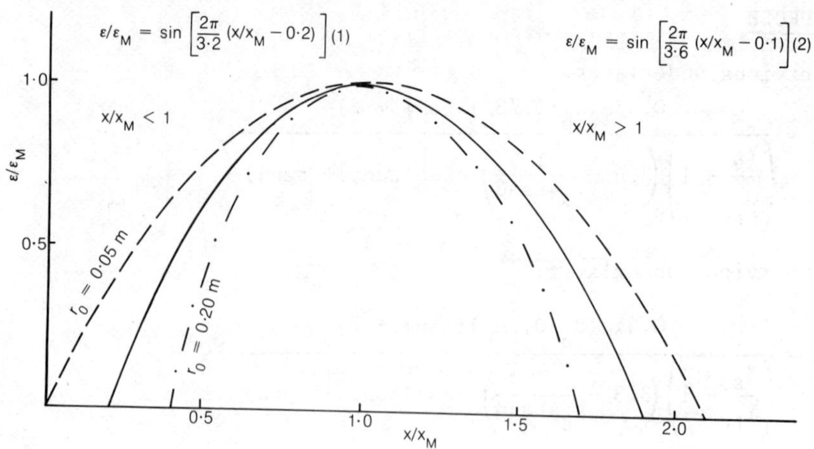

Fig. 7. Dimensionless scour profiles

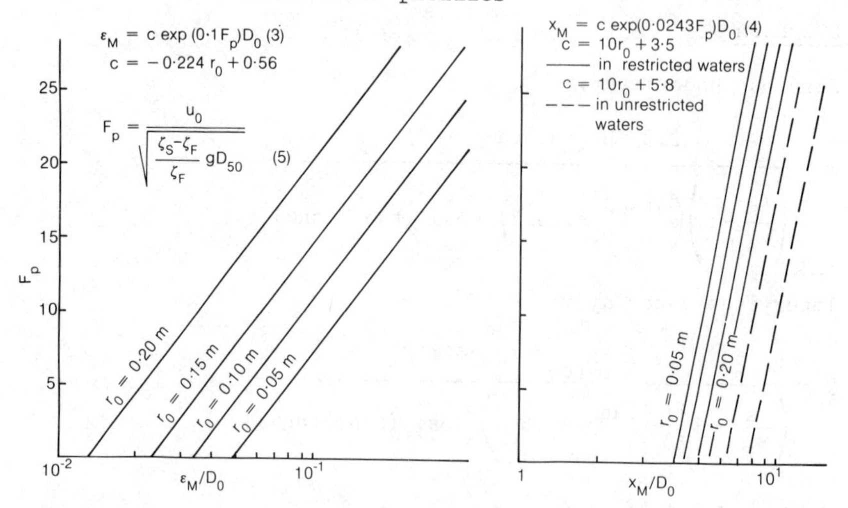

Fig. 8. Maximum scour depth ε_M and location x_M

$r_o = z - D_o/2$, and the sediment specific parameter D_{50}. Although it is known that the time dependent increase of the scour depth follows a logarithmic law it has not been possible to unite all these parameters, which have equal influence, into only one equation.

From the similarity of jet-induced scours, which show the same contours as computed curves of equal shear stresses in propeller jets, it can be anticipated that a dimensionless presentation of scour profiles, taken in the centre line of the propeller jet, should fit into one sinoidal curve. The result of an evaluation of all test series is presented in Fig. 7 (equations 1 and 2). It demonstrates a very good agreement as far as scour depths are concerned, but remarkable deviations regarding the beginning and the end of the erosions, due to the influence of the distance of the propeller from the soil surface.

Table 1. Mattress characteristics

	Type	Weight: g/mm^2	Mesh 0 90	Permeability
Top fabric	Woven PP + PP felt	520	300 μm	$i_1 \simeq 10$
Base fabric	Woven PP	520	350 μm	$i_1 = 11.4$

Fill material: coarse sand, D_{50} = 400 μm
Mattress height average: 0.20 m
Mattress weight average: 3.20 kN/m^2 (dry weight)
Level of pressure meters under mattress: NAP − 0.273 m
NAP + 0.101 m -A-
NAP + 0.567 m

Table 2. Some preliminary results

Run No.	Type of barge	Speed: m/s	Water level ref. NAP:m	Water level depression:m	Uplift pressure -A- (MWC)
8814	Pushed	4.50	+ 1.54	0.57	0.02
8826	Pushed	4.38	+ 0.50	0.34	0.14
4528	Tug	5.49	+ 0.55	0.45	0.30

Table 3. Factors of safety to uplift

Run No.	Weight of mattress: kN/m^2	FOS
8814	4.0	20
8826	2.0	1.42
4528	4.0	1.33

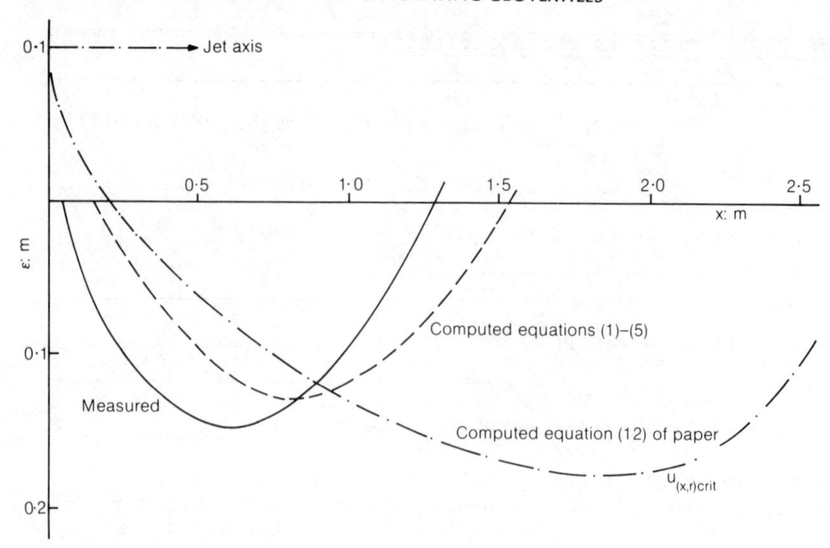

Fig. 9. Measured and computed erosions

The maximum scour depth ε_m (equation 3) and its location off the propeller plane x_m (equation 4) as functions of a specific densimetric Froude number F_p are given in Fig. 8. Together with equations 1 and 2 they now allow the computation of the form, location and maximum depth of propeller jet induced scours. Fig. 9 shows an example for a typical measured erosion, its computation after the equations 1 to 5 and a computation after equation 12 of the paper. It demonstrates a good harmony of measured and computed scour depths in all three cases, and a fairly good one concerning measured erosion profiles and those computed according to the empirically derived equations, but an unexpected failure of the computation of the profiles after equation 12, although here the additional forces at the scour slopes due to the percolation and the gravity forces had also been considered. This means that the hydrodynamic processes and conditions at the moment of contact of a propeller jet with the bottom of waterways are not yet quite understood.

5 Geotextiles as filters beneath revetments

T. S. INGOLD, BSc, MSc, PhD, DIC, FICE, FIHE, FASCE, MSocIS (France), FGS, Ground Engineering Division, Laing Design and Development Centre

SYNOPSIS. In the design of geotextile filters to prevent excessive loss of soil fines under conditions of alternating turbulent flow there are two schools of thought. The first prescribes the positive retention of the smallest particle size by the geotextile while the second attributes the soil with a self-filtering capability. This implies that only the larger size soil particles need be retained. These design concepts are explored and filter design criteria are developed. In addition to the problem of erosion and suffusion caused by flow of water through the bank soil there is the hazard of pumping which may be induced in the bank soil through dynamic hydraulic loading of the revetment. This problem is defined and the results of recent research work are presented.

INTRODUCTION

To give adequate protection to banks in soils susceptible to erosion, it is vital that the revetment incorporates a suitable filter system. Traditionally such filters have been constructed using aggregates with specific gradings. For problem soils it is often necessary to employ a multilayer filter system. Over the last decade aggregate filter systems have been largely superseded by geotextile filters. Although the mechanism and design of geotextile filters under steady-state flow conditions is well researched and documented, there is a dearth of information relating to geotextile filter performance under the alternating and turbulent flow conditions often prevailing beneath a revetment. Filter requirements are investigated and a theory is developed to relate bank soil grading and coefficient of uniformity, U, to the geotextile pore size O_{90}.

DESIGN CONCEPTS

When geotextiles are employed beneath revetments they must operate as effective filters under conditions of alternating or turbulent flow caused by vessel wash and propeller action. One school of thought prescribes to the notion that even under severe hydraulic conditions it is adequate to retain only the coarse soil fraction and that this will then act in association with the geotextile to retain finer particles in the main body of the soil mass. A second school maintains that complete retention of soil fines can only be achieved if the geotextile pores are fine enough to retain the smallest particles of the soil to be protected. Both concepts can be quantified by determining the relation between the characteristic pore size of the geotextile, defined as O_{90}, and other particle sizes of the soil defined as d_n.

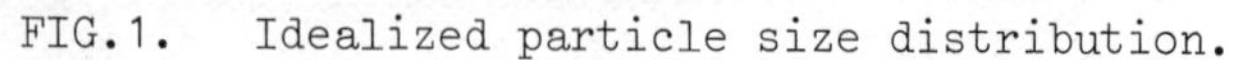
FIG.1. Idealized particle size distribution.

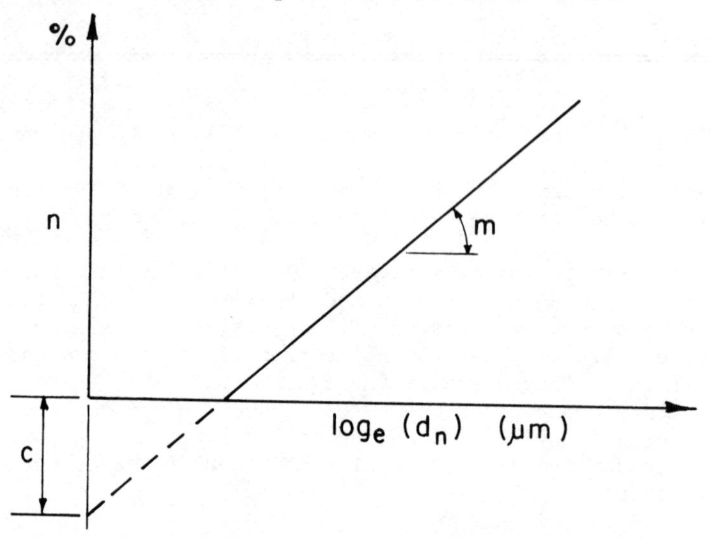

As an idealisation the particle size distribution curve of a soil can be considered as a plot of percentage passing, n, on an arithmetic scale against particle size d_n on a logarithmic scale. In the simplest form a particle size distribution may be taken to be a straight line. If it is assumed that this line has a slope m and an intercept c, Fig.1, then the equation of the simplified particle distribution is

$$n = m \ln (d_n) + c \qquad .....(1)$$

The slope m can be defined in terms of increments of n and ln (d_n) :-

$$m = \Delta n / \Delta \ln (d_n)$$

Taking Δn from n = 10% to n = 60% gives Δn = 50%. Similarly, taking $\Delta \ln (d_n)$ from $\ln (d_{10})$ to $\ln (d_{60})$ gives $\Delta \ln (d_n) = \ln (d_{60}/d_{10})$ which by definition equals ln (U) where U is the conventional coefficient of uniformity. From this it follows that $m = 50/\ln(U)$ whence equation (1) becomes:-

$$n = 50 \ln(d_n)/\ln(U) + c \qquad .....(2)$$

The constant c may be evaluated by substituting a value of n. In this case take n = 50. On evaluating the intercept constant c and rearranging equation (2), an expression is obtained for d_n.

$$d_n = \exp\left[(n/50-1)\ln(U)+\ln(d_{50})\right] \qquad .....(3)$$

THE NOTION OF POSITIVE RETENTION

Now to constrain a certain particle size d_n it has been suggested, (1), that the particle should be larger than the characteristic pore size of the geotextile, O_{90}. Expressed in mathematical terms $d_n \geq O_{90}$. In the limiting case the particle size to be retained would be equal to the pore size, $d_n = O_{90}$. This pore size can be related to the d_{50} of the soil to define a coefficient O_{90}/d_{50} which can be expressed in general terms by using equation (3).

$$O_{90}/d_{50} = (d_{50})^{-1} \exp\left[(n/50-1)\ln(U)+\ln(d_{50})\right] \qquad .....(4)$$

The variation of O_{90}/d_{50} with the coefficient of uniformity U is shown in Fig.2 for a range of values of particle size, d_n, by Teindl [1] during his investigation of alternating turbulent flow.

FIG.2. Positive retention criteria.

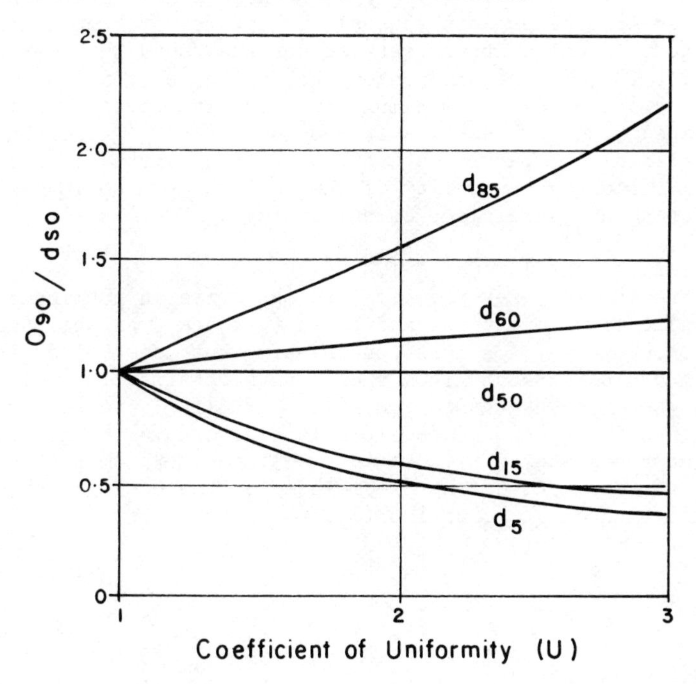

As can be seen use of the criteria in Fig.2 is very onerous from two points of view. First it leaves the designer to select what particle size is to be retained. Secondly, it is found that unrealistic O_{90} sizes are obtained. There is some guidance on the first point from Ogink [2] who recommends $O_{90} \not> d_{15}$. So suppose, for example, that a silt with d_{50} = 10μm and U = 3 is to be protected such that the d_{15} particle size is to be positively retained. Reference to Fig.2 gives $O_{90}/d_{50} = 0.46$ hence $O_{90} = 4.6\mu m$. In practice the smallest O_{90} available is at least an order of magnitude higher than this value. Additionally, a geotextile with such a low O_{90} is likely to have a low permeability which would allow destructive hydraulic over pressures to generate.

It is apparent that in the formulation of such a theory of positive restraint the coefficient of uniformity is used as a mathematical ploy and does not represent the likely behaviour of the soil. This is borne out by inspection of Fig.2 which implies that as a soil becomes more well graded, that is as U increases, there is need for a smaller O_{90} to retain a given d_n when $d_n < d_{50}$. Although it is true, according to the mathematical model, based on a given d_{50} the absolute value of d_n, ($d_n < d_{50}$), will become smaller as U increases, the self-filtering capability of the soil will also increase. What is thought to be a more realistic assessment comes from alternating turbulent flow research on slotted and geotextile well screens carried out at Ground Engineering Limited, [3, 4]. This indicates that as the coefficient of uniformity increases the maximum particle size to be positively retained also increases. It was found that significant loss of finer soil particles was prevented by virtue of the internal filtering capability of the soil which also increases with increasing coefficient of uniformity.

A PRACTICAL FILTRATION CRITERION

There are very few practical design criteria published for dynamic flow conditions and those that are available tend to be conservative. A less conservative but nonetheless safe approach can be extended from the results obtained for well screens where the hydraulic conditions are likely to be more severe than those encountered in inland waterways. The crux of the approach stems from the idea that the maximum particle size to be retained, d_n, can be related to the coefficient of uniformity and d_{50} by equation (5).

$$d_n = 2\, d_{50}\, U^{(n/50-1)} \qquad \text{....(5a)}$$

where

$$n = 100\,(1 - 1/\sqrt{2U}) \qquad \text{....(5b)}$$

Combination of equations (5) leads to equation (6).

$$d_n = 2\, d_{50}\, U^{(1-\sqrt{2/U})} \qquad \text{.....(6)}$$

Now if d_n is the maximum particle size to be retained, then $d_n \geqq O_{90}$ which leads to equation (7).

$$O_{90}/d_{50} = 2\,U^{(1-\sqrt{2/U})} \qquad .....(7)$$

As always with any soil retention problem a check must be made on the relationship between the O_{90} pore size and a large particle size such as d_{90}. In this case the relationship between O_{90} and d_{90} can be expressed in the form of equation (8).

$$O_{90}/d_{90} = 2U^{(0.2 - \sqrt{2/U})} \qquad .....(8)$$

Equation (8) is valid for U>5 and the resulting variation of O_{90}/d_{90} with U is shown in Fig.3. For U<5 the value of O_{90}/d_{90} is taken to be nominally unity. For cohesionless soils containing more than 50% by weight of silt the O_{90} value is limited to 200µm as prescribed by Calhoun [5].

FIG.3. Variation of O_{90}/d_{90} with U

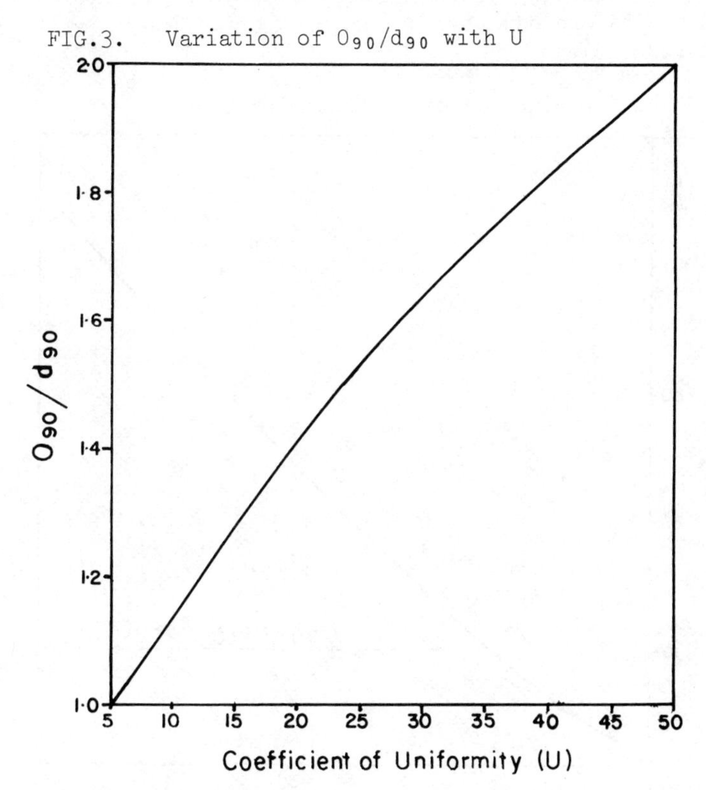

COMPARISON OF FILTER CRITERIA

The filter criteria published by Heerten (6) state that for cohesive soils

$$O_{90} < 10\,d_{50} \qquad ....(9a)$$

and

$$O_{90} \leqq d_{90} \qquad \text{.....(9b)}$$

and

$$O_{90} \leqq 100\mu m \qquad \text{....(9c)}$$

Criteria 9a and 9b can be represented by equation (10).

$$O_{90}/d_{50} = U^{0 \cdot 8} \qquad \text{....(10)}$$

This is plotted in broken line in Fig.4 up to the cut-off of $O_{90}/d_{50} \not> 10$. Without this restriction O_{90}/d_{50} would increase as depicted by the dotted line. The corresponding relationship defined by equation (7) is shown in solid line in Fig.4. Concerning the limitation imposed by criterion 9b, that is $O_{90}/d_{90} \not> 1$, it can be seen from Fig.3 that this limitation is exceeded by the proposed criterion represented by equation (8).

FIG.4. Variations of O_{90}/d_{50} with U

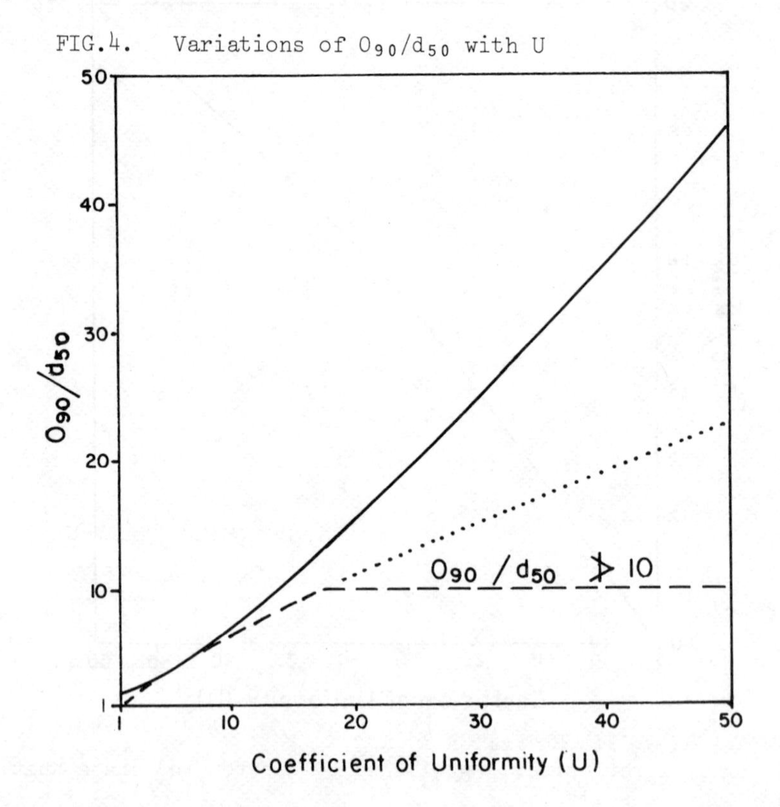

To compare the criteria defined by equations (7) and (9) it is useful to plot out O_{90} against U for a specific value of d_{50}. This is shown in Fig.5 for $d_{50} = 10\mu m$. As can be seen Heerten prescribed smaller values of O_{90} than equation (7) with there being a cut-off at O_{90}= 100μm. Equation (7) renders much higher values of O_{90} which increase steadily until U = 50 which is a cut-off point. This follows from limiting O_{90}/d_{90} to 2 which is a value consistent with minimal loss of fines. As such this is twice the value prescribed by Heerten. However, a limiting value of O_{90}/d_{90}= 2 is prescribed by the Nederlandse Vereniging Kust-en Oeverwerken (Netherlands Coastal Works Association) for bank protection works. A limiting value of O_{90}/d_{90} = 2 has also been recommended by Tan et al [7] who have suggested that the U.S. Army Corps of Engineers adopt this value in lieu of the more conservative value of unity currently employed. If this limitation is applied to equation (8) it is found that for O_{90}/d_{90} = 2 the coefficient of uniformity must take a value of 50. When this value of U is substituted in equation (5b) it confirms that the particle size to be retained is indeed d_{90}. Also this limiting value of U indicates that O_{90}/d_{50} from equation (7) must be limited to a value of 45.

FIG.5. Variation of O_{90} with U for d_{50} = 10μm.

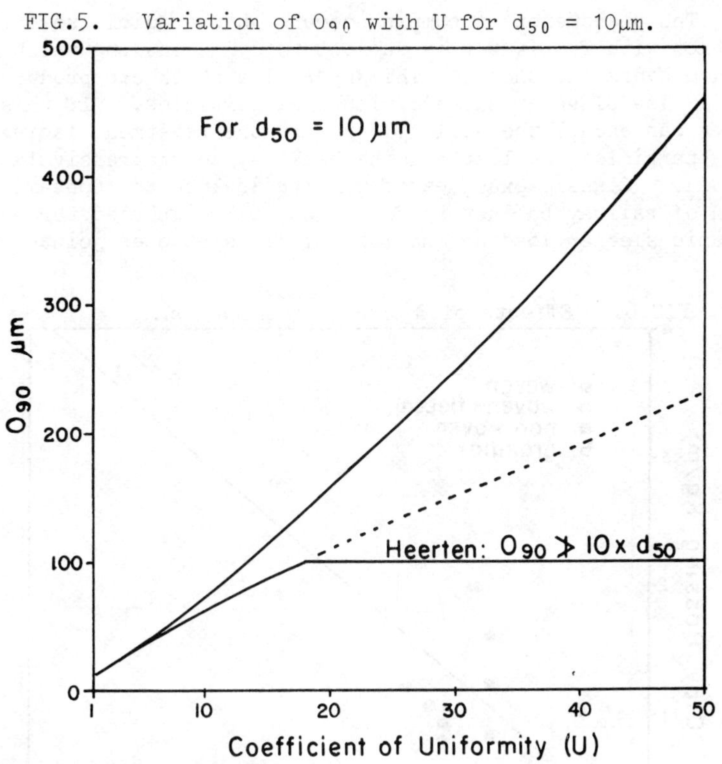

SUMMARY OF FILTER CRITERIA

The suggested filter criteria can be summarised as follows:-

(a) For $1 \leqq U \leqq 50$

$$O_{90}/d_{50} = 2\,U^{(1-\sqrt{2/U})}$$

but

$O_{90}/d_{50} \not> 45$

(b) For U<5

$O_{90}/d_{90} \not> 1$

(c) For U>5

$$O_{90}/d_{90} = 2\ U^{(0.2-\sqrt{2/U})}$$

but

$O_{90}/d_{90} \not> 2$

(d) For non-cohesive soils containing more than 50% by weight of silt.

$O_{90} \not> 200\mu m$

PUMPING

The phenomenon of pumping occurs when erodible soil in contact with free water is subject to cylic loading which can induce hydraulic shock. This hydraulic shock can produce rapid flow of water charged with soil particles. If this water can escape the soil filters and the revetment facing, soil particles are lost and the bank may be progressively eroded. Classic examples of pumping include the contamination of railway ballast by formation soil displaced through dynamic sleeper loading and loss of fines at open joints in

FIG.6. Effects of O_{90} on contamination.

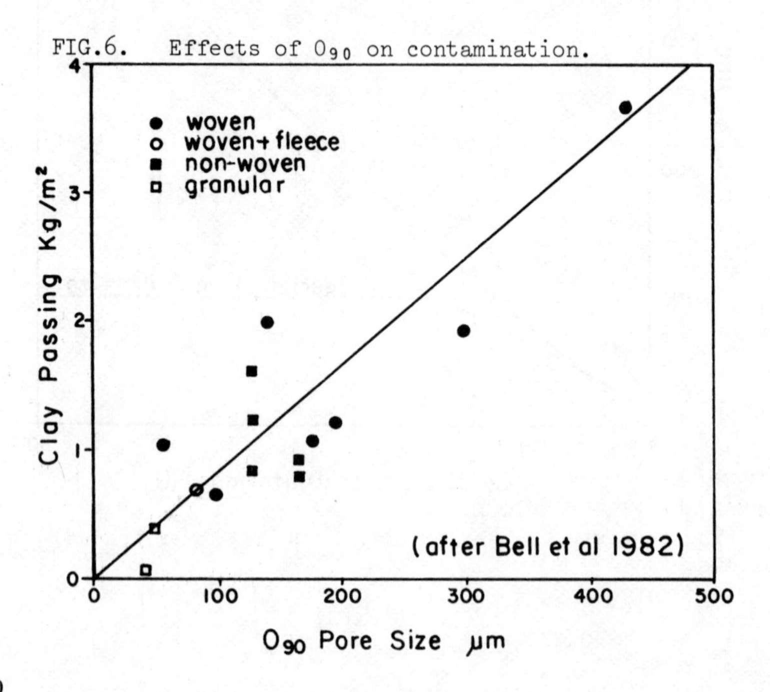

concrete pavements due to dynamic wheel loading. Clearly this phenomenon could be of importance in flexible revetment design where significant loss of fines could lead to distress in the armour. The role of geotextiles in reducing loss of soil fines from cohesive formations through pumping is currently a topic of research at the Queen's University of Belfast. A series of tests have been carried out involding the application of a cyclic load to a bed of coarse sub-base aggregate over a bed of cohesive soil with the aggregate underlain by a geotextile filter. The latest results published indicate that, for a given applied load frequency and intensity, soil loss increases as the O_{90} size increases, (7). This is depicted in Fig.6, which shows the effect of O_{90} on clay contamination passing through the geotextile and suggests that the nature of the geotextile, that is whether it is woven or non-woven, has no marked effect on the retention ability. Work at Birmingham University by Hoare [9] suggests that loss of fines increases as the cyclic stress level increases and as the number of applied loading cycles increases. The frequency of applied loading seemed to have little effect in the range 2.5 to 10 Hz.

Although the mechanism of pumping is not well understood it appears that under certain circumstances it may involve rapid extrusion of cohesive soil. This notion follows from the results of the work at the Queen's University where in the case of woven geotextiles there appears to be some abatement in the quantity of soil fines passing the geotextile as the open area ratio of the geotextile is reduced. Research work at Ground Engineering Limited [10] showed that as the open area ratio of woven fabric decreased the static pressure, P, required to extrude clay through the fabric increased. The results obtained are plotted in Fig.7 with the extrusion pressure plotted in the dimensionless form P/2Cu where Cu is

FIG.7. P/2Cu versus Open Area Ratio.

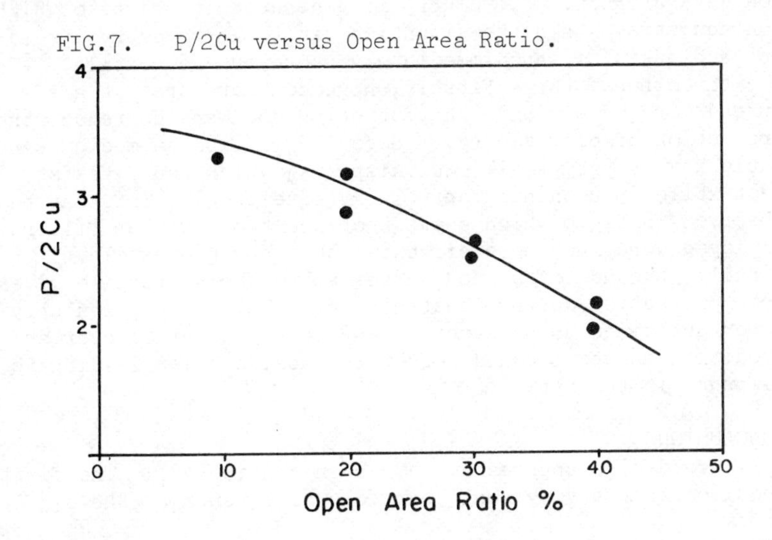

FIG.8. Effects of moisture content on contamination.

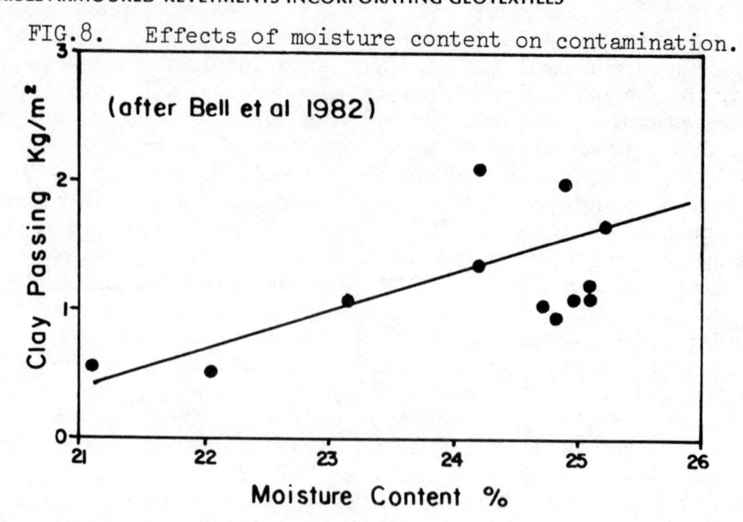

the undrained shear strength of the soil. Since it is the quotient P/2Cu that governs extrusion, it follows that for a fabric of a given open area ratio extrusion will occur at a progressively lower pressure as the undrained shear strength of the soil decreases.

This possibility is clearly reflected in Fig.8 which shows that as moisture content increases and, therefore, undrained shear strength decreases, there is an increase in the rate at which clay passes the geotextile.

Clearly in a revetment it is not feasible to attempt to control the moisture content of the bank soil. However, other tactics might be adopted. These are suggested by the observations (7, 8) that the clay slurry tended to form on the surface of the clay formation at the points of contact of the coarse aggregate. If slurry is generated at points of high contact stress then the situation can be improved by spreading a blanket of sand immediately above the geotextile. This has three benefits. Firstly, high contact stresses are spread more evenly onto the formation soil and so reduce the production of clay slurry. Secondly, a fine to medium sand would have a very small equivalent O_{90} which would assist the geotextile in retaining any slurry developed. This can be observed in Fig.6 which shows that the two granular filters employed were much more effective than any geotextile. Finally, the use of a sand carpet above the geotextile gives some protection during construction. This is particularly important since an undetected tear in a geotextile over an erodible bank soil could result in local erosion leading to slippage in the bank.

CONCLUSIONS

Two design approaches have been considered. The first requires that a very small particle size, such as the d_{15}

size, should be positively retained. This can give rise to unrealistically small pore sizes. Although such small pores would retain the soil they would be likely to be associated with low geotextile permeability that could allow the development of destructive overpressures. In the second design approach which is considered to be the more practical of the two, it is assumed that only the larger particles sizes, such as d_{90}, need be positively retained. With a reasonable grading bank soils can be well compacted such that a maximum pore size equal to twice d_{90} can be employed without significant loss of soil particles. For uniformly graded soils the soil particles, even after compaction, are not so densly packed thus the geotextile pore size defined by O_{90} should not exceed d_{90}. In addition to the problem of erosion or piping there is the hazard of pumping which may be induced in the bank soil through dynamic hydraulic loading of the revetment. It has been shown that the magnitude of soil loss due to pumping decreases as the O_{90} pore size decreases. However, geotextile filters seem to be less efficient than granular filters in the form of a sand carpet. It is suggested, therefore, that a sand carpet be employed in association with a geotextile since as well as enhancing filtration the sand acts as a protective layer during construction.

REFERENCES

[1] TEINDL , H. Filter Kriterien von geotextilien. Doctoral Thesis 1979. Insbruck University.

[2] OGINK, H.J.M. Investigations of the hydraulic characteristics of synthetic fabrics. Publication No. 146, WaterloopKundig Laboratorium. Delft Hydraulics Labratory 1975.

[3] GROUND ENGINEERING LIMITED. An introduction to ground water abstraction with particular reference to the development of Hydrotec. Confidential Report No. 2H/3100 1980. (Unpublished).

[4] GROUND ENGINEERING LIMITED. A research investigation into the operational characteristics of Hydrotec. Confidential Report No. 929800 1981. (Unpublished).

[5] CALHOUN, C.C. Development of design criteria and acceptance specifications for plastic filter cloths. Technical Report F-72-7.U.S. Army Engineers Waterways Experiment Station Vicksburg 1972.

[6] HEERTEN, G. Dimensioning the filtration properties of geotextiles considering long-term conditions. Proceedings Second International Conference on Geotextiles, Las Vegas, 1982. Vol.1. 115-120.

[7] TAN, H.H. et al. Hydraulic function and performance of various geotextiles in drainage and related applications. Proceedings Second International Conference Geotextiles, Las Vegas, 1982. Vol.1. 155-160.

[8] BELL, A.L., McCULLOUGH, L.M., and SNAITH, M.D. An experimental investigation of sub-base protection using geotextiles. Proceedings Second International Conference on Geotextiles, Las Vegas, 1982. Vol.2. 435-440.

[9] HOARE, D.J. A laboratory study into pumping clay through geotextiles under dynamic loading. Proceedings Second International Conference on Geotextiles, Las Vegas, 1982. Vol.1. 423-428.

[10] GROUND ENGINEERING LIMITED. The use of clay as hydraulic fill. Confidential Report No. 3E/1204, 1978. (Unpublished).

T1 Influence of the filtration opening size on soil retention capacity of geotextiles

Y. FAURE, J. P. GOURC, and E. SUNDIAS, Université de Grenoble, France

SYNOPSIS. Investigations are developped at the University of Grenoble (IRIGM) to determine filter criteria of woven and non woven geotextiles. The particule size retention of the geotextile is obviously depending of its filtration opening size O_f but also of site and use conditions.

Two testing procedures were performed for "soil-geotextile - granular drain" samples :

- either dynamic loading tests, simulation of setting on conditions ;
- or static loading tests with steady flow of water during 24 hours.

The laboratory tests results illustrate not only the influence of O_f/D^s_{85} ratio value (D^s_{85} : greatest grain size diameter of the soil) but also the solicitation effect (dynamic compaction energy, static load, flow gradient).

INTRODUCTION

In civil engineering, granular filters are more and more often replaced by synthetic filters with polymer fibres (geotextiles). A filter has to satisfy two criterias : a retention criterion and a permeability criterion. In this paper, we limit our study to the first one.

FILTRATION PARAMETER OF A GEOTEXTILE

The determination of a retention criterion for granular or fibrous filters requires parameters characterizing the soil to be filtered on the one hand and the filter on the other hand.

In the case of granular filters, Terzaghi's retention criterion use the D^s_{85} of the soil (initially an uniform soil) and designs the filter by its pore mean diameter $\overline{d}$:

$$\overline{d} < D^s_{85}$$

$\overline{d}$ of a granular medium is depending of grain diameter and may be estimated by the relation :

$$\overline{d} \simeq D_{15}/4$$

Thus the D_{15} of a granular filter : D^f_{15} is given by the wellknown relation :

$$D^f_{15} < 4D^s_{85}$$

In the case of geotextiles, the pore mean diameter is independent of the fibre diameter and consequently this last one is not sufficient to characterize the fibrous medium. $\bar{a}$ have to be defined by others structural parameters of the geotextile.

1. Porometry

Rollin (ref. 1) measures, by mean of a quantimeter, distances between fibres on cross sections of encapsulated geotextiles and obtains the histogram of pore diameter (fig.1).

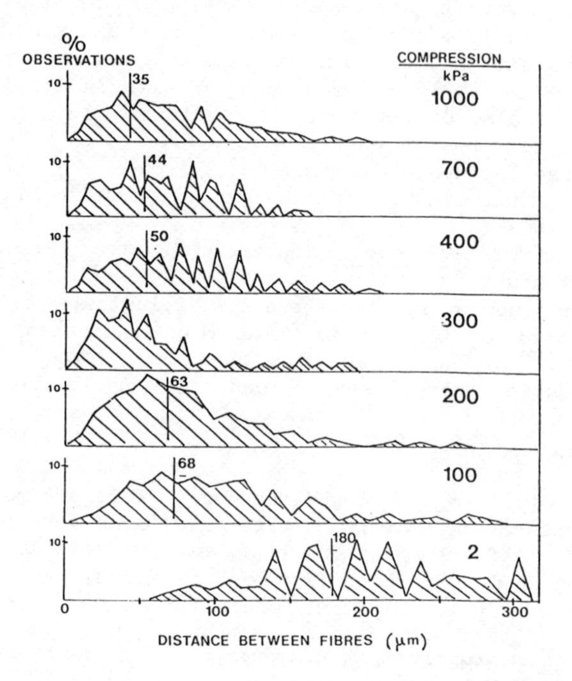

Fig. 1 : pore histogram of a needle-punched non woven obtained with a quantimeter under various compression σ_N (Rollin, ref. 1).

Gourc (ref. 2) and Rollin using morphometric analysis describe the fibrous medium by a porometric curve which agree with the previous results (fig. 2). (Note that capillary methods give different values).

But the porometric curve is not enough to characterize the filtration behavior of a geotextile : its thickness will modify it. Indeed, the probability of blocking a soil particle increases when the pore size decreases but also when increases the length of the way through the geotextile of constant porometry.

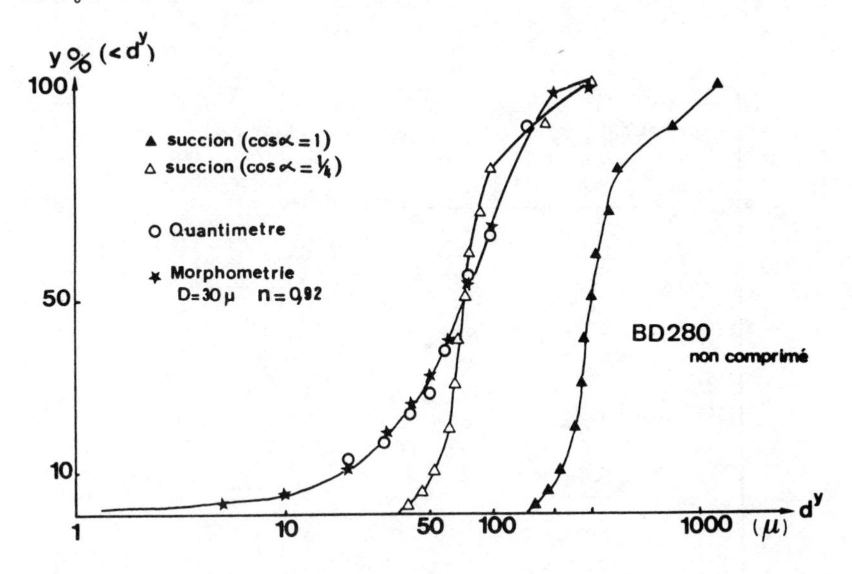

Fig. 2 : comparison between teoretical pore distribution and measures with quantimeter or capillary method for a needle-punched non woven.

2. Filtration opening size

For this reason, a sieving experiment would give complementary informations about retention capacity of geotextile. A filtration opening size O_f can be determined by the greatest grain size of the passing soil through the geotextile : D_{90}, D_{95} or D_{98} according to the authors.

Various sieving methods exist : wet or dry sieving of glass beads or of soil with uniform or spread grain size distribution. The French Comittee of Geotextiles recommends an hydrodynamic sieving like this presented by Fayoux (ref. 3) : sieving of a well graded soil with alternating flows. The filtration opening size is taken equal to the D_{95} of the passing soil. This method creates critical conditions with hydraulic forces which prevent clogging.

3. Retention criterion of geotextiles

The most elementary retention criterion could mean that the geotextile is able to block the greatest particles of the soil.

The criteria found in the litterature are formulated from filtration opening size O_f measured by different methods and correlate O_f to the D_{85}^{s} or D_{50}^{s} of the soil. Few of them consider the soil uniformity coefficient U_s , the geotextile structure (woven or non woven) and the mechanic or hydraulic forces.

The diversity of those criteria and some contradictions may be noted on the figure 3 where a synthesis is presented.

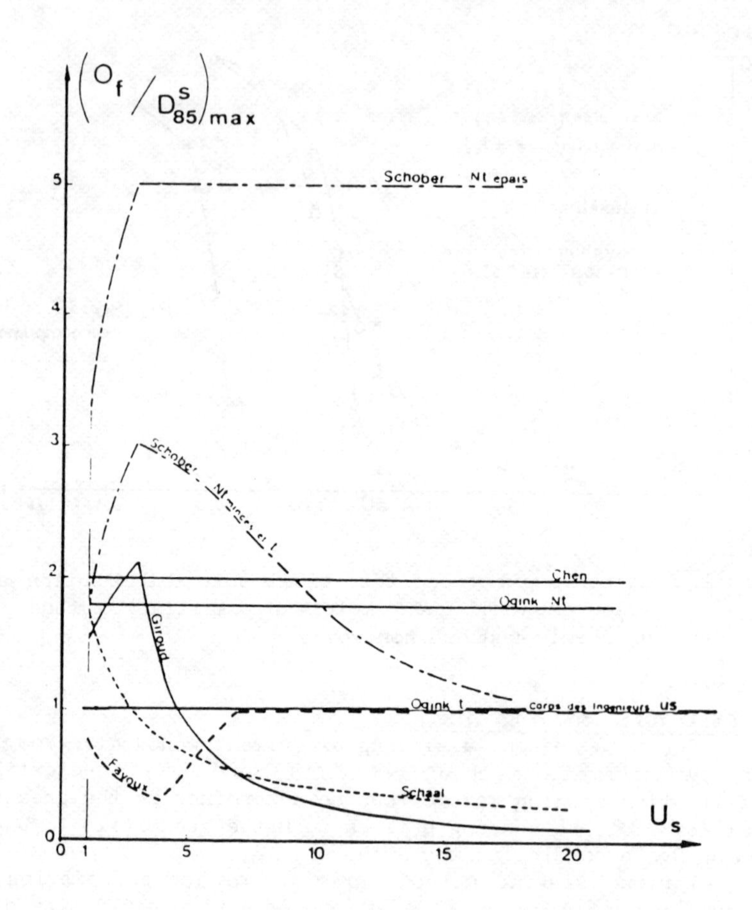

Fig. 3 : retention criteria proposed by various authors for geotextiles.

EXPERIMENTAL STUDY

Investigations are developped at the Institut de Recherches Interdisciplinaires de Géologie et de Mécanique, University of Grenoble (France) to show off the influence of site conditions (flow gradient, static or dynamic loading) and the influence of the geotextile (by its filtration opening size) on its soil retaining ability.

Two testing proceduces were performed on "soil-geotextile - granular drain" samples :

- either static loading tests with steady flow of water during 24 hours ;

- or <u>dynamic loading tests</u> to simulate setting on conditions and to illustrate the geotextile filtration behavior in function of the applied energy.

- <u>The tested soils</u> are quartz powders (cohesionless soil) fig. 4 :

. soil 1 : D^s_{85} 110 µm, U_s = 25

. soil 2 : D^s_{85} 105 µm, U_s = 1,5

. monoclass soil : 63 - 80 µm, 80 - 100 µm, 100 - 125 µm...

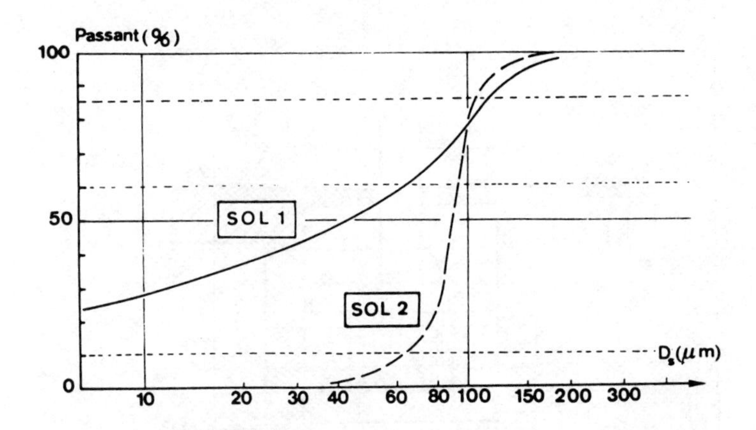

Fig. 4 : Grain size distribution of the soils used for filtration tests.

- <u>The tested geotextiles</u> :

. a needlepunched non woven :
BD 280 g/m^2 (Bidim U34), O_f = 108 µm.

. a spunbonded non woven :
TP 270 g/m^2 (Typar 3807), O_f = 40 µm.

. some wovens, square mesh : (D^f = fibre diameter, d = mesh opening)

tPt 54 g/m^2 (Polytrame) D^f = 103 µm, d = 338 µm
tPt 48 g/m^2 (Polytrame) D^f = 80 µm, d = 169 µm
tFy 25 g/m^2 (Fyltis) D^f = 42 µm, d = 69 µm

For non wovens O_f is determined by hydrodynamic sieving (Fayoux, ref. 3); for wovens, O_f is taken equal to d.

- <u>The drain</u> : glass beads of 10 mm diameter

The stability of the "soil-geotextile filter" sample is characterized by the mass μ_F of the passing soil through an unit surface of geotextile during 24 hours. In the case of wovens, it is easy to calculate the passing soil mass μ^*_F per unit <u>pore</u> surface :

$$\mu^*_F = \mu_F \cdot \left[\frac{d + D^f}{d}\right]^2$$

For non wovens, the pore surface is not accessible. It would be possible to estimate μ_F^* by : $\mu_F^* = \mu_F/n$, n : porosity but the volumic porosity do not represent the pore area offered to water flow.

1. Static tests

An oedometric cell specially designed for "soil-geotextile - drain" sample is used (fig. 5). The compression may be up to 1000 kPa. A retention paper is placed below the beads in order to recover passing soil in the drain after a flow of 24 hours.

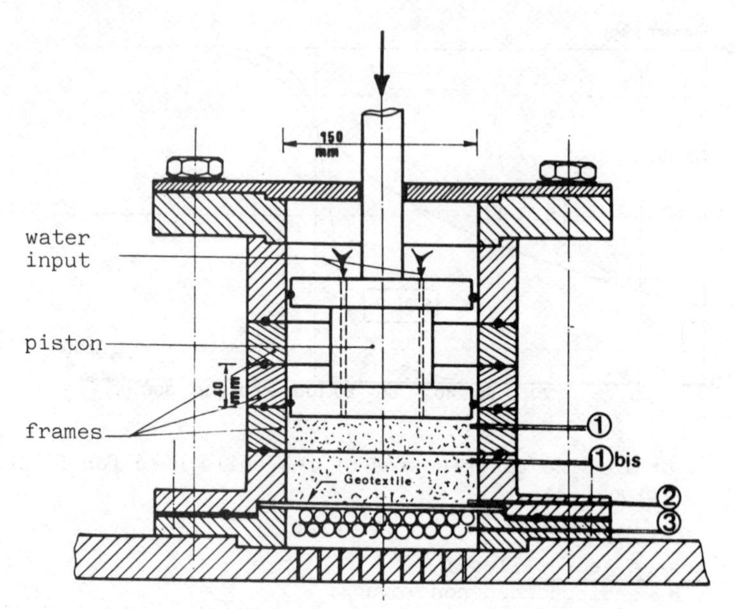

Fig. 5 : Filtration cell for static loading tests.

A first lot of tests, performed on non wovens, showed the influence of the gradient ratio i (fig. 6) and of the compression σ_N (fig. 7).

- The mass of passing soil μ_F increases with the flow gradient i but not in great ratio : from 50 to 80 g/m^2 while i varies from 1 to 30 for a soil with many fine particles (50 % < 40 µm) under 10 kPa. Note that here $O_f/D^s_{85} < 1$.
- Static loading influence is more important (fig. 7) : the passing soil decreases a great deal even with a high flow gradient ratio but the pressure have two roles :
  . modifies the pore size distribution due to the geotextile compressibility and also the filtration opening size ;
  . increases soil confining pressure and the grains friction.

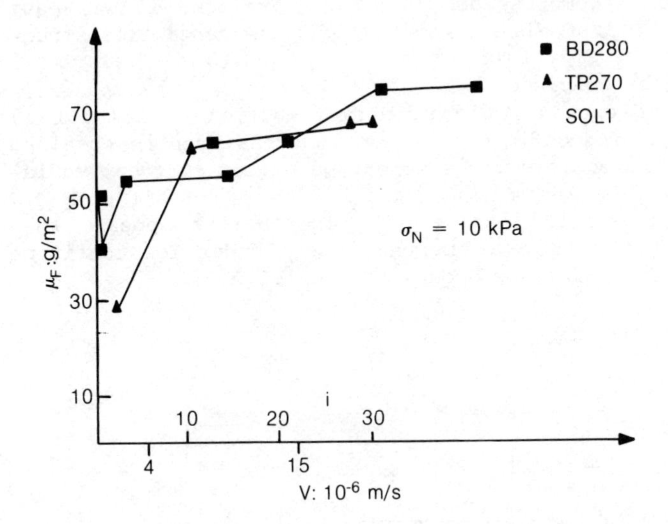

Fig. 6 : Static test. Influence of flow gradient ratio i on passing soil mass μ_F.

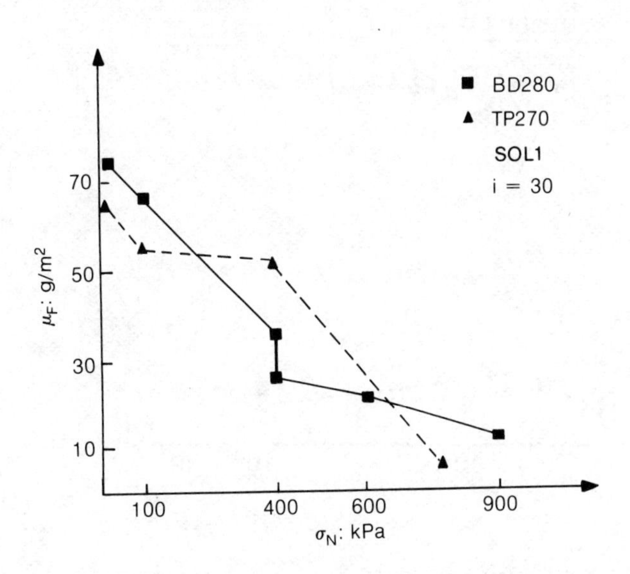

Fig. 7 : Static test. Influence of compression σ_N on mass μ_F

To illustrate the influence of the confining pressure, we have performed various tests with woven textiles, considered as incompressible (fig. 8). Besides, it was easy to vary the ratio O_f/D^s_{85} without changing the geotextile structure (fig. 9 and fig. 10).

It may be concluded :

. μ^*_F of spread grain size distribution soil (soil 1) is influenced by confining pressure whereas results obtained with an uniform soil (soil 2) do not point out so great variations.

. Under static loading soil instability appears when $O_f/D^s_{85} > 2$ but, for a well graded soil under low confining pressure, instability may occur if $O_f/D^s_{85} > 1$.

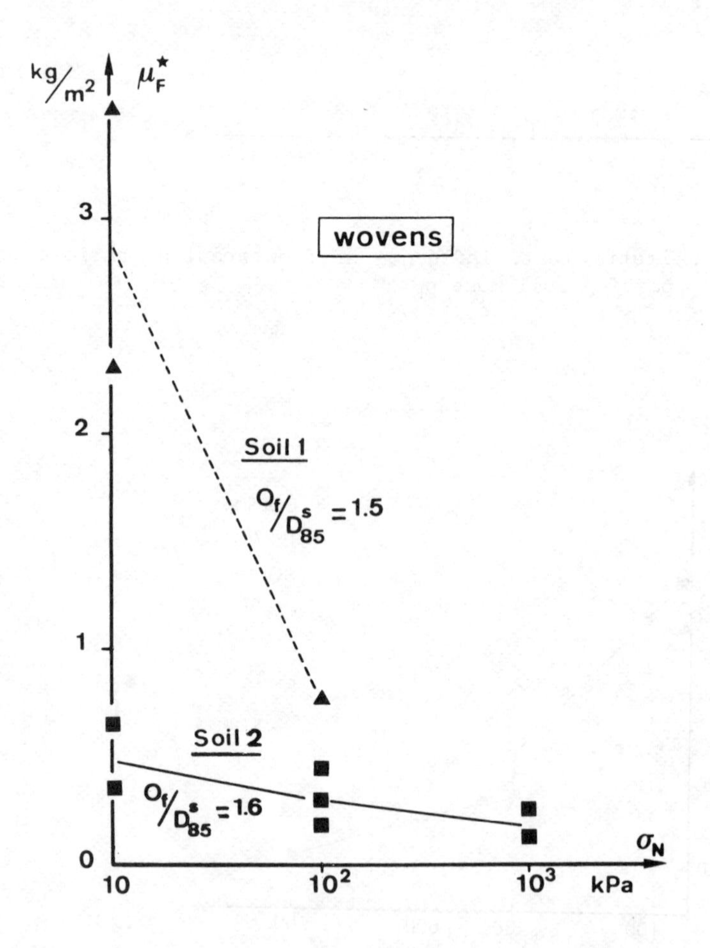

Fig. 8 : Static test. Influence of the coefficient of uniformity U_s and of compression σ_N.

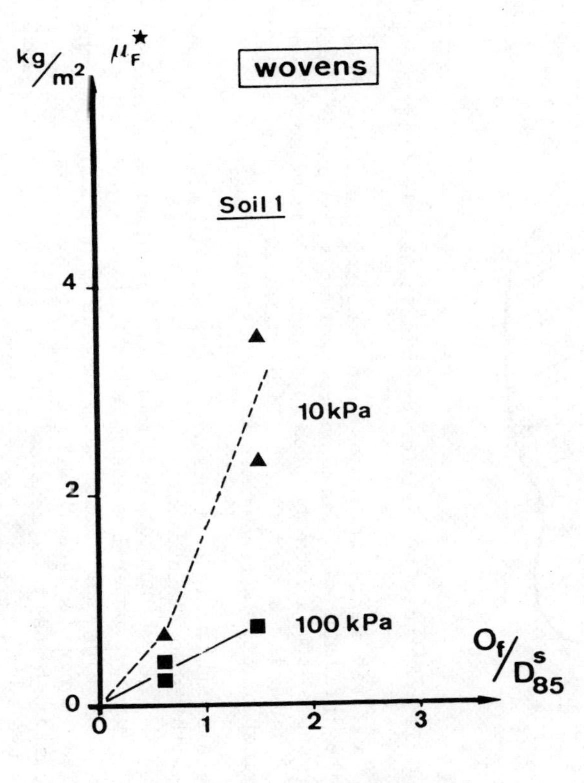

Fig. 9 : Static test. Role of O_f/D^S_{85} (soil 1).

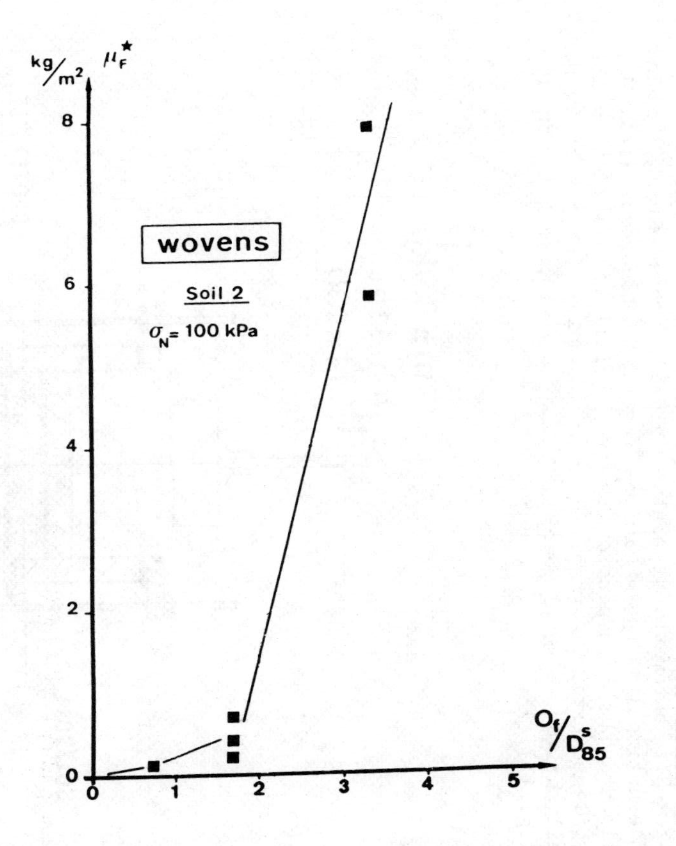

Fig. 10 : Static test. Role of O_f/D^S_{85} (soil 2).

2. Dynamic tests

The automatic apparatus, already presented at the 2nd International Congress at Las Vegas, by Loubinoux (ref. 4), allows dynamic compaction by cyclic fall of a weight on the "soil-geotextile - granular drain" sample : weight of 90 N, fall height 1 m, number of falls N = 10, 20, 50 or 100 (fig. 11).

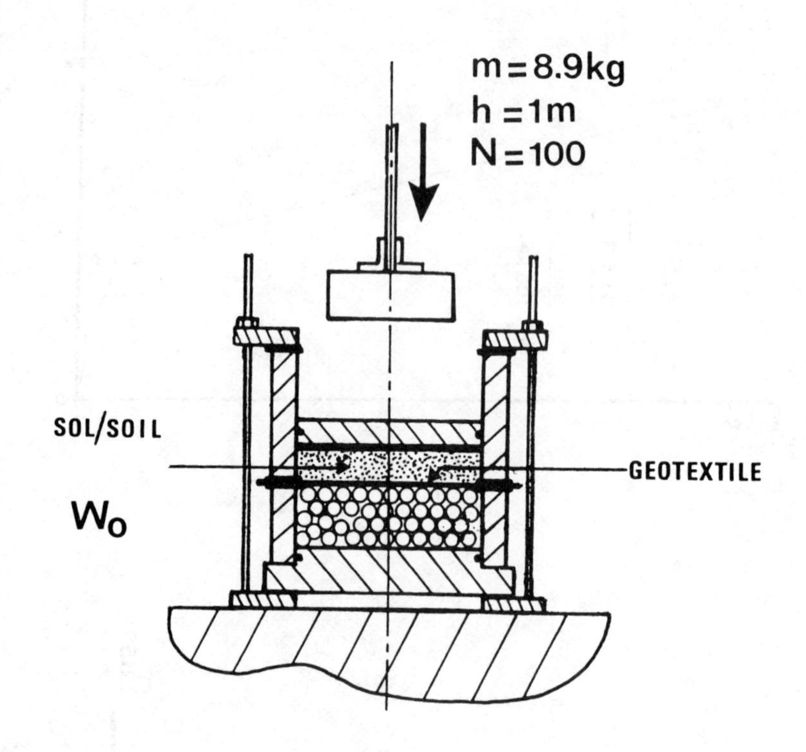

fig. 11 : Filtration cell for dynamic loading test.

A first serie of tests on non wovens indicates (fig. 12) the influence of initial water content W_o on passing soil mass and a comparison between passing soil from soil 1 or soil 2 shows the uniformity coefficient influence (fig. 13) : soil instability increases with W_o and also with U_s.

A second serie of tests were performed on wovens in contact with "mono class" soil :

. Fig. 14 carries out that soil stability is ensured when $O_f/D^s_{85} < 2$ under low energy N < 20.

. Passing soil mass seems to be proportional to the applied energy for great ratio O_f/D^s_{85} (3.5 or 4.4, fig. 15) whereas it tends to a constant level, i.e. to a stable soil structure for $O_f/D^s_{85} < 2$.

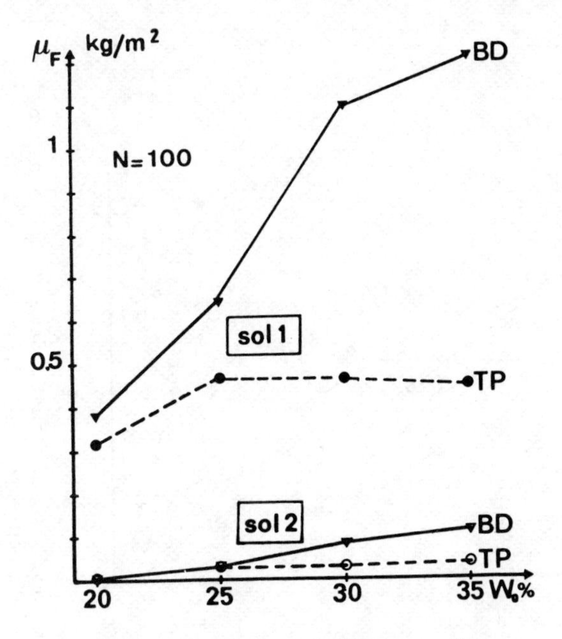

Fig. 12 : Dynamic test. Influence of w_o on mass μ_F.

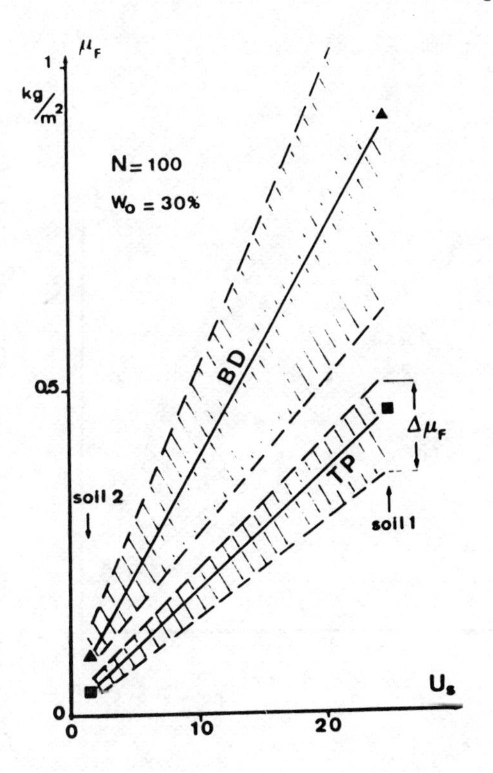

Fig. 13 : Dynamic test. Role of soil uniformity coefficient U_s.

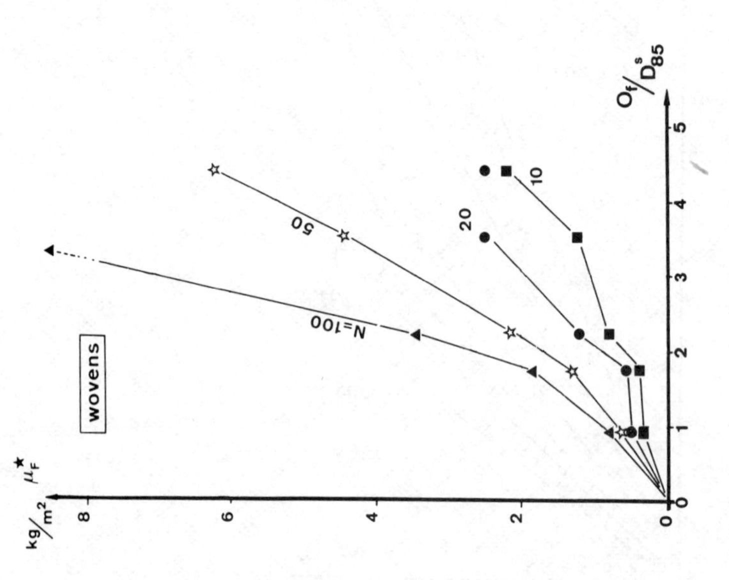

Fig. 14. Dynamic test. Influence of relative filtration opening size O_f/D^s_{85} and of applied energy (N falls)

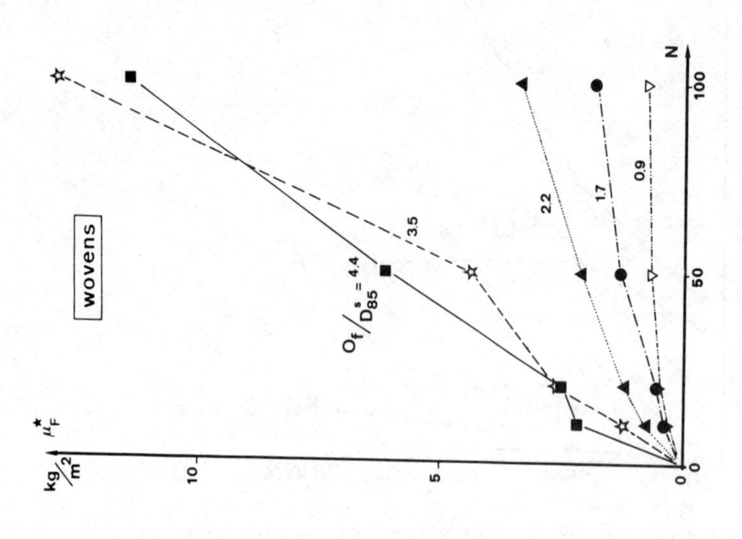

Fig. 15. Dynamic test. Same results as on Fig. 14 but with soil instability when O_f/D^s_{85} equal to 3.5 or 4.4 versus compaction energy

CONCLUSION

. Confining pressure furthers stability of the soil to be filtred.

. Flow gradient ratio increases a little the passing soil if $O_f/D^s_{85} < 1$.

. A spread grain size distribution soil is more unstable than a uniform soil, particulary under a low confining pressure.

. $O_f/D^s_{85} < 2$ ensures stability of an uniform soil under the considered conditions : static or dynamic.

REFERENCES :

1. ROLLIN A.L. Measurement of permeability of geotextiles under compression. ASTM Meeting at Kansas City, June 1983.

2. GOURC J.P. Quelques aspects du comportement des géotextiles en Mécanique des Sols. Thèse de Docteur-ès-Sciences. Université de Grenoble, 1982.

3. FAYOUX D. Filtration hydrodynamique des sols par des textiles. Colloque International sur l'emploi des textiles en Géotechnique. Paris 1977.

4. LOUBINOUX D., FAURE Y., GOURC J.P., MACHIZAUD Ch. Conservationde la fonction filtre des géotextiles sous sollicitations dynamiques et statiques. 2e Congrès International des géotexiles. Las Vegas 1982.

LIST OF SYMBOLS :

$d^f = d$: mesh opening of a woven filter

$\bar{d}_f$: pore mean diameter of a granular filter

D^f : fibre diameter of a textile filter

D^f_{15} : D_{15} of a granular filter

D^s_{85} : D_{85} of the soil to be filtered

O_f : filtration opening size of a textile filter

U_s : uniformity coefficient of the soil to be filtered

μ_F : mass of soil passing through the geotextile per unit surface during 24 hours

μ^*_F : μ_F per pore unit surface of geotextile

i : flow gradient ratio

σ_N : normal compression stress on soil-geotextile-drain system

Discussion on Session 2: Geotechnical aspects

Mr A. I. Woestenenk, Bitumarin B.V., The Netherlands

When thinking of the use of asphalt for flexible armoured revetments one is inclined to remember a long history of try-outs and tests up to half a century ago involving impermeable asphalt concrete or asphalt mastic mattresses, for instance on the Missisipi River. Some of these tests were successful; others were not, the impermeability of asphalt being the main cause for failure.

A permeable asphalt has existed for some 16 years, even before geotextiles were used as a filter for bank protection. An aluminium matting was used as a filter and armour layer. The system was soon modernized and used for larger projects, including the Dutch Delta Works, and also for smaller river bank revetments (Figs 1 and 2). The aluminium matting was changed to a polypropylene geotextile, common in Holland, and for the large Fixtone mattresses a steel cable reinforcement was used for laying mattresses up to 40 m in depth.

Extensive wave flume tests have shown that the permeable structure of Fixtone does not allow uplift pressures to develop, such as can be measured under impermeable elements forming an articulated mattress. The open joints in that type of construction, which take care of pressure release, at the same time induce uplift pressures under the impermeable elements. Therefore a uniform permeability, as in the Fixtone ballast layer, will always have the lowest level of uplift pressures.

Just after installation the black colour is somewhat protruding, but as Fixtone seems to stimulate vegetation it soon amalgamates with the surrounding environment (Fig. 3). Damage to the revetment due to vegetation has never been found. It is assumed that the reason for this is the thick coating between the individual stones that allows plant and root growth without damage.

Mr M. H. Goldsworthy, Golder Associates, Berkshire

I would like to comment on the interaction between soil and

Fig. 1.

Fig. 2.

Fig. 3

revetment permeabilities by reference to a particular case.

The revetment, consisting of concrete blocks on a woven fabric layer, was constructed in the tidal arm of a river where the dominant influence is probably tidal fluctuation. The soil is a gap graded mixture of fine to medium sand and medium gravel. The slope was cut at a slope of 1:2 and the revetment mats were laid on this slope. Normal tidal variation is from +4 m to -1 m (Fig. 4).

After installation it was noted that the lower mats were moving downslope, buckling at the toe and forming a gap in the protection around mid height. This movement appeared to be slow and progressive.

In order to study the problem piezometers were installed on cross-sections. These piezometers enabled the fluctuations of pore water pressures over a tidal cycle to be examined, and thus permitted an estimate to be made of the in situ permeability.

Laboratory tests were also undertaken both of permeability, on the finer component of the soil, and of friction between the fabric and the soil. Tests were also made of the permeability on the revetment system as installed.

Consideration of the flow rate out of the slope during the ebb tide and of the head required beneath the fabric to permit sliding enables a relationship to be developed between the soil and revetment permeability to produce a limiting equilibrium.

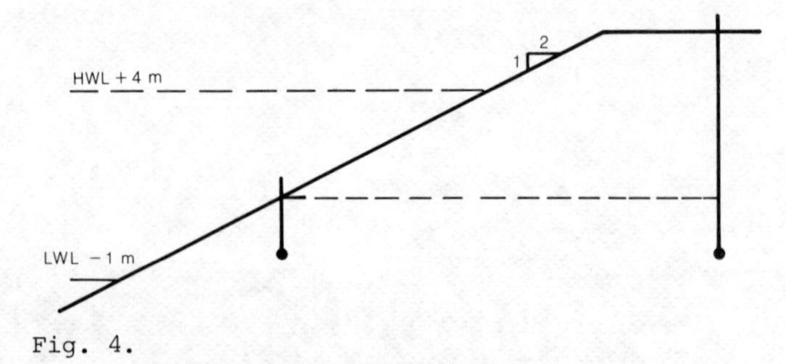

Fig. 4.

When this relationship is compared with the available data it shows that for a normal tidal range the revetment should be safe, even though the revetment permeability is apparently as much as one hundred times less than the soil permeability (Fig. 5).

The observed failure is thought to have resulted, initially at least, from exceptional tidal conditions, possibly combined with additional flow from the land side. It may also be that the effective revetment permeability is somewhat lower than measured due to the head losses at entry noted in a previous contribution.

It is interesting to note that if the revetment thickness is increased to, say, the full block depth by siltation and weed growth then the likelihood of lift off and downward seep at the main outflow zone is increased markedly.

These observations illustrate the importance of considering both the thickness and the permeability of the revetment system and how the thickness may be affected by environmental factors. They also suggest that advantages would be gained by anchoring the mats to prevent local instability affecting the overall performance.

Dr G. Heerten, Naue - Fasertechnik, Germany

Professor Ingold mentioned in his paper the filtration rules I gave at the 2nd International Conference on Geotextiles at Las Vegas in 1982. These filtration rules are based on extensive field studies, laboratory tests and questionnaires considering mainly practical experience. The filtration rules are being discussed by the German committees which are involved in geotextile recommendations at this time. The latest, but perhaps not final, version is:

1. For soils with $d_{50} \geqslant 0.06$ mm

(a) static load conditions and non-turbulent flow:

$D_w < 2.5 \cdot d_{50}$ and $D_w \leqslant d_{90}$

in case of high uniformity number $U > 5$

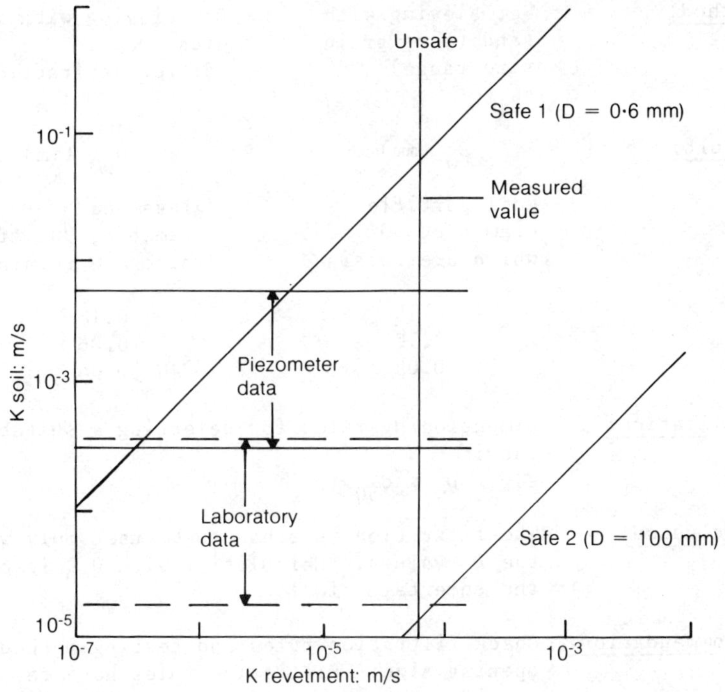

Fig. 5.

$d_{50} < D_w < d_{90}$

(b) dynamic load conditions:

$D_w < 1.0\ d_{50}$

2. For soils with $d_{50} < 0.06$ mm and all load conditions

$D_w < 10\ .\ d_{50}$ and $D_w < d_{90}$ and $D_w < 0.1$ mm

D_w is the effective opening size as estimated by the wet sieving method of the Franzius-Institut for Hydraulic Research and Coastal Engineering of the University of Hannover, West Germany. It is very important to know the test method which has been used for the estimation of a fabric opening size. Results of different opening size testing methods can differ in a wide range, as given in the following example for non-woven fabrics being tested using the wet sieving method and the glass bead method.

Problem: Estimation of soil particle sizes, which are passing a geotextile in the ground-system of water and soil.

Method:	Wet sieving with sand (wide grain size range)	Dry sieving with glass beads (different fractions)
Result:	D_w (mm)	O_{90} (mm)
	soil particle diameter, 90% of which are retained	glass bead diameter, 90% of which are retained
	0.15	0.18
	0.12	0.06
	0.09	0.04

Application: Dimensioning rules for selecting a suitable fabric
e.g. $D_w < d_{50}$

Conclusion: The filtration rule has to be used only with the D_w values. Calculating with O_{90} is on the uncertain side!

Recommendation: Check filtration rules and testing method of opening size. Filtration rules have to consider the method of opening size testing and definition of opening size.

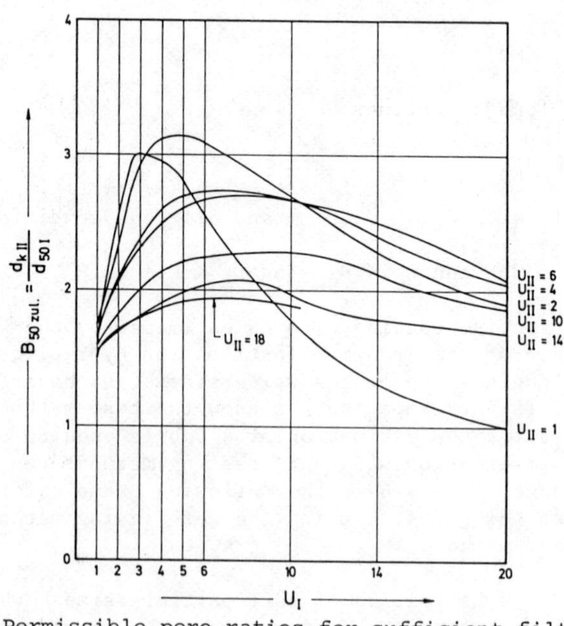

Fig. 6. Permissible pore ratios for sufficient filter conditions after TEINDL

A comparison of grain filters and fabric filters has to consider the pores of the filter medium, because the filtering properties of the filter medium are mainly determined by the dimensions and distribution of the pores. For grain filters TEINDL has transformed the CISTIN/ZIEMS determination diagram into the permissible pore ratio equation (Fig 6):

$$B_{50\ perm.} = \frac{d_{pII}}{d_{50I}} = f(U_I/U_{II})$$

For all the cases that have been investigated (U_I and U_{II} = 20), the evaluation shows that the average pore sizes (d_{pII}) of the filter medium may exceed d_{50I} of the base earth by a factor of between 1 and 3:

$$d_{pII} = 1 \text{ to } 3\ d_{50I}$$

This is in good accordance to the filtration rules for geotextiles given above.

Professor T. S. Ingold

A problem that currently besets geotextiles technology is the lack of application of international nomenclature and test methods. As defined in the appendix to Volume III of the proceedings of the second International Conference on Geotextiles, there is a move towards using the effective opening size defined by O_{95}. However, there is of yet no internationally accepted standard for determining this pore size. In consequence the magnitude of the measured pore size, whatever it is called, will tend to vary according to the test procedure employed. This is highlighted by the results given by Dr Heerten for nonwoven fabrics. Similar variations are found in woven fabrics, particularly those with woven tape yarn structures. I agree wholeheartedly with the recommendation that both filtration rules and test methods should be checked before application to ensure that they are consistent.

Mr I. R. Whittle, MAFF, London

In what ways can geotextiles be modified to increase the frictional resistance against sliding?

Mr P. R. Rankilor, Manstock Geotechnical Consultants, Manchester

I do not look to a geotextile to have any frictional support effect on a slope design.

I design my slopes to be stable in their own right, and use

the geotextile solely for its separation and filtration functions.

Can Professor Ingold comment on the following two points:

(a) Blocking of fabrics with uniform pore distributions

(b) The effects of alternating hydraulic gradients on particle loss through geotextile filters.

Professor T. S. Ingold

In response to the first part of the question it is worth reconsidering briefly the definitions of 'clogging' and 'blocking'. Clogging can be considered to be the lodging of soil particles within the pores and structure of the fabric. In contrast, blocking might be considered to involve the total or partial obstruction of a fabric pore by a soil particle resting on the surface of the fabric. A measure of the degree of blocking in a soil-geotextile system is generally assessed in the laboratory by observing the change in flow rate with time through the system. Under constant head conditions creating unidirectional flow, the flow rate should settle down fairly rapidly to a constant value. This flow rate will generally be lower than that at the start of the test and this reduction can be ascribed to partial blocking and the generation of a filtering network in the soil close to the geotextile. In the case of woven geotextiles, which tend to have a more uniform pore size distribution than nonwovens, the initial drop in flow rate tends to be less than for a nonwoven, provided that the geotextile is compatible as a filter to the soil. If the flow rate through the soil-geotextile system does reach a steady state there is no problem, since at the design stage the permeability of the geotextile can be selected to be several times higher than that of the soil so that even allowing for some blocking the geotextile is still of adequate permeability. Where a steady state flow rate is not reached it can only be assumed that there is continuing blocking and clogging which could ultimately reduce the permeability of the system to an unacceptably low value. Clearly such a system would not comply with the essence of any sound filtration criteria and should therefore be avoided. An entirely different situation arises when flow changes from unidirectional to alternating, since there appears to be no opportunity for the formation of a stable particle bridging network or filter cake. The magnitude of loss of soil fines is found to be strongly influenced by the magnitude of the alternating hydraulic gradient operating. Experimental work carried out to test the validity of equations 7 and 8 of my Paper has since shown that large losses of fines are encountered (1). While this might be acceptable for well screens, and it was from experimental work on well screen filtration that these equations were

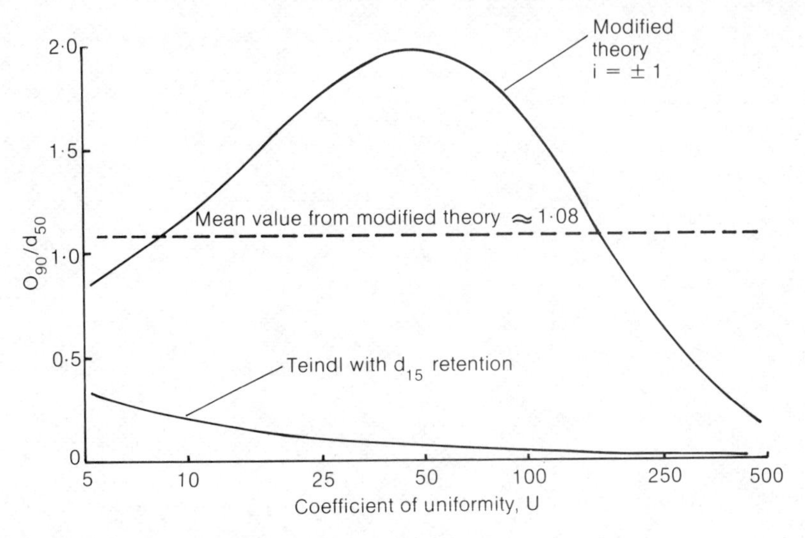

Fig. 7. Variation of O_{90}/d_{50} with U

derived, it was thought that such losses would be unacceptable beneath revetments. In the light of this more recent experimental work equations 7 and 8 have been modified to include a term relating to the magnitude of alternating hydraulic gradient and these are given in equations 11 and 12 below

$$O_{90}/d_{50} = \alpha U^{(1-\sqrt{2/U})} \qquad (11)$$

$$O_{90}/d_{90} = \alpha U^{(0.2-\sqrt{2/U})} \qquad (12)$$

where $\alpha = (|i|-0.5)^{(U/4)}$ valid for $\alpha > 0$ only

Experimental work based on these modified ratios of O_{90}/d_{90} and O_{90}/d_{50} gave losses of fines, by weight, of one percent or less. The variation of O_{90}/d_{50} with U from equation 11 is plotted in Fig. 7 and compared with positive retention theory which appears to render conservatively low values of O_{90}/d_{50}. It is interesting to note that the mean value of O_{90}/d_{50} from equation 11 is approximately unity over a range of coefficients of uniformity from 5 to 500, thus giving a threshold value of $O_{90}/d_{50} \simeq 1$. Since the method of measurement of O_{90} in the more recent experimental work is similar to that employed by Dr Heerten for determination of D_w it can be seen that allowance for the effects of hydraulic gradient in equation 11 gives a mean value of O_{90}/d_{50} which is in accord with the value of $D_w/d_{50} \ngtr 1$ recommended by BAW for soils with $d_{50} > 60\mu m$ subject to dynamic hydraulic loading.

1. INGOLD T.S. A theoretical and laboratory investigation of alternating flow filtration criteria for woven structures. Geotextiles and geomembranes, Vol. 2, No.1, 1985, Elsevier Applied Science Publishers, London (in press).

6 Development parameters for integrated flexible revetment systems

E. G. WISE, Consultant

SYNOPSIS. The Paper discusses the principal engineering and economic factors which have stimulated the development of single-layer armoured revetment structures, orthognally integrated from pre-assembled flexible panels of castable armour blocks, in association with underlying geotextiles. Cost effectiveness of such revetment structures is reviewed in relation to conservation of materials, labour, installation time and maintenance; typical components and their production methods are described; and design aspects, affecting stability and durability are considered particularly in relation to the two-dimensional cable restraint of the armour units and the anchoring of completed structures.

INTRODUCTION

1. The principal concerns of this Paper will be with the development and behaviour of revetment systems provided by an inter-connected single layer of armour units, overlying geotextile filters in alternating flow situations. However it is perhaps appropriate as a prelude, to briefly review the general catagories into which all forms of revetment may be grouped: And thereafter to establish a theme by defining the characteristics peculiar to such integrated systems.

2. It is clear that all methods of erosion control provide protected surfaces which may be either rigid or non-rigid, installed either as single or multiple layers, and by means which result in either a uniform or a random surface finish.

3. Beyond these conceptions however there are significant differences between armour surfaces which are rigid or merely non-rigid or which are intrinsically flexible.

4. Although no definition of a rigid revetment is necessary it may be noted that such works are predominantly fabricated in situ; that they involve relatively slow and labour intensive methods; that construction standards are vulnerable to climate, staff capability and access to the works and particularly so in the execution and supervision

of all underwater installations: Furthermore rigid revetments do not adapt to weaknesses or movements which may develop within any protected earth mass and which, if indicated, signal the need of maintenance effort to prevent a progressive deterioration of that mass.

5. The components of non-rigid revetments, on the other hand, may be pre-formed and assembled under carefully controlled production conditions, and with the close degree of supervision inherent in factory processes. This manufacturing characteristic is one which, in effect, extends from the screening of a specified rock size, through gabion or mattress construction, and to the machine casting of concrete blocks or other armouring units which have been devised as alternatives to rock.

6. A non-rigid revetment comprising such pre-formed armouring elements may be classified either as one whose stability is conditioned by the weights of its discreet units:

(a) In conjunction with an indeterminate interlock:

or (b) Supplemented by some inbuilt degree of bonding between units.

7. As neither class of revetment can be assembled with degrees of either interlock or bonding which will ensure a three-dimensional stability of armoured surfaces, and since both classes readily respond to any ground movement, there is an inherent probability of random displacement of individual armour units, or of assemblies, at times of extreme distress.

8. Where random-rubble or pre-formed systems of discreet armour units are used to protect an earth embankment against surface wave attack such armour must necessarily extend downwards to the embankment base, not withstanding that a stable underwater unarmoured zone might otherwise have developed above that base.

9. It may also be that such total embankment revetment will generate difficulties in routine maintenance dredging operations.

10. It is considerations such as these which logically support a system of flexible two-dimensionally integrated armour units and which suggest the following definition:-

"An integrated flexible revetment structure is one of any extent comprising an arrangement of castable armour units whose movements relative to adjacent units and to the armoured surface are restrained by an inter-connecting two-dimensional mesh of flexible cables or the like, and by the anchoring of the integrated structure through the protected earth surface."

11. Although the concept of integrated armour relates primarily to improved stability, the following are supple-

mentary economic advantages also inherent in the system:-

ECONOMICAL ADVANTAGES

12. Ever-increasing labour, material and transportation rates stimulate a continuing need to minimise the costs of engineering projects. While notable successes have been achieved through the developments of new materials, computer-aided production, and improved constructional techniques, it is unfortunate, to say the least, that resulting economies should frequently have attracted a "low cost" tag: For it remains a fact of life that just as a dog may be doomed by a bad name so too can "low cost" come to be interpreted as "low grade" or synonimous with "cut price" or "second best" or worse.

13. In reality a "low cost" project is any which seeks to maximise the efficient installation of a minimum amount of economic material which is capable of performing a specified function; and in the field of erosion control a sound "low cost" case is surely emerging for a very close association between flexible integrated armour and carefully selected types of geotextiles, both of which are intrinsically "low cost" but certainly neither "low grade" or "second best".

14. An absolute justification of any inovation will always be difficult for the very simple reason that newness is always difficult to justify. Certainly any attempted comparison between recently developed flexible revetments and earlier, more widely proven rigid or non-rigid alternatives can only be invidious since both must necessarily be related to past effectiveness, maintenance and eventual replacement costs.

15. Nevertheless the following do seem to provide significant present day economic arguments in favour of typical single-layer flexible revetments:-

a) U.K. installation rates of the order of £20-£30/m^2 for embankment protection against 1-1.5m surface wave attack.

b) Installation teams of 4-5 men, having only relatively simple handling procedures to follow and whose most skilled member is a crane operator.

c) Revetment laying rates of up to 75 m^2 per hour which are primarily influenced by site preparation and access and material deliveries.

d) The reduction to a minimum, or in some situations the total elimination, of diving effort for revetment installation below tidal ranges.

e) The recovery, for future re-use, of components of a revetment in the event of accidental damage to the structure or unforseen distress in its underlying earth mass.

DEVELOPMENT CONCEPTS

16. The attributes of non-rigid armouring systems have generally been claimed under the following headings:-

a) Production standardisation of armouring units under controlled conditions.
b) The simplicity of their bulk handling and/or assembly into easily transportable loads.
c) A variety of available installation techniques to suit differing site conditions.
d) Accurate underwater laying and anchoring of prefabricated assemblies.
e) The facility to retrieve and to re-use non-rigid panels in the event of accidental damage to the protected earth mass.
f) The ease with which geotextiles may be laid in association with non-rigid armour, particularly in underwater situations.

17. While the above attributes are also endemic to flexible revetment systems the following additional factors, introduced by integration, merit further consideration:-

g) The ease with which variably contoured and planometrically curved ground surfaces may be armoured without an attendant risk of subsequent individual armour displacement especially from vertical convex underwater surfaces.
h) An improved stability of armour units resulting from their integration.
i) Means by which pre-assembled panels of armour units may, themselves, be inter-connected and pre-tensioned to provide integration of an entire revetment.
j) Finite determination of continuous long-term anchoring restraint.

COMPONENTS OF FLEXIBLE REVETMENTS

18. A schedule of the components of an armoured flexible revetment, including its handling and installation equipment, is a quite modest one amounting to no more than:-

a) Precast concrete armour units.
b) Anti-abrasive liners.
c) Cables, having anti-abrasion sheaths.
d) Cable connectors and associated tools.
e) Geotextile filters.
f) Handling equipments.
g) Anchors.

Armour Blocks or Units

19. Of flexible revetment components the armour block is certainly the simplest, and therefore probably requires the least said about it.

20. Generally, but not necessarily, blocks are cast with a cellular configuration; in each there are at least two cable tunnels, penetrating the block and connecting opposing sides, in mutually transverse directions. Preferably the tunnels are sleeved with anti-abrasion liners to provide a second-line of defence for the anti-abrasion sheaths of the cables themselves, which are free to move within the liners.

21. There is a wide range of block-making machinery, incorporating computerised batch monitoring censors, and which are therefore capable of production rates, with multiple moulds, of up to 180-200 blocks per hour, each with an assured 28 day strength of not less than 50N/mm^2.

Cables

22. The degree of stability of an armoured flexible revetment is necessarily closely related to the durability of its intergrating cabling mesh. Of the several types of cable employed in the past to inter-connect armour units, among the earliest were high-tensile steel aircraft tendons, used in revetment projects in the Southern States of America. Within the draw-down range of waters having high chemical concentrations, surface corrosion of the cables became evident after 5-7 years, followed by initial strand failure at 10-12 years: In cable zones, adjacent to connections, the rates of corrosion were, of course, frequently accelerated. Stainless steel cables have also been tried; but their cost, allied to a comparative inflexibility and considerable handling difficulties, make their future general use appear somewhat unlikely.

23. During recent years there have been very significant developments in the production of cables from extruded plastic filaments, prominent among which are parallel-laid cables of high-tensile polyester filaments, contained within braided sheaths. In the absence of severe ultra-violet light such filaments are inert and seemingly resistant to almost all forms of degenerative attack within wide temperature ranges; while cable sheaths, in addition to providing high abrasion resistance also act as UVL shields. In accordance with normal practice, manufacturers of synthetic fibre cables provide Guaranteed Minimum Breaking Loads (GMBL) for new cables: And while strength fall-off, with time, is notoriously difficult to predict owing to the several factors detailed at paragraphs 28-31 it seems that a working life in excess of half a century is quite realistic.

Principal Quantifiable Cable Properties

26. Table 1. sets out Average Breaking Strengths, and Guaranteed Minimum Breaking Loads (GMBL) for typical overall cable diameters when measured at BS 5053 check loading; also included are corresponding Safe Working Loads (SWL) for Factors of Safety (FS) of 3, 4 and 5, given by:-

$$SWL = (GMBL \times M \times K) \div FS$$

where M is a "mode of lift" factor being 1.4 for angled slings; and K, a splicing factor of 0.85 for continuous slings.

Table 1

Cable Diameter	8.5mm (5/16")		12.0mm (½")	
	Tonnes	lbs.	Tonnes	lbs.
Av. Break Strength	3.2	7053	6.7	14,777
GMBL	2.9	6392	6.0	13,224
SWL for....FS=3 (MK=1.19)		2535		5246
=4		1901		3934
=5		1522		3147

27. The total stretch of any cable, under load, is made up of an immediate elastic stretch, a delayed elastic stretch and creep. As Table 2. illustrates, on repeated cyclic loading some delayed elastic stretch and creep are taken out of the cable which is then said to have become "harder" or to have acquired a state of lower stretch.

Table 2

Cable Diameter	8.5mm	12.0mm
Cycle 1 Loading		
Stretch from check load to 10% GMBL	1.6%	1.6%
25% GMBL	3.2%	3.2%
30% GMBL	5.0%	5.0%
Cycles 2-5 Loadings to 50% GMBL, followed by a one hour rest, followed by:-		
Cycle 6 Loading: ...... to 10% GMBL	1.2%	1.2%
25% GMBL	2.6%	2.6%
30% GMBL	4.0%	4.0%

Fatigue

28. Cable fatigue conditions may involve a combined variable loading and deflections of the longitudinal axis of the cable. In general it seems unlikely that a cyclic cable loading due, for instance to wave or draw-down conditions, will exceed 20% of the cable GMBL (equivalent to a panel-handling Factor of Safety of 5). At such low loads it is believed that fatigue-free life should exceed 10^8 cycles, or 25 years at maximum wave frequency, before any appreciable strength fall-off occurs.
29. If the solid angle through which the longitudinal axis of the cable may deflect during this cyclic-time does not exceed 3° the mutually sliding capability of the parallel cable fibres will accommodate an infinitesimal cable transverse deformation without reduction of full-strength cable life.

Exposure to UVL

30. A completely clean and unsheathed cable, exposed to sunlight, in European latitudes, for one average month per year would lose 19% of its initial strength after 20 years; although it is highly unlikely in practice either that the cable would be unsheathed, or remain clean enough for this fall-off rate to proceed.

Unquantifiable Cable Properties

31. The following are among the hostile modes of cable attack which are either difficult or impossible to quantify:-

a) Chemical: This is dependent upon concentration and water temperature. Although some alkaline build-up could occur adjacent to revetment concrete components, experience suggests a maximum cable deterioration of 3%-5% in 10 years

b) Micro Biological: Is negligible on polyester.

c) Fish Bite: Limited to unprotected cables in tropical waters or to attack by warm water Gribbon who live in power station effluent.

d) Abrasion: Although abrasion cannot be quantified, the rate at which it occurs can be rigorously inhibited; for example by a passive flexibility of cable sheaths working in bell-mouth soft-skin liners.

e) Vandalism: What one man is seen to build another will always be able to demolish; and there is little doubt that the best defence against a determined vandalism lies in revetment cable concealment either beneath established vegetation or within the revetment itself. At the same time, as with all social ills, the symptoms of vandalism may be exaggerated out of all proportion to the complaint.

Cable Connection

32. Following the installation of separate flexible panels, the cables threading together the panel armour units, must themselves be inter-connected both laterally and longitudinally by cable connections to form an integrated revetment structure. Such cable connections may either be "static", which type includes all forms of sleeves, permanently crimped onto cable ends; or "dynamic", and capable of responding to increasing cable loading by a progressively firmer grip upon the connected cables.

33. Because "static connectors" cannot counter an inevitable small, but progressive, cable "wasting" with increasing load, they are, as a type, structurally inferior to their dynamic alternative. However, the cost and comparative complexity of the latter type have lead to telling arguments in favour of well-known knots in conjunction with simple tensioning tools; notwithstanding that such knots cause an approximate 50% reduction of unbroken cable strength, unless techniques, involving friction devices, are employed to minimise such strength reduction.

GEOTEXTILES

34. Nothing can be added in this Paper to the already fully documented characteristics of geotextile filters whose effectiveness is widely established and whose uses are very well known. However it may not be inappropriate to pay tribute to the professional advisory service, provided by manufacturers, concerning the correct geotextiles for specific applications.

35. In calm-weather conditions the handling and laying of geotextiles above water, well in advance of covering armour, presents no particular problem. However because of low specific gravity, the fabrics usually require either pre-weighting or similar restraint when laid, alone, below tidal ranges; and particularly so when wave or current conditions prevail. This may entail added expensive diving effort and it has been found advantageous in several underwater revetment projects to pre-attach the geotextile underlay, having a leading valence or skirt, to the underside of a flexible armour panel with the skirt temporarily secured to its upper surface, and to thereafter install the combined revetment unit in a single operation; following which the leading edge of the skirt may be automatically detached from the upper surface of the panel and be temporarily restrained on the earth bed pending the placing of the following revetment unit.

36. Fig. 1 illustrates one such assembly suspended from a spreader beam prior to its installation in deep tidal water.

Fig. 1

HANDLING EQUIPMENTS

37. Generous handling safety factors make it feasible to crane-lift flexible revetment panels from one end; and to thereafter install them simply by an outward lowering of the crane jib. However there are advantages in suspending a flexible panel by its opposing ends from the cross-heads of either a fixed or variable-length spreader beam. Apart from requiring a lower safety factor, and a shorter crane jib, a spreader beam facilitates a remote release of the panel following its underwater installation.

38. Fig. 1 also shows one form of fixed spreader beam having cross-heads with fixed bollards around which the panel cable slings are looped to lie upon a horizontal gate at the base of the bollards. Following the installation of the panel, the gate is free to be raised, by remote means, and to disengage the untensioned slings from their bollards.

39. Equipments for use with fixed spreader beams and on restricted sites are shown at Figs. 2 & 3 respectively.

40. Finally space limitations permit only a passing reference to flotation methods of panel installation, currently under investigation.

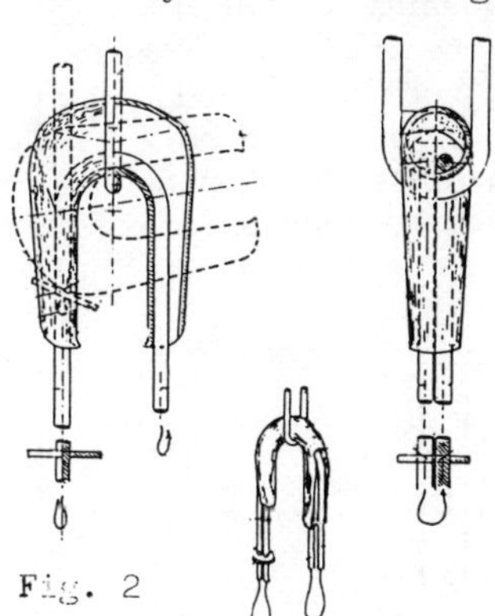

Fig. 2

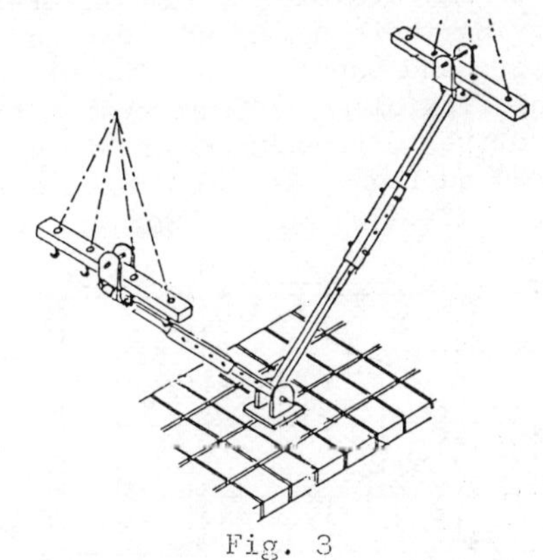

Fig. 3

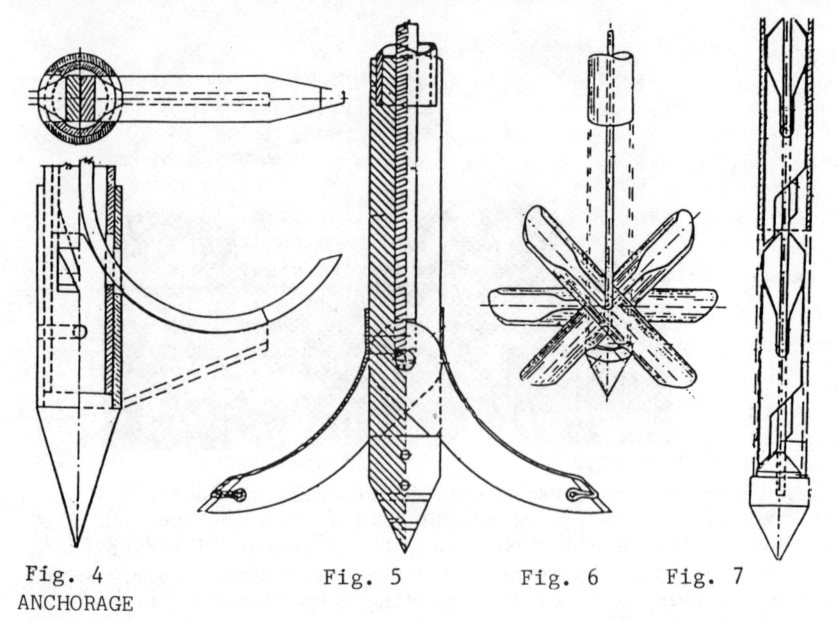

Fig. 4 Fig. 5 Fig. 6 Fig. 7

ANCHORAGE

41. The history of earth anchorage is both brief and recent, and accordingly descriptions of revetment anchors have generally lacked depth. Even today, methods available for insitu testing of multiple earth layers admit no prediction of their joint behaviour under variable physical conditions.

42. Nevertheless recent field tests have shown that anchoring mechanisms, driven less than 2m through wide spectrums of soils, can develop displacement resistances which exceed 3 tonnes at extractive angles of less than 45°. While further investigation of anchor behaviour is necessary, enough is known to identify basic requirements and to include Figs. 4-7 of mechanical anchors which embody principles of minimum cohesive disturbance and variable resistive surfaces.

43. While flexible revetments are frequently anchored by burying their upper or lower ends in trenches, a precast anchoring and panel-connecting block has been developed for one flexible revetment system for use, either as an alternative to trenching, or at intermediate points on an armoured surface. The block is restrained by appropriate anchors and is shown in Fig. 8.

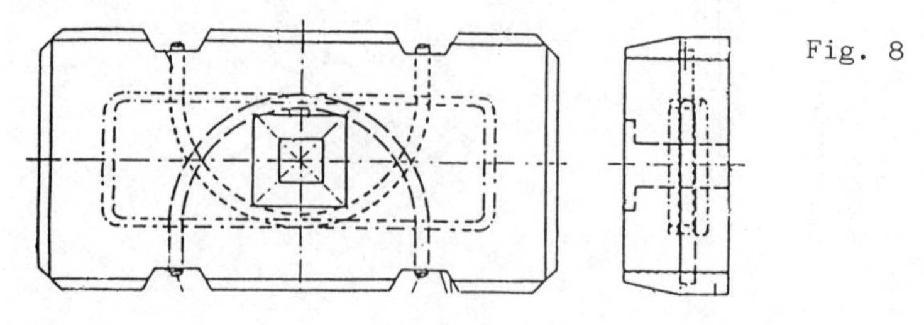

Fig. 8

DESIGN ASPECTS OF FLEXIBLE REVETMENTS

44. It has been said that engineering developments, borne in theory, often die in practice. This certainly cannot be true of all non-rigid revetment systems, many of which seem to have evolved primarily from precedence and with only slight technical justification. A glance at non-dimensional Stability Coefficient (Kd) values, in the well known Hudson Equation, for weight determination of irregular rock armour, will support this view.

45. The mean values of all Kd Coefficients, published in 1975 for "breaking" and "non-breaking" wave conditions, are 5.6 and 6.3 respectively. Since then an increasing use of precast non-rigid armour has introduced a trend towards much higher Kd values: In 1981, for example, a suggested value of 42.5 was published for one such system; more recently the concept of total armour integration, as a possible prime factor of revetment, prompted Kd guide-line values of 80+, subject, of course, to test verification. Finally initial notification of full-scale hydraulic tests carried out in America on integrated flexible armour in 1983 include references to modified Kd values of 230. Some time must necessarily elapse, however, before such early suggestions can be substantiated. In the meantime, since fully integrated flexible revetments exist further consideration of them seems necessary now.

46. To this end, since a GMBL, and therefore an SWL, is assignable to any cable size, and because armour blocks weighing up to 800 lbs. (360 Kgs.) can be economically produced, it is proposed, in the first instance, to relate SWL's to block weights, by considering an arbitary square panel of side "l", comprising l^2 unit-sized blocks each weighing W and integrated by lateral and longitudinal cables, freely passing through "n" orthognally arranged tunnels in each block, and having their ends secured at the periphery of the panel. Then if the panel, as a single flexible revetment component, be suspended by the one block at its centre, the maximum value of W will be given by:-

$$n(SWL) = W\ (l^2 - 1)$$
$$\text{and } W = n(SWL)/(l^2 - 1):-$$

or for any assigned SWL, a maximum block weight will vary directly with the number of cables and inversely as the square of the panel side.

47. By way of example, since the permissable road transportable width of a revetment panel is 8ft. (2.44m), then an 8' x 8' square panel, of unit-sized blocks, integrated by a 1/1 mode of 12.0mm cables, each having a SWL of 3147 lbs. (vidc. Table 1), results in a maximum block weight W = 2 x 3147/63 = 100 lbs. (45 Kgs.).

48. If the panel corners were to be anchored, or otherwise restrained, it will be evident that for a constant value of "l", other cable modes such as 2/1 or 2/2 would

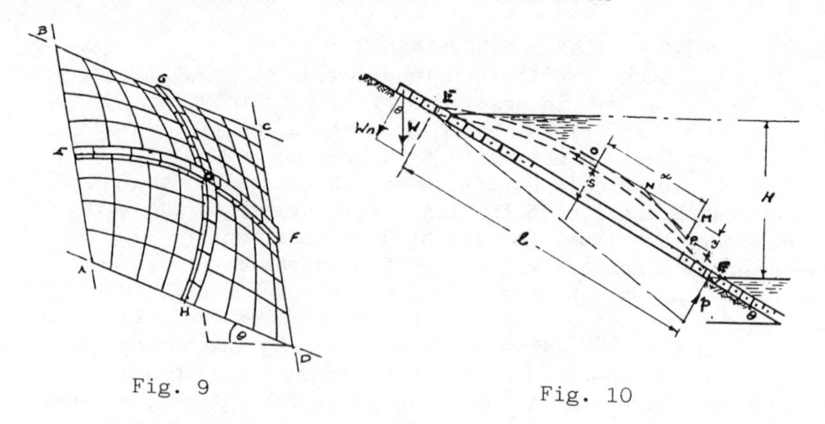

Fig. 9

Fig. 10

permit proportionately differing maximum block weights.

49. The concept of a flexible revetment component ABCD is pursued in Fig. 9, when anchored either at its corners, or at mid-points EFGH, through a porous earth surface at θ to the horizontal; and subject to a hydraulic max. pressure "p" normal to that surface, induced by a periodic water-level draw-down H: Then if "w" be the unit weight of water, p will tend towards ($wl \sin\theta$) as the earth inter-granular pressure falls, with a corresponding increase in the hydraulic gradient, towards its critical, or "quick" state at final earth-surface collapse.

50. Considering the equilibrium of the row of blocks EF in Fig.10, restrained by cables in the two planes EF and GH (Fig.9), each with a maximum deflection angle α and tension T: The weight W of each block will have a resolved component Wn normal to the surface of ($W \cos\theta$) and when subject to an instantaneous mean hydraulic pressure of ($wl \sin\theta/2$), the net upward force on the row or column becomes:-

$$(wl \sin\theta/2)-((W \cos\theta)+(T \tan\alpha + nT \tan\alpha/2))$$

51. It is interesting to note that since α is small, and if p/2 be considered as uniform over l, the cable, if displaced, would approximate to a parabola, having "O" as origin. Hence if "s" were the maximum rise of the cable at times of extreme embankment distress:

let R= total Restraint over length l,
$$= Wnl + T \tan\alpha + lt \tan\alpha/2$$
and P= total instantaneous pressure uplift
$$= p/2 \times l \sin\theta$$

Then net unit uplift $= \dfrac{R-P}{l}$

Considering the cable stability of a length "x" (Fig.10)

$$MP = (R-P)x/l \quad \text{and} \quad MN=T$$

$$\text{Then } \frac{MP}{MN} = \frac{(R-P)x}{lT} \text{ and this } = \frac{2y}{x}$$

or $y=(R-P)x^2/2lT$.
When $y= s$, then $x= l/2$

and $s=\frac{(R-P)l}{8T}$

52. As an extension of the generalised example at Para. 47:-

For $l= 8$; $\theta = 19°(1/3$ approx.); $\alpha = 2°$
$W= 100$; $T= 1000$ (⅓ SWL) and
Tt= Max. transverse cable normal component

Then $H= 8 \sin 19° = 2.6'$
$p/2= 64 \times 1.3 = 83$lbs./ft²
and $P= 83 \times 8 = 664$ (lbs./ft. width of panel)
$Wn= 100 \cos 19° = 95$lbs.
$Tt= 1000 \tan 2° = 35$lbs.
$\therefore$ $R= (8 \times 95)+(8 \times 35)/2 = 900$ approx.
and $s= \frac{(900-664)8}{8 \times 1000} = 0.326'$ or 2.8"(71mm)

53. It is not envisaged that the armour unit displacement discussed in paragraph 49-52, however small and contained such movement may be, should in fact occur when a reveted surface is subject only to predicted attack, and in the absence of sub-surface failure: Neither is it suggested that the stability of a flexible revetment should depend only upon its anchorage. Nevertheless it is evident that an anchored integrated revetment may be rationalised and designed to provide increased structural stability in the event of extreme distress.

CONCLUSIONS

54. a) Single-layer flexible armour integration will result in improved structural stability with increased economic benefits.

b) Much experimental and observational effort remains to be undertaken in order to prove recent developments in flexible revetment techniques.

c) As with much engineering progress, there is a latent danger of market forces exerting powerful influences, upon early technical developments. Were such influences to operate it could well be that a cutting of commercial corners would contribute to premature revetment failures either of non-rigid or flexible revetment systems.

THE WAY AHEAD

55. If the above conclusions are acceptable in principle then there are seemingly good reasons for further research into the whole concept of anchored flexible revetments for incorporation into a Code of Practice.

REFERENCES

1. BAIRD W. The Erosion of River Banks in the St. Lawrence Estuary. Vol. 1 Canadian Soc.Civ.Engng.1977.
2. BAIRD W. The Erosion of River Banks of the St. Lawrence. Can. Journal Civ. Engng. Vol.5 1978.
3. BURSALI S. Economic Revetments for Protecting Banks Against Wave Erosion. Paper A19 Proc. IAHR INT. Symp. on River Mechanics, Bangkok 1973.
4. HYDRO DELFT No. 41-1975. Protected Canal Banks Against the Effects of Passing Ships.
5. MEYER H. The Construction of Permeable Protective Embankment Surfaces Underwater in Shipping Canals. Vol. 4/72 Zeitschrift fur Binnenschiffahrt und Wasserstraben
6. Patent References:-

U.S. Patents No. 4,227,829,

U.K. " Nos. 1,599,312, 1,577,708, 2,031,973
2,063,327, 2,089,862,

U.K. Patent
Application Nos. 2,085,943, 8,300,113, 8,303,931,
8,325,056, 8,320,123, 8,308,646,
8,303,931, 8,326,701

7. SHUGAER R.R. An Expedient System for the Design of Large Earth Canals. Gidrotekh. Strut. No.1 1979.
8. Shore Protection Manual: Dept. of the U.S. Army Corps of Engrs. 1975.

7 Technical and economical design of modern revetments

Dipl.Ing. H.-U. ABROMEIT and Dr Ing. H.-G. KNIESS, Bundesanstalt für Wasserbau, Karlsruhe, West Germany

SYNOPSIS. By the investigation of the rentability of permeable revetments in inland waterways the probability design is useful in which the main forces are determined in category, intensity and frequency. The main determining forces are indicated as spectra which yield the dimensions and out of these the prospective total costs of investment and maintenance as a function of the design level and finally an optimization of the design in technical and economical way. The most benefits result from bonded and flexible covering layers on geotextiles.

Prefaces

1. The revetments of the nearly 5000 km waterways of W.Germany represent an important capital of investment and maintenance. A lot of investigations resulted that the determining values of design, investment and maintenance of permeable revetments of inland waterways are correlated with the intensity and frequency of the interaction between sailing ship and waterway. The design engineer becomes more and more involved with the economic decision problem in which the benefits of a higher design criterion must be weighted against the total costs. The present paper gives statements and criteria for the optimization of the total costs involved based on the inland canal with the cross-section of the European category no. IV. - Ref. 1.

Historical review

1. When the inland shipping was determined by tugboats and dumb barges the attack of the banks and beds of waterways was modest so that even easy revetments were sufficient. The increasing dimensions and engine power of ships increased the loadings since the late fifties and caused damages which made former proofed constructions of no use. To improve the stability of bank protections various constructions of fixed ripraps bonded with asphalt or concrete, concrete slabs and

Fig. 1 : Stern transverse wave, Mittelland-Canal, W.-Germany

some mats or pavements were developed. Thes constructions became stable in technical way, but even more and more expensive.

Forces

1. A sailing ship causes in a waterway with limited width and depth forces by return flow and water-level depression as primary waves and stern transverse waves as secondary waves. The essential forces are determined as probability values in the first step. Based on results of measurements in nature the forces F are indicated as spectra of return flow, water-level depression and secondary ship waves. The spectra correlate the intensity and the frequency of interactions with the level BL of interacting forces. Fig. 2 shows the procedure.

2. Return flow. The return flow is marked by the current velocity V_x and the water-level depression z_A. Both parameters are determined and combined as spectra in height and frequency in Fig. 3.

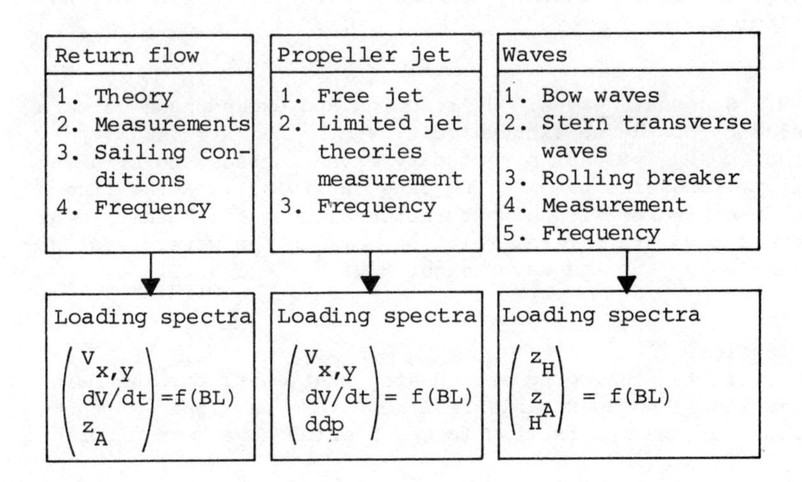

Fig. 2 : Procedure to state loading spectra

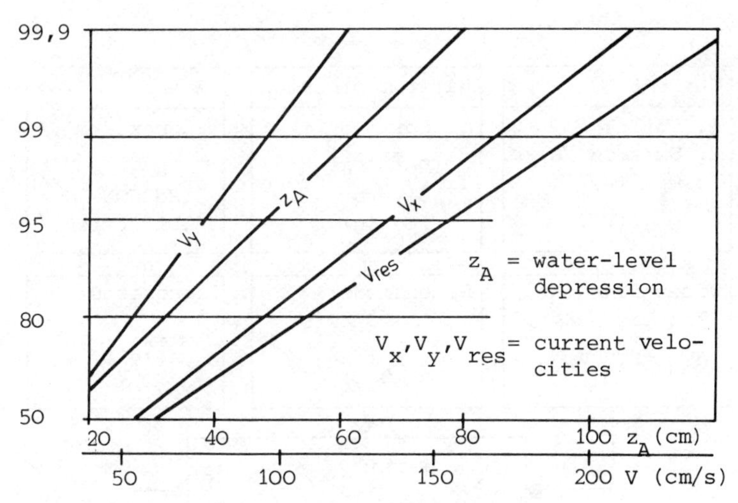

Fig. 3 : Spectra for loadings by return flow

3. Propeller race. In accordance with investigations in theory and model-tests measurements in the Mittellandcanal showed that the propeller yet hits the bottom of a canal only if a ship starts and accelerates out of the normal sailing. The bottom-loading decreases with increasing sailing velocity.

4. Secondary waves. By reaching and exceeding a certain relation of the sailing velocity V_s to the wave-velocity V_c of $V_s/V_c = 0.5$ the secondary waves increase to breaking stern transverse waves which cause a violant loading like a rolling breaker with high turbulence. Long-time measurements gave a good statistical relation between the water-level depression z_A and the wave height H with $H = 1.5 \cdot z_A$.

Dimensioning

1. In the following second step statements for the main dimensions B of permeable revetments are developed as functions of the interacting forces F which have been discribed in spectra before.

2. The basic assumptions of the further procedure are :
   1. The bank is stable in the case of soil mechanics
   2. The soil of bank and bottom consists of a silty fine sand with a permeability of $k = 6 \cdot 10^{-5}$ m/s

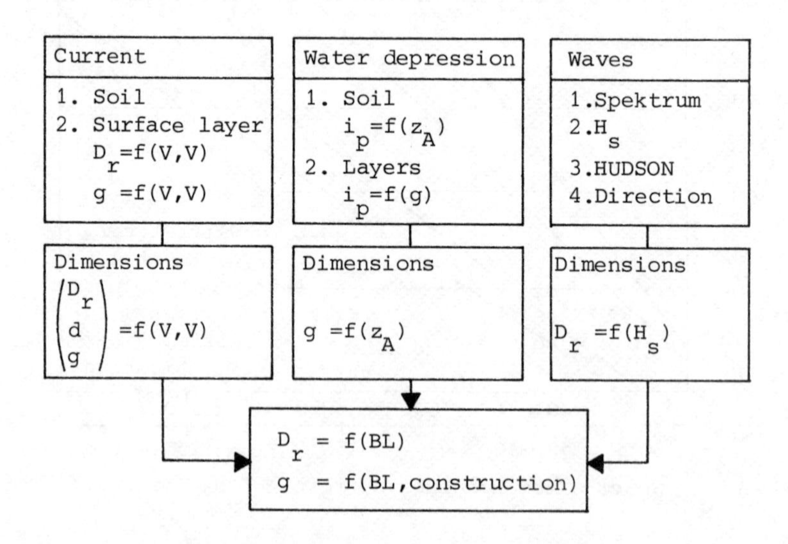

Fig. 4 : Procedure of dimensioning permeable revetments

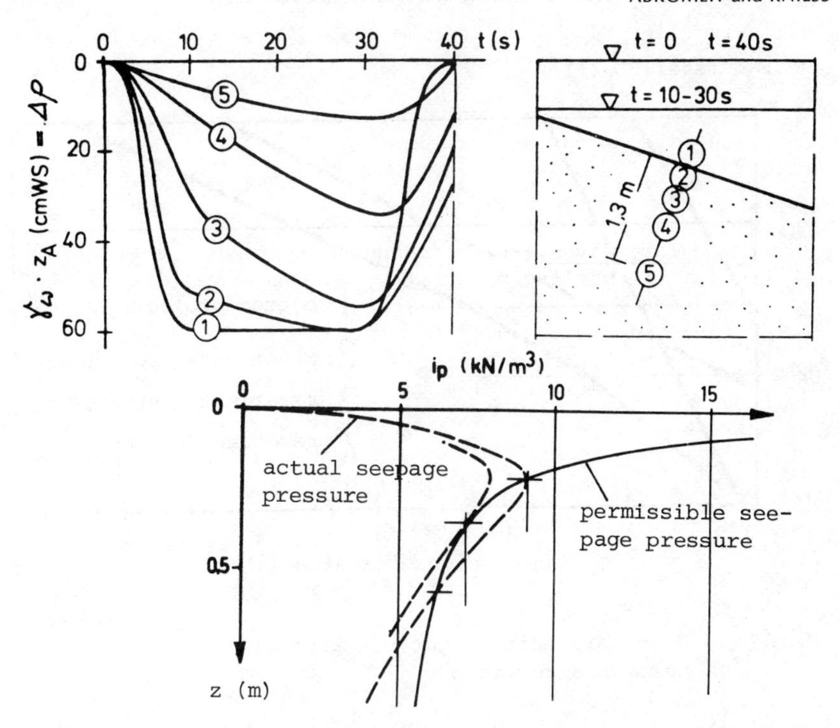

Fig. 5 : Seepage pressure induced by a quick water-level depression

3. Loading by current and waves. For the loading condition "current" the representative sice of rubbles D_r and the weight per unit area g of bonded layers are developed and obtained from known and adapted solutions considering drag, lift and even inertial forces.
For the loading condition "waves" the criterion of HUDSON is well sufficient if the sloping direction is considered.

4. Water-level depression. The quick water-level depression z_A causes a pressure gradient and a seepage pressure i_p inside soil and permeable revetment on slope and bottom which vary with depth z and time. As the permissible seepage pressure can be developed from stability conditions at the critical time the necessary weight per unit area g can be calculated as a function of the water-level depression z_A

5. Standard values. The three procedures discribed above result in technical necessary standard values D_r and g of permeable revetments. As the intensity of the forces F depends on the frequency level which is equal the intensity level or the design level BL the standard values depend on the design level BL, too - Fig. 6 + 7.

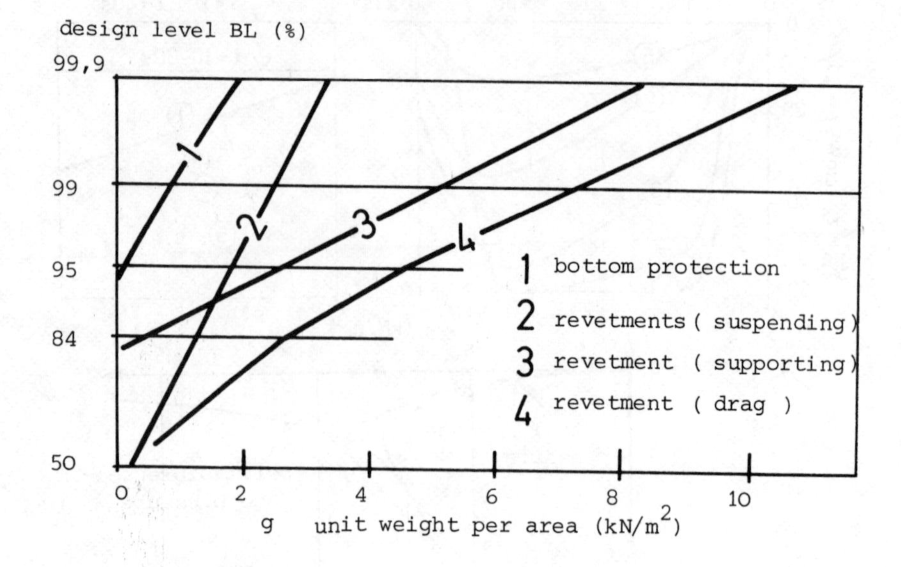

Fig. 6 : Necessary unit weight per area of permeable revetments

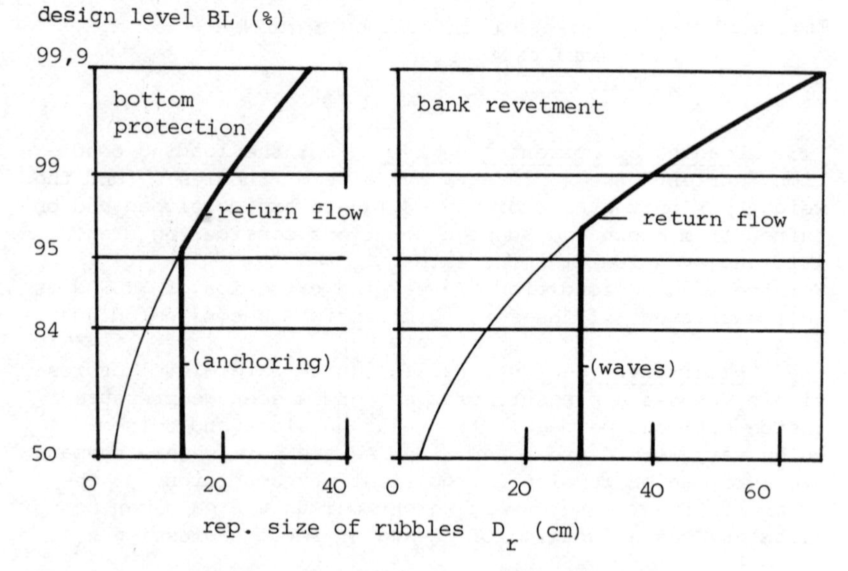

Fig. 7 : Necessary size of rubbles of unbonded ripraps

parts of construction		constructions
surface layer	riprap	oooooo
	riprap part.grouted	oooooo
	riprap tot. grouted	oooooo
	concrete slabs	oooooo
	pavements/mats	oooooo
filter layer	unbond.grain filter	o o o o o o o o o o
	bonded grain filter	o o o o o o o o o o
	geotextiles	o o o o o o o o o o
statics	drag	ooo ooo ooo ooo ooo
	support	ooo ooo ooo ooo
	suspending	ooo ooo
selected constructions		o o o o o o o o

Fig. 8 : Possible constructions of permeable revetments

Constructions

1. Permeable revetments can be built up with a lot of proofed constructions above and under water during still going shipping. Including the statical features fig. 8 shows a matrix of today used, proofed and appropriated constructions.

Rentability

1. In the third step the prospective costs of investment and maintenance of eight selected constructions are calculated with the discounting method within an estimated period of 50 years. The interdependence between dimensions B and the design level BL on the one hand and between dimensions B and total costs C on the other hand yields to an interdependence between total costs C and the design level BL, too.

2.The costs of investment of the eight selected constructions are calculated with mean cost-values of 1981/82 from waterways in W.Germany. The costs of the prospective maintenance are calculated with statistical mean values of the usual effort of maintenance together with assumed strategy models demonstrating the sequence and the quantity of maintenance-work. The strategy-model in fig. 9 shows that the quantity of maintenance of unbonded riprap layers depends essentially on the thickness of the surface layer and on the frequency of ship passages. Bonded surface layers let expect less maintenance and a longer period of using but in all the same tendency of wearing out.

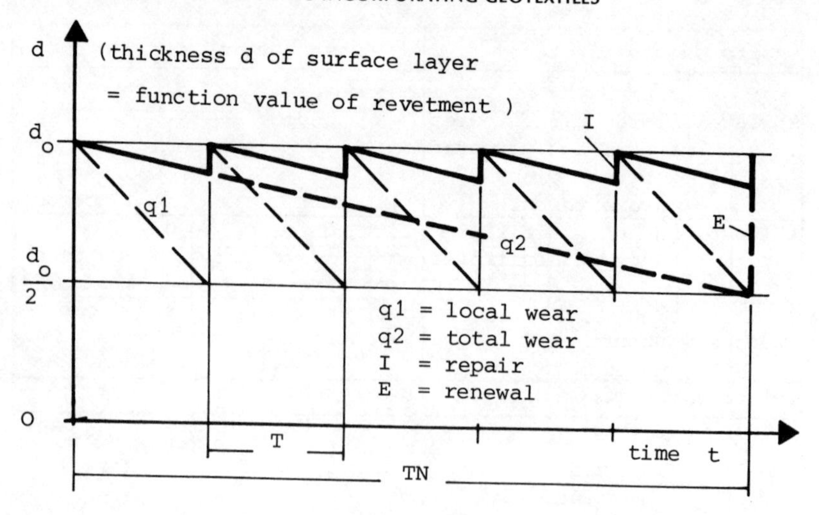

Fig. 9 : Strategy-model for usual maintenance

3. Optimization. To get the optimal design level BL the in technical way necessary dimensions (D, g) and the corresponding total costs C can be calculated now. Fig. 10 shows the simplified results of the investigated constructions. The best rentabilities could be expected by design levels between 50 and 90 percent. This demonstrates that not the construction will be the best one which is dimensioned for the strongest loading but that one which ist dimensioned für lower loadings.
The results show further on that the most benefits can be expected today from constructions with bonded layers of rubbles grouted with partial filling by concrete and with geotextiles as filter layer. But it must be remarked that these constructions are only satisfactory if the bonded layer is either flexible or the bank let expect no worth mintioning setteling.

Property test

1. To get a mostly objective knowledge about the specific properties of bonded revetment constructions with different materials of grouting we make property tests nowadays that enclosure three main measurements :

- flexibility test
- permeability test
- stability of grouting

Fig. 11 shows the equipment to test the flexibility with water filled cushions which can be changed in pressure to simulate the support conditions.

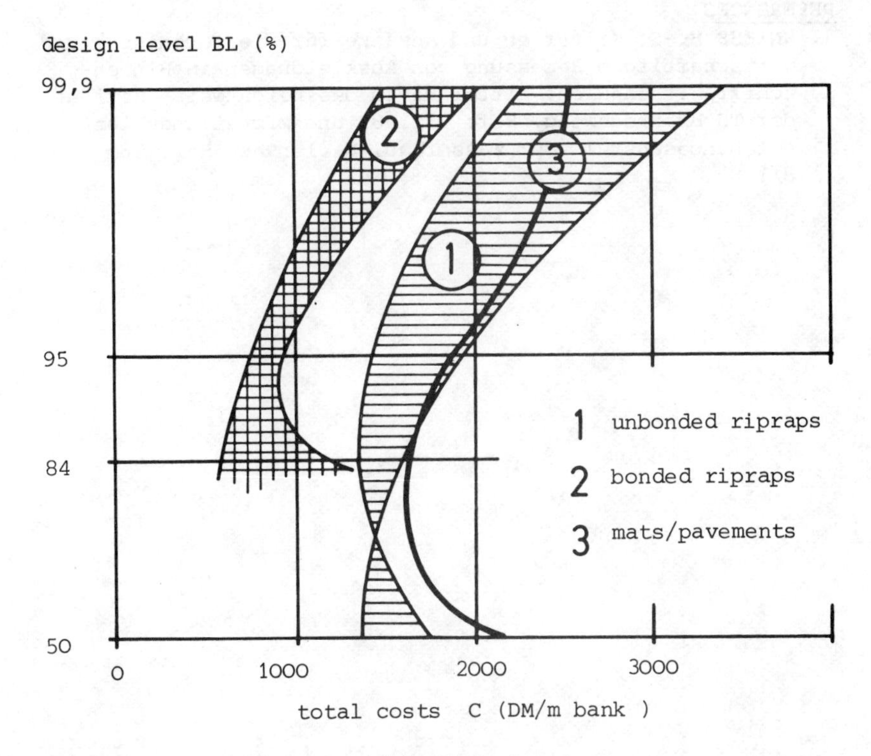

Fig. 10: Total costs of bank protection as a function of design level BL

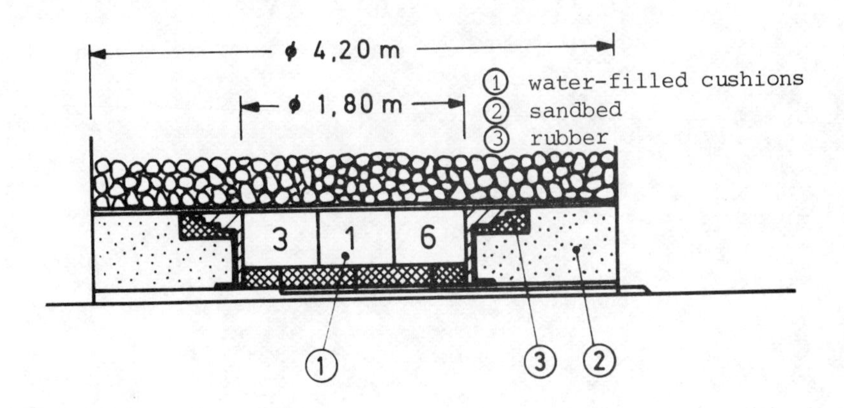

Fig. 11: Equipment for testing the flexibilty of revetments

REFERENCES

1. KNIESS H.-G. Kriterien und Ansätze für die technische und wirtschaftliche Bemessung von Auskleidungen in Binnenschiffahrtskanälen, Mitteilungen des Leichtweiß-Instituts der TU Braunschweig, Heft 77/1983 und Mitteilungsblatt der Bundesanstalt für Wasserbau, Karlsruhe, Heft 53, 8/1983

8 Stability of Armorflex revetment system under wave attack

Ir. C. van den BERG, Nicolon BV, and Ir. J. LINDENBERG, Delft Soil Mechanics Laboratory

SYNOPSIS. For a safe application of Armorflex concrete block slope protection mats it is necessary to know the stability under wave attack. For this reason 1:1 scale tests were done in the Delta Flume of the Delft Hydraulics Laboratory. The 1:3 slope consisted of a sandcore covered by 30 cm coarse gravel and was protected against wave attack by an Armorflex 180 concrete block system. The weight of the Armorflex system was about 180 kg/m2, the height of the blocks 11,5 cm. The system had an open area of 17%.
The Armorflex system and the Delta Flume test program including the results are described.

INTRODUCTION

1. In July 1982 Nicolon B.V. commissioned the Delft Hydraulics Laboratory to carry out a research program on the stability of Armorflex concrete block slope protection mats under wave attack (ref. 1).
2. To determine the stability under wave attack the research program involved model tests in the Delta Flume of the Delft Hydraulics Laboratory, and the Delft Soil Mechanics Laboratory, including the execution of a number of measurements.
3. The objective of the Delta Flume investigation primarily was the determination of the maximum hydraulic conditions during which the Armorflex revetment system with a realistic structure, inclination 1:3, and subsoil can be applied without damage.
4. Following a preliminary analysis it appeared that the dimensions and capabilities of the Delta Flume were suitable to carry out the investigation at "actual size", i.e. by using a prototype Armorflex 180 concrete block mat.

DESCRIPTION ARMORFLEX

5. The prefabricated Armorflex revetment system is constructed of interlocking concrete blocks with

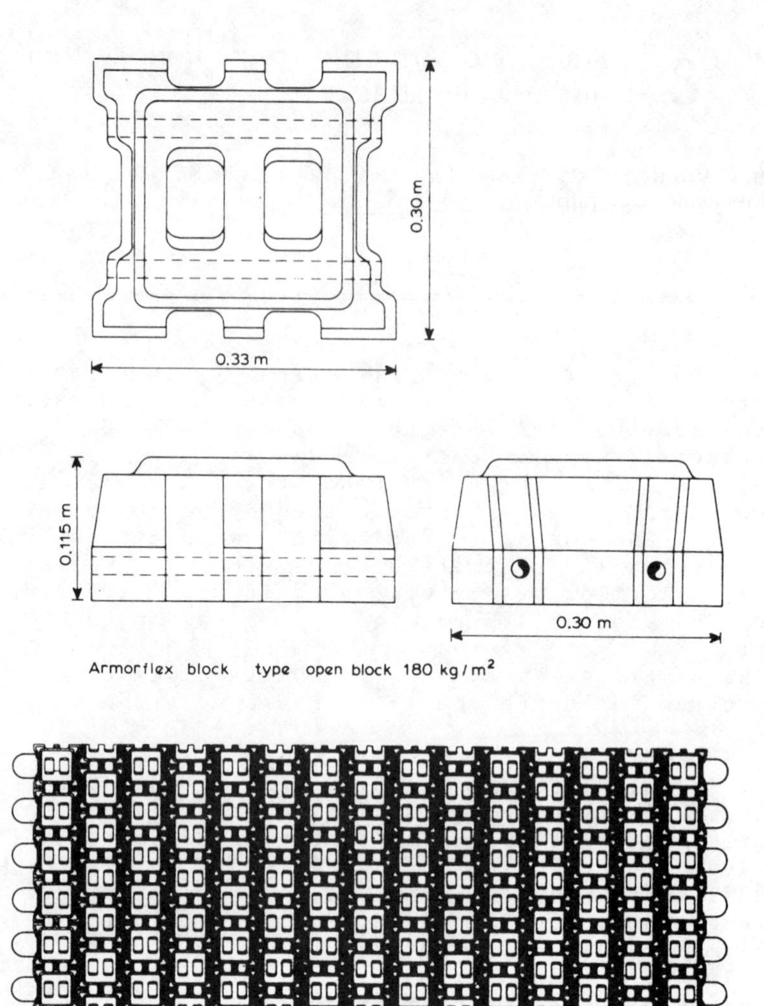

ARMORFLEX BLOCK AND ARMORFLEX MAT

Fig. 1

specific hydraulic properties (fig. 1). The Armorflex mat as such is assembled by connecting blocks by means of cables. The blocks are placed in stretcher bond. The cables run through special ducts in each block.

6. Block heights vary from 0,10 - 0,25 m for blocks with holes as well as for blocks without holes in the middle. Even blocks of 0,60 m are considered at the moment. The maximum length of an Armorflex mat is in principle determined by the available handling equipment.

7. The Armorflex mat system can be used for protection against various types of hydraulic loads along sea, river and harbour slope defences, e.g. storm wave loads, ship waves, river currents and ice flow.

DEFINITION STABILITY

8. As a lower limit the Armorflex revetment system is considered to be stable under wave attack as long as all the single blocks, of which the system is assembled, are stable.

9. In general the concrete block-cable systems or concrete block-geotextile systems need a certain vertical movement over a distance "d" before the dead weight of the surrounding blocks will be mobilised. This distance "d" may be more than 2 cm and is depending a.o. on the diameter of the ducts through which the cables run, the cable diameter and stiffness and the elongation of the geotextile to which the blocks are attached. Although, depending on sublayer properties (e.g. grain size) a vertical movement of this order of magnitude is often not acceptable for normal design circumstances. However, under exeptional hydraulic conditions this might be acceptable. In such cases the cables can be introduced as an additional safety. If a certain movement is accepted serious attention must be given to the stability of the sublayer material underneath the slope protection.

10. A previous model study on Armorflex mats on a 1:10 scale showed these mat uplift phenomena very clearly (ref. 2). This model study was carried out some years ago for Nicolon by Tetra Tech Inc., California, as a first evaluation of the behaviour of Armorflex concrete block mats when exposed to a range of wave conditions. In these tests the cables were not removed and could act as a reinforcement. Parts of the mat consisting of several blocks were lifted up and down under wave action apparently heavier than could be withstood by loose blocks. During a number of these tests deformation of the sublayer as a result of erosion could be observed.

MODEL BOUNDARY CONDITIONS

11. The model layout has mainly been determined based on the next 4 items.

a. Permeability of the slope protection system. From previous research (ref. 3 and 4) it was known that the permeability of the slope protection system is an important parameter for the stability under wave attack. The more permeable the protection system the higher the wave height at which damage occurs. For this reason an open Armorflex block and a very permeable filter fabric, placed between the Armorflex system and the sublayer, were chosen. The open area of the chosen Armorflex revetment, block thickness 11,5 cm, system was 17%. The filter fabric was Nicolon 66447.

b. Permeability of the sublayer. Further it was known that the permeability of the subgrade material is also an important parameter. The more permeable the subgrade material the lower the wave height at which damage occurs (ref. 4 and 5). Because a lower limit for the equilibrium conditions of the Armorflex blocks was wanted by Nicolon, a very permeable subgrade material (silex) was used. This material has a D50 of 37 mm and is sufficiently resistant to internal erosion.

c. Definition stability. As explained before for the stability of the Armorflex revetment system the stability of the single Armorflex blocks, not connected by cables, is considered. So in the area where damage could be exptected loose blocks were placed on the slope.

d. Slope inclination in practice. A slope of 1:3 was chosen because this is a reasonable average of the slopes on which Armorflex revetment systems are used. The parameter analysis makes it possible to convert the results to other slope inclinations within a range 1:2 - 1:4.

12. Figure 2 presents 2 longitudinal sections of the model layout in the Delta Flume, scale 1:1000 and scale 1:200 respectively. The slope inclination 1:3 was present over a vertical height of 8.75 m. About 5.5 m of this height was protected by 3 Armorflex mats with a total length along the slope of 18 m. During the tests the mean water level was kept at 5 m above the flume bottom.

TEST PROGRAM

13. A test programme was planned, with which the

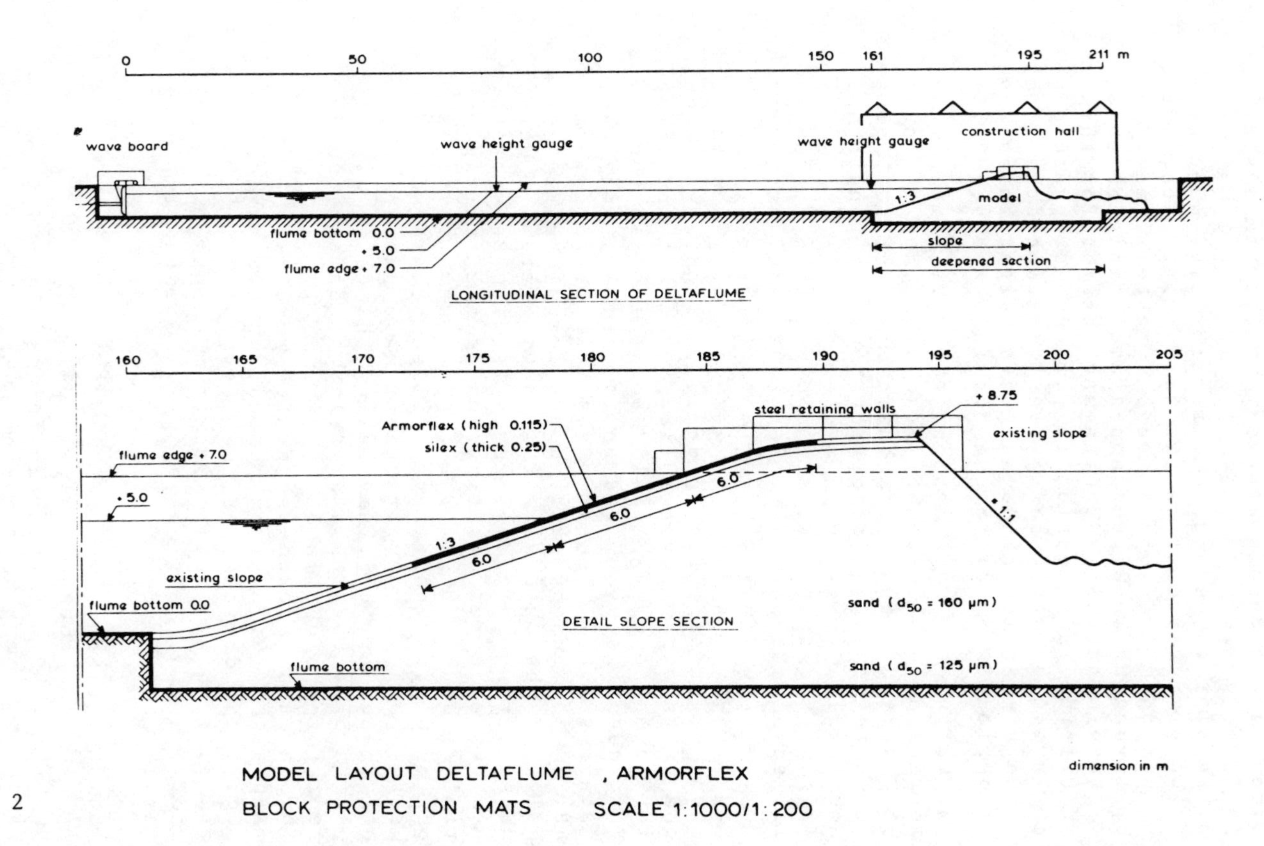
0
50
100
150
161
195
211 m
wave board
wave height gauge
wave height gauge
construction hall
1:3
model
flume bottom 0.0
+ 5.0
flume edge + 7.0
slope
deepened section
LONGITUDINAL SECTION OF DELTAFLUME
160
165
170
175
180
185
190
195
200
205
steel retaining walls
+ 8.75
Armorflex (high 0.115)
silex (thick 0.25)
existing slope
flume edge + 7.0
+ 5.0
6.0
6.0
± 1:1
1:3
6.0
existing slope
flume bottom 0.0
sand (d50 = 160 μm)
DETAIL SLOPE SECTION
flume bottom
sand (d50 = 125 μm)
dimension in m
MODEL LAYOUT DELTAFLUME , ARMORFLEX
BLOCK PROTECTION MATS SCALE 1:1000/1:200

Fig. 2

stability of both non-gravel filled concrete blocks and mats with gravel filled narrow tapered joints between and holes in the blocks, could be determined. The sequence of the test program was as follows:

a. Regular waves and non-gravel filled blocks. Determination of damage wave height for 3 wave periods, T=3 s, 4 s and 5 s respectively.
b. Irregular waves and non-gravel filled blocks. Determination of damage wave height for the critical wave period $T=T_p$, found in phase a.
c. Irregular waves and gravel filled blocks. Determination of damage wave height for a wave period T_p equal to the one in phase b.

14. During the wave action, the wave height was increased in steps, until the first damage could be observed visually. A summary of the test program including the most important boundary conditions and results is given in table 1.

15. For the test phases a and b with non-gravel filled blocks only data for 2 wave heights for each wave period are included in table 1, viz. the highest wave at which no damage took place and the wave height at which no damage was observed. During the tests with gravel filled revetment and irregular waves no damage could be obtained in the flume. For these tests, 57 and 58 in table 1, only the data for the 2 highest waves are presented. Tests 48, 50, 52 and 54 carried out immediately after the holes in the mats were filled with gravel, were added during the model investigation and meant to achieve a better distribution of the filling material without damage. The regular wave tests with period 6 s and non-gravel filled blocks were added because of the very irregular wave pattern that originated in the flume during the tests with wave period 5 s. For T=5 s no relevant damage could therefore be obtained.

16. In each test the generated wave height was maintained for a given period. When damage did not occur, the duration of the test was 20 minutes minimum, and 40 minutes maximum (during test 58). The wave generator was stopped immediately, once damage was observed.

TEST RESULTS

17. In this chapter only a summary is given of the results obtained from the full scale model investigation. A much more comprehensive description of the test results is presented in ref. 1.

18. Damage wave.

In figure 3a and 3b the dimensionless significant

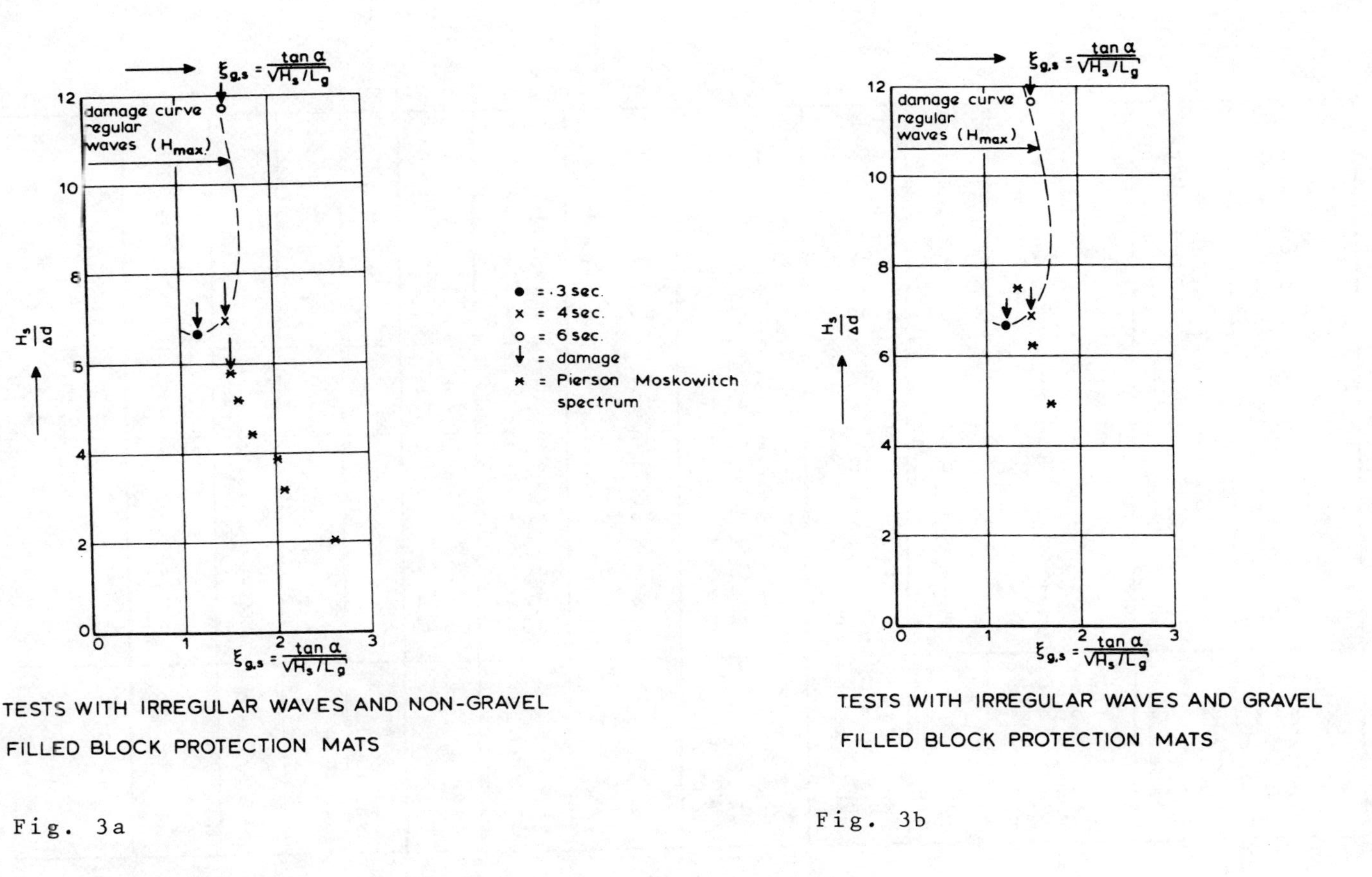

TESTS WITH IRREGULAR WAVES AND NON-GRAVEL FILLED BLOCK PROTECTION MATS

Fig. 3a

TESTS WITH IRREGULAR WAVES AND GRAVEL FILLED BLOCK PROTECTION MATS

Fig. 3b

test series	test no	T or T_p (s)	regular waves	irregular waves	H_i or H_s (m)	H_{max} (m)	L_g (m)	$\xi_{g,max}$ or $\xi_{g,s}$ (-)	$\frac{H_{max}}{\Delta d}$ or $\frac{H_s}{\Delta d}$ (-)	r (%)	damage? damage level (i.r.o. flume bottom) (m)	wave runup vertically i.r.o. mean water level: runup Z (m)	Z/H_{max} (-)
regular waves non-gravel filled	10	3	+		0,96	1,00	13,76	1,24	6,15	4	no	1,21	1,21
	11	3	+		1,02	1,08	13,76	1,19	6,64	6	4.62+ 4.85+	-	-
	19	5	+		0,34	0,56	30,31	2,45	3,44	58	no	-	-
	20	5	+		0,58	0,80	30,31	2,05	4,92	38	no	-	-
	33	6	+		1,40	1,88	38,09	1,50	1,6	34	no	3,04	1,62
	34	6	+		1,44	1,90	38,09	1,49	1,7	32	3,98+	3,29	1,73
	39	4	+		0,89	0,98	22,20	1,59	6,03	10	no	1,54	1,57
	40	4	+		1,06	1,12	22,20	1,48	6,89	6	3.98+ 4.41+	-	-
irregular non-gravel filled	45	3,75		B	0,84	-	20,11	1,63	5,17	-	no	-	-
	46	3,75		B	0,93	-	20,11	1,55	5,72	-	4.19+ 4.41+	-	-
regular gravel filled	48	3	+		0,57	0,66	13,76	1,52	4,06	16	no	0,96	1,45
	50	4	+		0,68	0,86	22,20	1,69	5,29	26	no	1,37	1,60
	52	5	+		0,35	0,54	30,31	2,50	3,32	57	no	0,71	1,31
	54	6	+		0,61	1,00	38,09	2,06	6,15	64	no	1,46	1,46
irregular gravel filled	57	3,70		B	1,02	-	19,69	1,46	6,27	-	no	-	-
	58	3,68		B	1,22	-	19,52	1,33	7,50	-	no	-	-

Table 1. Summary of boundary conditions and test results

wave height $H_s/\Delta d$ for the irregular waves has been plotted as a function of the breaker parameter $\xi_{g,s} = \frac{\tan\alpha}{\sqrt{H_s/L_g}}$ (see table 1). In figure 3a the results from the tests with non-gravel filled blocks are shown. The assumed damage line for the regular wave test series for wave periods 3, 4 and 6 s is added. Damage tests are indicated by an arrow (↓). The damage line clearly illustrates the relatively great stability of the blocks at T=6 s compared with those at the smaller wave periods of 3 s and 4 s. Based on the results of the tests with regular waves and a non-gravel filled slope, the average period T_p=3.75 s was selected for the tests with irregular waves. It was decided not to use the period of 3 s (smallest $H/\Delta d$ at which damage occurred) for this test, as it was expected that wave breaking would too soon impose a limitation, during increasing of the wave height. The damage point for the test with irregular waves and non-gravel filled slope lies below the damage line for regular waves. Figure 3b shows the results of the three tests with irregular waves, after that the holes in and between the stones were filled with gravel. The damage line of the tests with regular waves and non-gravel filled slope based upon H_{max} and L_g has been added in this figure too. In the case of a gravel filled slope damage was not observed. In the last test 58 at H_s = 1.22 m, much wave breaking occurred between the wave generator and the foot of the slope. This means that the physical boundaries with irregular waves have been reached at T_p = 3.75 s. A further increase of the wave generator capacity then will not lead to an increase of H_s. The significant wave height H_s = 1.22 m in test 58 has been maintained during approximately 40 minutes. In this test the wave height exceeded by 1% of the total number of waves, was approximately 1.80 m.

19. Damage location.

Damage of the revetment was found at the end of 4 tests with non-gravel filled blocks. As expected, in each case damage occurred at a level below the mean water level in the flume.

The Armorflex investigation confirmed the findings from previous research on various slope revetment systems that first damage takes place in a zone between the mean water level and the level corresponding with one wave height below the mean water level (see table 1).

20. Maximum wave runup.

In figure 4 the dimensionsless wave runup Z/H_{max} is plotted as a function of both breaker parameter $\xi_{g,max}$ for all tests with regular waves. A compari-

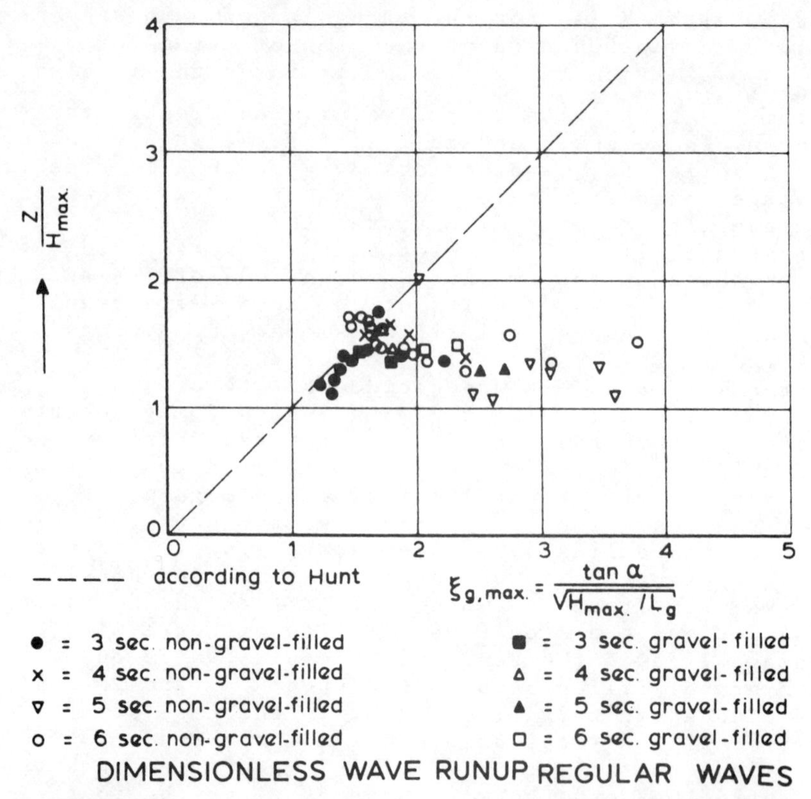

Fig. 4

Photo 1. Breaking wave on Armorflex slope protection system.

son with the wave runup relation according to Hunt (1) shows that the measured wave runup at the relatively low waves, lies below the line of Hunt. This particularly applies to the greater wave periods T = 5 and T = 6 s. At higher wave heights, the measured runup is about the same as that according to Hunt.
In figure 5 a distinction is made between wave runup for a non-gravel filled and a gravel filled slope. In general the dimensionless wave runup for the gravel filled slope is slightly higher than for the non-gravel filled slope. However, the data are rather scattered, so a consistant conclusion could not be drawn.
21. Wave reflection from the slope.
For regular waves the reflection from the slope is determined according to:

$$r = \frac{H_{max} - H_{min}}{H_{max} + H_{min}} = 100\%$$

in which H_{max} is the through-crest value of the wave height, at that point in the flume, where this is greatest. H_{min} represents the smallest through-crest. A comparison between the reflection percentage for a non-gravel filled and a gravel filled slope shows that there is no clear difference. Filling the holes in and between the Armorflex blocks with gravel hence does not cause an increase in reflection. For irregular waves the incident and the reflected wave spectrum are determined from the registrations of two pressure sensors on the bottom of the Delta Flume via a correlation of the measuring signal of a wave height gauge directly above one of the pressure sensors.
22. During the full scale tests only minor settlement and deformation of the slope took place (maximum settlement approximately 0.03 m). It was concluded that this settlement did not influence the performance of the slope revetment during the tests.
23. Behaviour of block filling material during wave action. After completion of test 46, the holes in the concrete blocks and the narrow tapered openings between the blocks were filled with a mixture of concrete gravel and fine gravel. In practice normally fine gravel, D50 = 3 to 6 mm, is used. All openings were filled completely. After wave action it appears that almost half of the gravel material has disappeared from the holes in the heavily loaded zone. This material was deposited at the bottom of the slope. In the narrow tapered seams between the blocks fine gravel was still present. The observed greater stability

of the gravel filled concrete block mats was most likely caused by this material.

CONCLUSION

24. For the circumstances during the investigation, the following conclusions can be formulated, on the basis of observations during the tests.

1. For the damage wave parameters, as in a previous research, the dimensionless quantities $H/\Delta d$ and $\xi = \frac{\tan\alpha}{\sqrt{H/L}}$ were selected.
   The smallest damage wave height found during the tests with regular waves and non-gravel filled Armorflex concrete block mats, without steel cables, occurred at a wave period T = 3 s and also the lowest ξ-value took place. The damage wave height of the block mat in tests with T = 6 s was substantially larger than that at T = 3 s and T = 4 s (H_i = 1.44 m, compared with 1.02 m and 1.06 m).
2. During the damage test with irregular waves, T_p = 3.75 s, and non-gravel filled mats H_s appeared to be slightly smaller than the incident damage wave height, derived from the tests with regular waves and the corresponding wave steepness parameter ξ.
3. During the tests with irregular waves, T_p = 3.75 s, with the holes filled with gravel, damage could not be brought about. A greater wave height than H_s = 1.22 m (maintained during 40 min.), could not be realized in the Delta Flume, because of the breaking of the waves at this T_p and H_s.
4. The stability of the Armorflex concrete block mats with gravel filled holes and tapered joints between the mats, is considerably larger than that of the non-gravel filled mats.
5. Filling of the holes in and between the Armorflex blocks with gravel results in much erosion during wave action. After a total of approximately 5 hours of wave action, about half of the gravel turned out to be eroded from these holes, over a total length of 4 m, measured along the slope.
6. After completion of the tests, most of the finer material (coarse sand and fine gravel) which was present in the narrow tapered joints between the blocks, appeared to be adhered. It can reasonably be assumed that the adhesion and interlocking effect of this material has led to a greater stability of the gravel filled block mat.
7. The stability of separate, non-gravel filled blocks on a slope being submitted to wave

attack, depends on a great number of parameters. Amongst other factors, permeability of the mat and filter layer, the relative permeability of the mat and the dynamic and quasi-static wave boundary conditions, play their part. The stability of the non-gravel filled Armorflex block mats appeared to be rather large during the Delta Flume research, e.g. larger than that of a slope protection consisting of closed square blocks, as found during a previous survey. This was probably caused by the relatively larger water permeability of the Armorflex block mats and of the interlocking of the blocks.

8. The results of the full scale model investigation can be used to determine the lower limit stability of the Armorflex slope revetment system. As mentioned before, under certain conditions, additional safety as a result of integrate mat behaviour, might be taken into account. In these cases the results of the earlier 1:10 scale model test (ref. 2) can be used to determine the total mat stability.

REFERENCES

1. Stability of Armorflex block slope protection mats under wave attack, Report on model investigation. Delft Hydraulics Laboratory and Delft Soil Mechanics Laboratory, M1910, May 1983.
2. J. Weckmann and J.M. Scales. Design Guidelines for cabled-block mat shore protection systems Proc. Coastal Structures '83, March 1983, Arlington, Virginia ASCE, pp 295-306.
3. A. Wevers, Stability of slope-revetment of paved concrete blocks under wave attack, M1057, 1970.
4. K. den Boer, C.J. Kenter, K.W. Pilarczyck. Large scale model tests on placed blocks revetment. Delft Hydraulics Laboratory, publication no. 288, January 1983.
5. Delft Hydraulics Laboratory. Slope revetment of paved concrete blocks under wave attack; the effect of the permeability of the foundation; M1410 - part II, April 1981.

LIST OF SYMBOLS AND PARAMETERS

Test number (chronological)

L_g (m) = wavelength in flume, according to linear wave theory

H_i (m) = waveheight of the incident wave

$$= \frac{H_{max} + H_{min}}{2}$$

H_{max} (m) = maximum trough-crest value of a combined standing and progressive wave

α(rad) = slope inclination

$$\xi_{g\,max}(-) = \frac{\tan\alpha}{\sqrt{H_{max}/L_g}}$$

r (%) = reflection coefficient

$$= \frac{H_{max} - H_{min}}{H_{max} + H_{min}} \cdot 100\%$$

T (s) = wave period (with irregular waves T_p)

$$\Delta\ (-) = \frac{\rho_s - \rho_w}{\rho_w}$$

ρ_s (kg/m³)= specific mass of block

ρ_w (kg/m³)= specific mass of water

d (m) = thickness of block

$\frac{H_{max}}{\Delta d}$ (-) = dimensionless height of the maximum wave

spectrum shape: B = Pierson Moskovicz spectrum

occurrence of damage + = yes

(or visible movement)- = no

area in which damage occurs (m), measured vertically i.r.o. flume bottom, limited by the centre of the upper surface of the highest and the lowest damage block

wave runup, measured along slope (m)

Z (-) = wave runup (m), measured vertically i.r.o. mean waterlevel

$\frac{Z}{H_{max}}$ (-) = dimensionless wave runup

9 Tubular gabions

C. D. HALL, MSc, MICE, MCIT, Netlon Ltd, Blackburn

SYNOPSIS. This paper describes the use of tubular gabions as a flexible armoured revetment. A design method is presented which enables the selection of a tubular gabion diameter for a specified bow wave height and current velocity. The design method is put into context by outlining the scope of its applicability in waterways. Two roles for geotextiles are discussed, an open textured grid for the gabion material and a filter fabric underlayer. A construction method appropriate to waterway working is described.

INTRODUCTION

1. A tubular gabion revetment comprises a battery of stone filled polymer grid tubes, forming a flexible armouring to the bank of a waterway.

2. The open textured, large aperture grids are members of the geotextile family (ref.1) ensuring containment of the stone whilst maintaining the revetment's porosity - a feature which is vital to the dissipation of wave and current energy and hence the protection of the bank from erosion.

3. Extruded polyethylene grids have been used in civil engineering for nearly 20 years with most of the early work undertaken in Japan (refs.2,3). These applications were mainly geotechnical in nature where the grid, in sheet form, was incorporated in weak soils as a reinforcing element. An exceptional case was the installation of rock-filled tubes providing a coastal defence on the island of Kyushu. This application revived the earliest use of gabion baskets when cylindrical wire mesh forms were used in breach repairs on the Reno River, Italy in 1894 (ref.4).

4. This paper takes a fresh look at tubular gabions by considering the features of interest to Engineers, namely a basis for the design of a tubular gabion revetment and the scope for geotextiles in providing a flexible armoured revetment suited to navigable waterways.

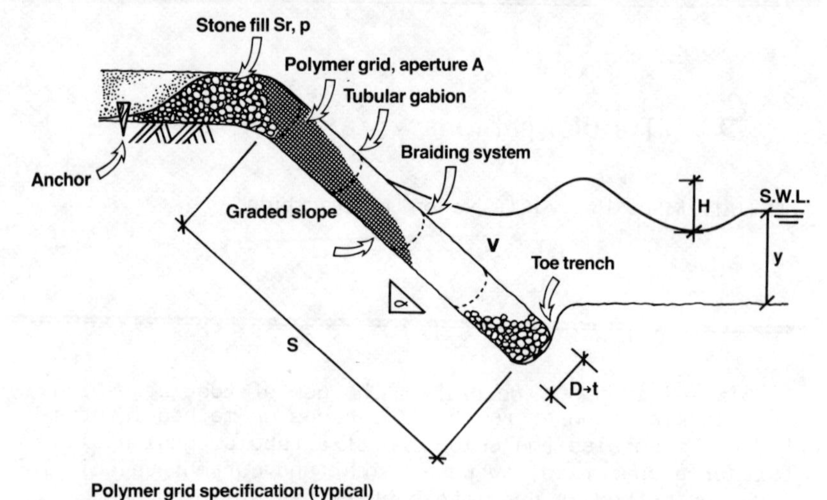

Polymer grid specification (typical)
Material: high density polyethylene
Characteristic tensile strength: 5.8 KN/M

Diameter D:	**0.63m**
Aperture A:	**27mm x 27mm**
Rib thickness:	**5mm**
Weight:	**660 g/m²**

Fig.1 Characteristics of tubular gabions

DESCRIPTION

5. Polymer grids are extruded from circular counter-rotating dies with the consequence that the tubular gabion is manufactured in a continuous form with no seams or joints. The variables in manufacture are the tube diameter, the grid aperture size and the rib thickness. The length is determined by the requirements of the site. Extrusion through counter-rotating dies offers the opportunity to produce complex rib/aperture configurations, but for tubular gabions, a simple rectangle is satisfactory for the retention of stone.

6. The characteristics of tubular gabions and a typical material specification are given in Fig.1.

DESIGN

7. A feature of tubular gabions that is pertinent to design is that the stone fill can be expected to migrate down the tube under current and wave action. To cater for this, a generous reservoir of stone is provided at the shoulder serving to 'top-up' the body of the gabion in these circumstances. Settlement of the gabion onto the slope can also be expected, such that the initial diameter (D) tends to a rectangle (Fig.2). The resulting thickness of revetment (t) is related to (D) by

$$t = 0.6D \qquad (1)$$

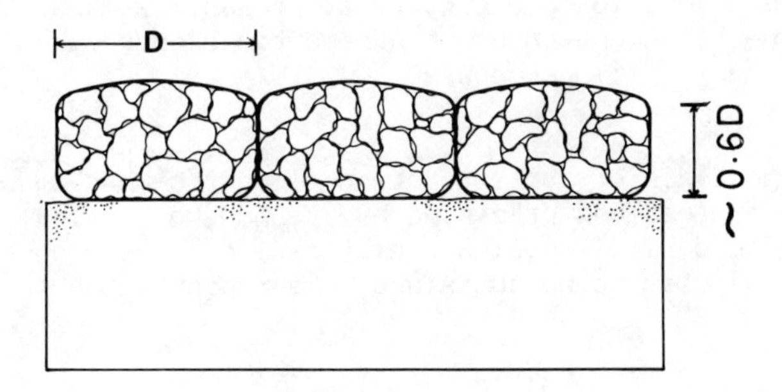

Fig.2. Ultimate settlement of gabions

8. The essential design of a tubular gabion revetment is to derive a diameter which ensures that a battery of gabions is stable when exposed to waves and currents. In waterway engineering, the design must cater for the hydraulic effects of vessels in passage, namely bow waves and return currents.

WAVES

9. For an analysis of porous revetments subjected to wave attack, the work of Brown (ref.5) is referred to here. This work provides a theoretical treatment of wave impact on a porous slope. It considers that the wave is in translation (i.e. physical propagation of the wave mass) and that its impact on the slope is akin to jet impulse. In such circumstances, the porosity and density of the revetment are of fundamental importance along with the slope's gradient. Two failure modes apply to tubular gabions - uplift (where the gabion is lifted from the slope) and buckling (where the gabion deforms locally). A third form of revetment failure, wholesale sliding, is considered not to apply due to the anchored shoulder and toe arrangements, as shown in Fig.1. The design criteria for the two failure modes are as follows.

i Uplift
$$\frac{H}{t} < Cbu\ (Sr-1)(1-p)\cot^{1/3}\alpha \qquad (2)$$

Cbu is an empirically derived coefficient with a value of 7.0 obtained from scaled laboratory testing.

Using equation (1), the uplift criterion becomes:

$$\frac{H}{D} < 4.2\ (Sr-1)(1-p)\cot^{1/3}\alpha \qquad (3)$$

ii Buckling $D > \frac{S}{35}$ (4)

This relationship is arbitrary at present: it ensures that the revetment does not become too slender when designing for mild exposures.

CURRENT

10. For an analysis of current action along the bank of a waterway, the expressions presented by Stephenson (ref.4) are used here to derive a gabion diameter. An expression for stable stone size to be contained in a gabion subjected to current is:

$$d > \frac{K_3.r.i}{(Sr-1)(1-p)\cos\alpha\sqrt{(\tan^2\phi - \tan^2\alpha)}} \quad (5)$$

where K_3 is a coefficient with a value of 8 (approx), r is the hydraulic radius of the waterway, i is the hydraulic gradient and $i = \frac{V^2n^2}{r^{4/3}}$ in Manning's equation.

If the hydraulic radius r is taken to approximate to the average water depth y in a 'wide' waterway and the value of Manning's n taken as 0.030, equation(5) becomes:

$$d > \frac{0.007V^2}{y^{\frac{1}{3}}(Sr-1)(1-p)\cos\alpha\sqrt{(\tan^2\emptyset - \tan^2\alpha)}} \quad (6)$$

11. A condition to be satisfied by the fill material is that individual stones are neither too large, which may cause difficulty during filling, nor too small such that they become excited by current action. Assuming, typically, that d=0.25t (7) then equations (1), (6) and (7) combine to give:

$$\frac{V^2}{D} < 21y^{\frac{1}{3}}(Sr-1)(1-p)\cos\alpha\sqrt{(\tan^2\emptyset - \tan^2\alpha)} \quad (8)$$

DESIGN SCOPE

12. For outline design purposes, the parameters in design criteria equations (3), (4) and (8) can adopt the following typical values.

$$Sr = 2.6$$
$$p = 40\%$$
$$\emptyset = 35^o - 40^o$$

13. The scope for these design criteria is greatly dependent on the slope of the waterway bank. For in-situ filling methods, the bank slope needs to be sufficiently steep to permit gravity feed of stone into the body of the gabion.

The range of bank gradients considered to be suitable for this application is:

$$(1:2.0) < \alpha < (1:1.0) \qquad (9)$$

Shallower than this range will require the adoption of a filling method other than that described in this paper. Steeper slopes limit the tolerable wave and current exposure somewhat and also requires the shoulder anchorage to come into permanent effect as the angle of repose of the fill material is exceeded. An examination of forces at the anchor suggests that slope length S should not exceed approximately 12 metres. With a maximum tubular gabion diameter of approximately 0.6 metres, the design criteria suggest that a revetment can be formed to withstand wave heights up to 2 metres and current velocities up to 3 metres/second.

BANK TREATMENT

14. Naturally, the geotechnical considerations of slope stability need to be satisfied as the tubular gabions revetment is purely a means of erosion control and contributes little to stability in limit equilibrium analyses for the slope.

15. The site preparation of the bank requires that a reasonably plane slope is provided with, preferably, a toe trench and a slightly rounded shoulder. A drag line excavator is highly suitable for this work.

16. To control erosion of the bank, the revetment must be sufficiently thick to dampen the energy of the water that is in motion in waves and currents. Tubular gabions are porous, however, and interstician water turbulence in the proximity of the slope surface may cause erosion. For many waterways, and particularly those with the steeper slopes required of tubular gabions, the indigenous or imported bank lining material is a stiff, homogeneous clay. To reduce the vulnerability of such banks to erosion through the revetment, the remedies include the introduction of a stone underlayer or lining the slope with a geotextile to serve as a filter membrane to protect and retain bank material.

17. The observed performance of mattress linings to waterways provides some empirical rules on a minimum thickness required to prevent bank and bed erosion under the action of currents (ref.6). For clay linings, a tubular gabion diameter of 500mm provides protection for current velocities up to approximately 3.0 metres/second; a value mentioned earlier as a likely upper limit for tubular gabion revetments.

18. A similar set of empirical revetment thicknesses criteria for wave action does not appear to have been formulated. The effect of waves is potentially more damaging than currents as first the bow wave crest and then, particularly, the wave trough passes along the revetment. The water level

drawdown associated with the presence of a wave trough on the revetment is the instant when erosion of bank material takes place. To overcome this problem, recourse is made to the Terzaghi filter criteria for the introduction of a filter medium to protect the bank. In these circumstances, a geotextile filter membrane is appropriate, the associated design and geotextile selection are well documented (ref.1).

CONSTRUCTION

19. An attribute of tubular gabions is that there is a choice of construction methods - the most suitable being site specific and commensurate with the technology available to the constructor. The following description applies to most waterway work, however.

i An 'A-frame' is erected on shore within which the tubular gabion is suspended and partially pre-filled with stone (Fig.3)

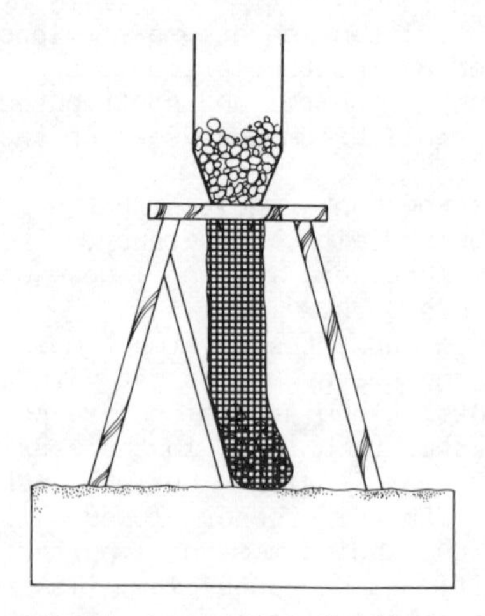

Fig.3. Pre-fill in 'A-frame'

ii The gabion is transported to site and lifted into position, the toe being keyed into a prepared trench (Fig.4)

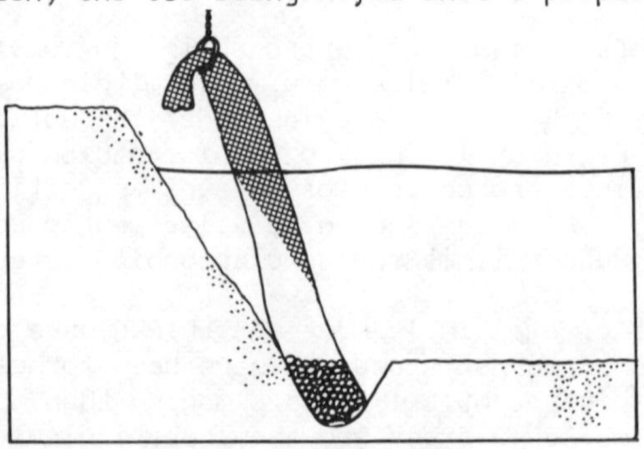

Fig.4 Lift in part filled gabion

iii The gabion is then filled with stone by the most appropriate method (Fig. 5)

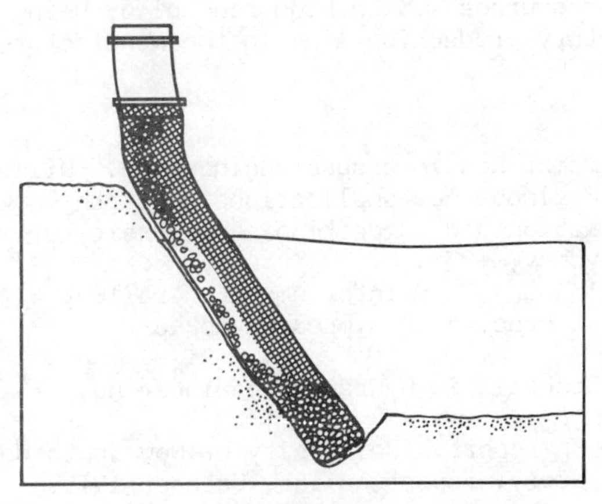

Fig.5. In-situ fill

iv The neck of the gabion is then closed using polyethylene braid and suitably anchored to the shoulder.

v As further gabion placement takes place alongside, polyethylene braid is interlaced to create a coherent battery of gabions for alongslope integrity (Fig. 6)

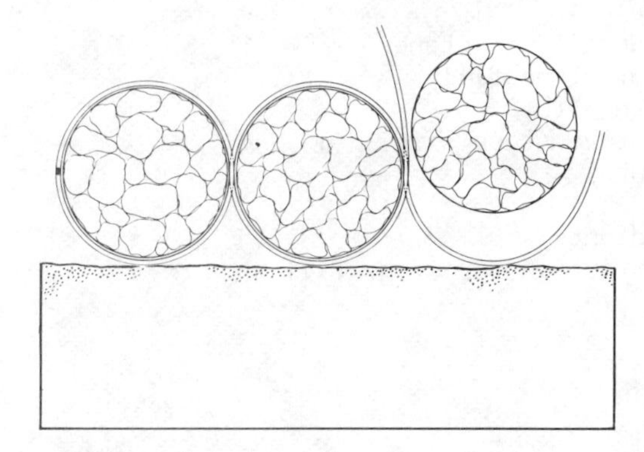

Fig.6. Braiding detail

20. Low technology filling methods have been used to date, as exemplified by the lining of Ulcinj-Solana reservoir in Yugoslavia, where stone filling took place manually. Here, the neck of the gabion was supported in a timber trestle and filled with hand implements. At a pre-construction trial for the lining of the Euphrates River, the tubular gabions were suspended by the neck with one crane whilst a second supplied stone from a hopper. The crane supporting the gabion laid it on the slope as stone filling proceeded. Ultimately, the most sophisticated and economical methods

may employ hydraulic fill using locally dredged gravels.

21. The wide range of site methods that may be adopted allows tubular gabion work to make a rational use of local plant and manpower resources - with high technology being confined to the factory production line in the manufacture of the geotextiles.

REFERENCES

1. RANKILOR P.R. Membranes in ground engineering. Wiley.
2. YAMANOUCHI T. Resinous net applications in earth works. Proc. conf. stabilisation and compaction. University of New South Wales. Sydney, 1975.
3. IWASAKI K. & WATANABE S. Reinforcement of railway embankment in Japan. Proc. ASCE symposium on earth reinforcement.
4. STEPHENSON D. Rockfill in hydraulic engineering. Elsevier.
5. BROWN C.T. Gabion report. University of New South Wales water research laboratory, report no 156, October 1979.
6. Commercial literature, Maccaferri Ltd.

NOTATION

D	tubular gabion diameter (m)
t	ultimate revetment thickness (m)
S	slope length (m)
A	grid aperture size (mm)
H	wave height (m)
V	current velocity (m/s)
Sr	relative density of stone
p	porosity of stone fill
y	water depth (m)
d	representative stone size (m)
Cbu	coefficient
α	slope gradient
ϕ	angle of friction for stone fill

10 Geotextiles for bank protection in relation to causes of erosion

F. G. CHARLTON, BSc(Eng.), MICE, FIWES, MASCE, Hydraulics Research Station Ltd

SYNOPSIS. Channel banks erode in different ways due to various causes. The type and cause of failure should be established before selecting a method of protection and the materials of construction. The paper summarises the methods of bank protection suitable for different situations and the forms of geotextiles available for use in river engineering works.

INTRODUCTION

1. The banks of natural and artificial channels which are unprotected many erode, and it is sometimes necessary to prevent the continuing loss of soil by constructing suitable works. There are different methods of protection and materials for construction, but no single method or material offers a sound, technical and economic solution to every erosion problem.
2. Banks erode in different ways for various reasons. It is essential therefore, to establish both the type and cause of the loss of bank material before choosing a method and materials for protection, if the result is to be economic and successful.
3. The use of geotextiles, particularly those which have come onto the market in various forms in recent years, makes protection cheaper and simpler in some cases. More attention, however, needs to be given to the potential of these materials, to their performance, to the problems of maintenance and to the difficulties which may arise when using them in the construction of bank protection works.

TYPES OF EROSION

4. The banks of natural and artificial channels erode in two ways (Ref 1,2):
   (a) abrasion, or the removal of material from the surface of the bank; and
   (b) slip, or the collapse of a mass of soil into the channel.
5. Abrasion may be caused by men and animals walking on the

face of a bank. The condition is often aggravated by rain-water flowing down the worn paths and washing soil into the channel. More usually, however, abrasion is caused by the movement of water in the channel, and is affected by high velocities, currents, local-eddies, waves and boat wash.

6. Slip is caused by a reduction in the internal soil strength or by an increase in the forces tending to cause the movement. The mass of soil which slips into the channel, breaks up and is carried away in suspension or as bed load. Factors producing a slip are seepage of water, cracking of the soil on drying followed by the entry of water along the potential surface of failure, or an increase in the load on the top of the bank.

PRINCIPLES OF PROTECTION

7. A careful examination of the bank, the morphology of the river and the flow characteristics in the channel should reveal both the mode of failure and its cause (Ref 3,4,5,6,7). The methods of protection which are technically sound may then be deduced and a solution developed which takes account of funds available, the extent of the eroded bank, the effect on the river upstream and downstream of any remedial works, the availability of labour and materials locally, and difficulties in obtaining manufactured materials (Ref 8,2,1).

8. Methods of protecting a river bank from the loss of material may be classified under two main headings depending on the type of failure.

(a) Protection against abrasion:
- (i) Armour face of bank.
- (ii) Retard the flow within the channel or near the bank.
- (iii) Deflect the flow away from the eroding bank.

(b) Prevention of bank slip:
- (i) Reduce seepage through the soil mass to increase intergranular pressure and decrease the forces causing failure.
- (ii) Drain the soil mass away from the face of the bank.
- (iii) Protect against surface cracking which allows the entry of moisture and the development of a lubricated potential slip surface.
- (iv) Increase the strength of the soil mass.
- (v) Reduce the external forces tending to cause sliding.

FORMS OF GEOTEXTILES

9. There is a great range of materials available for river training and bank protection. Those used in the past include:

Leaves (usually woven into mats)

Bamboo
Timber (piles, fences or woven)
Clay
Stone (loose, in crates or bonded with mortar)
Brick (loose, in crates or bonded with mortar)
Soil cement
Cement mortar
Concrete (precast slabs or pavements)
Bitumen
Rubber (natural and artificial)
Resins (for impregnating permeable soils)
Car tyres
Steel sheet (sheet piles and sheet from oil drums)
Asbestos sheet (sheet piles)

10. More recently engineers have begun to make greater use of the various geotextiles available (Ref 9,10). Without sub-dividing these into types of material (polyamide, polyester, polyvinyl chloride, polyolefine, etc), methods of processing (melted, woven, knitted) or the physical properties (tensile strength, resistance to ultra-violet light, etc), geotextiles are available in the forms listed in Table 1, some of which have been specially designed for use in river and coastal protection.

Table 1. Forms of Geotextile

Form	Permeability of finished works		Condition of Use
Sheet	Impermeable	(i)	Plain membrane (Plate 1)
		(ii)	Membrane with felt laminate uppermost onto which cement mortar is sprayed
		(iii)	Woven jute, reinforced with wire and coated with a synthetic rubber (Plate I and IV)
Cloth	Permeable	(i)	Filter fabric (needle punched, welded, knitted or woven) (Plate I and II)
		(ii)	Filter fabric with concrete blocks attached
Netting	Permeable	(i)	Plain net (plastic or jute) (Plate III)
		(ii)	Mats, deep openwork usually formed by welded threads. (Plate III)
		(iii)	Mats with a filter fabric backing (Plate IV)
	Impermeable	(i)	Net coated with asphaltic concrete

Basic Materials
(jute netting, impermeable sheet and filter cloth)
Plate I

Filter Cloth
(Needle punched, woven and welded)
Plate II

		(ii)	Mat fileld with asphaltic concrete (Plate IV)
Webbing	Permeable	(i)	Interwoven strips of impermeable material,
		(ii)	Interwoven strips of impermeable material backed by a filter fabric
	Impermeable	(i)	Interwoven strips of impermeable material backed by an impermeable sheet
Cloth with pockets	Permeable	(i)	Filter fabric with added pockets to retain ballast
Cloth with panels	Permeable	(i)	Two sheets of filter fabric joined at intervals to form tubes or bags into which cement mortar is pumped
Bags	Impermeable	(i)	Bags formed of impermeable sheet and which may be filled in situ with sand
		(ii)	Bags specially tailored of impermeable sheet which may be filled with water when required to form a temporary weir
	Permeable	(i)	Bags of cloth filled at site with dry sand-cement mix
Baskets	Permeable	(i)	Baskets of open mesh net packed in situ with stones (gabions)
	Impermeable	(i)	Baskets of open mesh net packed in situ with stones and filled with asphalt or sometimes with cement mortar
Strands		(i)	Cables which may be used to link concrete blocks to form a flexible or articulated set of blocks usually with a filter backing (Plate V)
		(ii)	Bundles of filaments attached to the bed (Plate V).

CHOICE OF MATERIALS

11. When methods of bank protection which appear to offer a solution for a particular erosion problem have been selected, the choice must be narrowed by an examination of available

Netting
(Openwork mat, woven and wire reinforced jute netting)
Plate III

Composite Materials
(Openwork mat with filter cloth, jute net with
synthetic rubber and openwork mat with asphalt)
Plate IV

materials. The factors to be considered include:

(a) Cost of materials and whether they have to be imported,

(b) transportation of materials to the site and whether this requires new roads, unusual or excessive quantities of transport, and

(c) construction which should take account of the type of equipment required, skill and cost of labour and particular difficulties in handling the materials.

12. When materials available locally are suitable there is often little justification for using items which may need particular skills or transport facilities. When construction and repair work could provide additional income for local inhabitants materials which need imported skilled labour should usually be avoided.

13. When it has been established that geotextiles are necessary the physical form (sheet, cloth net, etc) and its resistance to attack by ultra-violet light, insects, abrasion and heat should be carefully considered (Ref 10,9).

USE OF GEOTEXTILES IN BANK PROTECTION

Protection against Abrasion

14. Bank Armouring. A bank may be protected against the loss of material due to abrasion by armouring the sloping face, using:

(a) rigid revetments, or

(b) flexible revetments.

15. Rigid revetments are mainly impermeable being constructed of concrete (plain, reinforced or precast slabs), cement mortar, soil cement, sheet piles (steel, asbestos or timber), brickwork or stone and mortar. There is generally little scope for geotextiles here unless drainage through the revetment is provided when a filter fabric instead of a reverse filter, may be used to prevent the loss of bank material, and slabs may be laid on a filter fabric to minimise the loss of material through the joints. To increase the stability of a pre-cast block revetment, the blocks may be cast with longitudinal or lateral holes. When laid the blocks are linked together by ropes of polyester filaments to form an articulated mat.

16. Flexible impermeable revetments may be constructed of sheet (polyethylene, material or artificial rubber), webbing backed by impermeable sheet (polyesters and polyethylene), clay, bitumen, and asphaltic concrete (plain or reinforced by polypropylene netting). The impermeable revetments which are light in weight (sheet and webbing) must be secured to the bank either by concrete slabs or by staples into the bank. Generally, when sheeting is used in rivers and canals, it should be considered as a backing to reduce permeability rather than as a facing to protect against abrasion. It has insufficient structural strength and mass to resist the forces of flowing water and pressures due to seepage.

17. Flexible permeable revetments may be constructed of

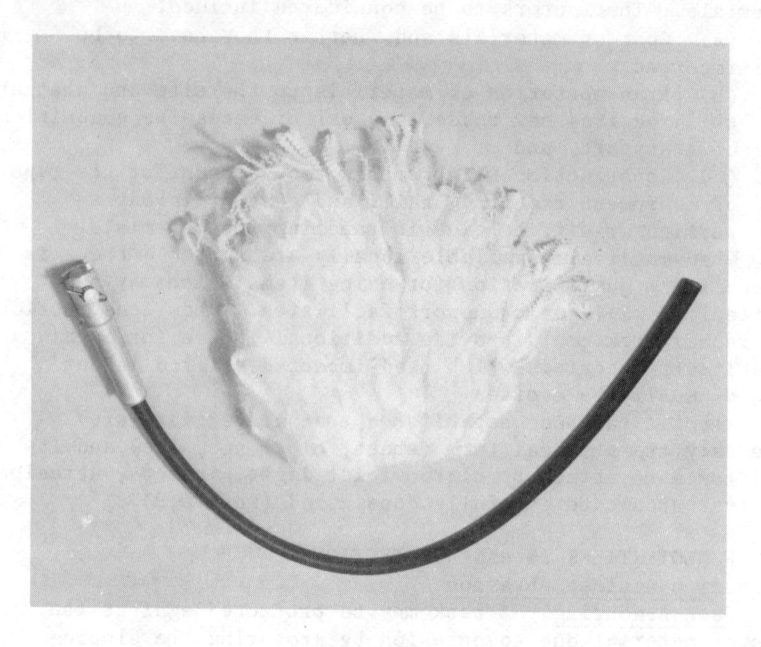

Strands
(Cable and bundle of filaments)
Plate V

brushwood or woven willow mattress, fibreglass and resin (sometimes in conjunction with vegetation), vegetation (protected by a layer of jute, wire or plastic netting), dumped or placed stone, gabions formed of stone in baskets of wire or plastic netting, bricks protected by a layer of wire or plastic netting, precast concrete blocks (sometimes bonded to a filter fabric (Ref 11), cloth with pockets to contain stones or panels into which cement mortar may be pumped, jute bags containing cement and sand, and plastic bags of sand.

18. Jute does not have a long life and soon decays. When used with veegetation its main purpose is to provide protection during the early stages of growth. Plastic or wire netting or jute netting reinforced with wire, may be used to protect vegetation for longer periods.

19. Gabions have usually been constructed of baskets made of wire mesh (Ref 12,13), woven willow, bamboo or timber. Plastic netting has, however, been used successfully and provided it is resistant to ultra-violet light, should give satisfactory service.

20. A relatively new method of revetment construction has emerged with the use of a permeable cloth to which pockets have been added. Those pockets are filled with stone or stone and mortar to give stability. Another variation is a cloth with panels into which cement mortar is pumped (Ref

14). This provides a stronger but less flexible revetment, but both have the advantage that they act as a filter and prevent loss of bank material.

21. Jute bags filled with cement and sand have long been used to form revetments; now bags formed of impermeable plastic sheet or permeable plastic cloth, and filled with sand may be used. (Ref 15).

22. Finally, except where the materials form an adequate filter membrane to prevent the loss of bank material due to seepage out of the embankment, the above revetments should be backed by a filter membrane of the required mesh size to protect the soil in the bank.

23. Flow Retardation. Methods of reducing the speed of flow near a channel bank may be achieved in three ways:

(a) Instal traiing fences
(b) By-pass the area under attack
(c) Raise water levels.

24. Training fences to reduce flow speeds may be constructed by driving into the bed closely spaced piles in rows running out at right angles from the bank; the piles may be of timber, bamboo, concrete, plastics, etc. Other methods include netting (steel, plastic or woven willow, bamboo or timber) attached to piles and projecting into the steam, or steel jacks linked to one another by wire or plastic cables, or bundles of filamentss attached to the bed. The principal object is to provide a set of closely spaced obstacles which retard the flow. Geotextiles which can be used for such works are jute, polyethylene netting polyester-polyethylene cables, and polypropylene filaments.

25. By-passing the area under attack by the excavation of a cut-off channel is not within the scope of this conference.

26. Raising water levels is not normally an economic solution to this type of problem as the cost of a weir, dam or barrage would probably be excessive. In some cases, however, a specially tailored plastic bag anchored in position across the channel and which could be filled with water as required to raise the crest level is a possibility (Ref 16).

27. Flow Deflection. Currents which impinge on a bank or flow at speeds sufficient to dislodge surface particles may be deflected away from the bank thus eliminating erosion by abrasion. There are three basic systems, although the first two are similar differing only in their height relative to the depth of flow.

(a) Spur or groyne
(b) Sill
(c) Vanes

28. A single spur or groups of spurs set at carefully chosen distances along a channel bank are often used. Rigid and impermeable spurs are constructed of steel, asbestos or timber sheet pile, reinforced concrete or stone and mortar. There is little scope for the use of geotextiles in such structures as they must be strong and rigid. Sills whose

crest is below water level are of similar construction (Ref 17).

29. Spurs required to be impermeable but flexible may be constructed of clay with a stone facing to protect the core from erosion. Plastic sheet or a filter cloth should be laid between the core and facing to preveent fines being washed through the stone.

30. Permeable and flexible spurs can also be constructed of brushwood, woven mats of locally available materials attached to piles, dumped or placed stone, concrete blocks, or gabions with baskets of woven bamboo, steel or plastic netting. There is considerable scope for using geotextiles in such structures. Mats may also be made of woven jute, plastic netting or plastic webbing, and they may be attached to piles of timber or plastic. Concrete blocks may be linked together by cables of plastic and plastic netting may be used for gabion baskets.

31. Finally, vanes to generate secondary currents which affect the positions at which scour and accretion occur in a channel may be constructed of timber, concrete, steel or plastic webbing with timber beams and struts.

Protection against bank slip

32. Reduction of Seepage. Erosion due to the seepage of water through the mass of soil in the bank may be reduced by:

(a) Controlling rate of drawdown in channel.
(b) Reducing permeability of bank.

33. The rate of drawdown may be controlled by constructing a dam, barrage or weir. This is not normally a practical approach unless the structures have a second purpose also. Such structures would normally be of concrete, brickwork, stone, gabions or earth, but an inflatale plastic bag tailored to suit the dimensions of the channel is a possibility for small rivers (Ref 16).

34. The permeability of the bank may be reduced by a cut-off wall of steel, concrete or plastic sheet. Reductions could also be achieved by injecting cement grout or resin to fill the pores of the soil mass.

35. Drainage. Improved drainage to reduce the quantity of water available to pass through the bank may be achieved by constructing a drain to lower the water table in the bank or by improving surface runoff. Subsoil drains constructed of stone surrounded by a filter cloth should be considered.

36. Protection against Surface Cracking. When the top surface of a bank containing clay is subjected alternately to wetting and drying, cracking may occur. The cracks weaken the bank and allow water to enter which aggravates the condition. Vegetation, bitumens and asphaltic concrete may be used to protect the top surface of the bank.

37. Increase of Soil Strength. The overall strength of a soil may be increased in two ways:

(a) Injection.
(b) Earth reinforcement.

38. Cement grout or resin and hardener (Ref 18) may be injected to strengthen a bank of permeable soil. This is an expensive method and would normally be reserved for short lengths of bank or for sites where the high cost could be justified (e.g. in urban areas) (Ref 19).

39. Reinforcement of the earth embankment may be achieved by laying ropes or high tensile strength plastic cloth across the bank at different elevations. Where the rope or cloth emerges on the face of the bank it is linked to a facing of metal, timber or plastic webbing.

40. Reduction of Sliding Force. The force tending to cause a soil mass to slide into a channel may be reduced by:

(a) Reducing an overburden.
(b) Increasing the load on the face of the bank.
(c) Reducing the slope of the face of the bank.
(d) Raising the water level in the channel.

41. Loading the face of the bank by the construction of revetments has been discussed above.

42. Reducing an overburden or the slope of the bank face are problems in earthwork, and the construction of a dam, barrage or weir to control water levels has been discussed above also.

CONCLUSIONS

43. Unprotected banks erode by abrasion and slip due to a variety of causes.

44. There are many systems of bank protection; a choice should only be made after the method and cause of failure have been established and suitable materials selected.

45. Systems of protection against erosion by abrasion include:

(i) revetments,
(ii) flow retardation, and
(iii) flow deflection.

46. Systems of protection against erosion by bank slip include:

(i) reduction of seepage,
(ii) drainage,
(iii) protection against surface cracking,
(iv) increase of soil strength, and
(v) reduction of sliding force.

47. Geotextiles are available in the form of:

(i) impermeable sheet,
(ii) permeable cloth,
(iii) netting,
(iv) cables,
(v) strands.

48. In addition there are many items fabricated from these geotextiles and used for river engineering. They include:

(i) webbing,
(ii) cloth with panels,
(iii) baskets,

(iv) netting and mat coated or filled with asphaltic concrete.

(v) bags of impermeable or permeable sheet.

49. Geotextiles are used in various ways:

(a) Impermeable sheet to prevent seepage. Used behind structural defence formed of concrete slabs, stones, gabions, cement mortar, etc.

(b) Permeable filter cloth to permit seepage but prevent loss of soil. Used behind structural defence formed of concrete slabs, stones, gabions, etc.

(c) Reinforcement to protect vegetation, to strengthen soil (earh reinforcement), to strengthen blocks (cables).

(d) Container to hold a heavier material which when confined forms the defence. This includes pocketed and panelled cloth, baskets for stone, bags for sand and cement mortar, netting and mat filled with asphaltic concrete.

(e) Structural defence in form of webbing, netting, or mat on the face of a bank or between piles.

(f) Flow retarding system in form of netting or bundles of strands.

ACKNOWLEDGEMENTS

50. This paper is published by the kind permission of the Managing Director, Hydraulics Research Limited, Wallingford.

REFERENCES

1. CHARLTON F.G. River stabilisation and training. Chapter XII. Gravel bed rivers (Ed Hey RD, Bathurst JC & Thorne CR) John Wiley and Sons, Chichester, 1982.

2. Hydraulics Research Station and C.H. Dobbie and Partners. Report on Bank Protection in Rivers and Canals. Hydraulics Research Station, Wallingford, July 1980.

3. CHARLTON F.G. The importance of river morphology in the design of river training works. Int. Comm. on Irrig and Drain, Eighth Congress, Varna, May 1972, Question 29.1, Report No. 11.

4. BLENCH T. Mobile bed fluviology. The University of Alberta Press, Edmonton, Alberta, 1969.

5. KELLERHALS R. Stable channels with gravel paved beds. Proc. A.S.C.E. Vol 93, WWI, Feb 1967, P63.

6. CHARLTON F.G., BROWN P.M. and BENSON R.W. The hydraulic geometry of some gravel rivers in Britain. Report IT 180, Hydraulics Research Station, Wallingford, July 1978.

7. JANSEN P.Ph, VAN BENDEGOM L., VAN DEN BERG J, DE VRIES M and ZANEN A. Principles of river engineering, the non-tidal river. Pitman Publishing Limited, London, 1979.

8. KEOWN M.P., OSWALT N.R., PERRY E.B. and DARDEAU E.A. Literature survey and preliminary evaluation of streambank protection methods. U.S. Army Engineer Waterways Experiment Station. Tech Report H-77-9, Vicksburg, May 1977.

9. RANKILOR P.R. Membranes in ground engineering. John Wiley and Sons Limited, Chichester, 1981.
10. CANNON E.W. et al. Fabrics in civil engineering. Civil Engineering and Public Works Review, March 1976, p.39.
11. Anon. A new Dutch method of sea and river bed protection by blankets weighted with concrete blocks. La Houille Blanche, 1976, p11-15.
12. STEPHENSON D. Rockfill in hydraulic engineering. Elsevier Scientific Publishing Co, Amsterdam, 1979.
13. STEPHENSON D. Rockfill and Gabions for Erosion Control Civil Engineer in South Africa, Vol 21, No 9, Sept 1979, p.203.
14. LAMBERTON B.A. Revetment construction by fabriform process. Proc. Am Soc. Civil Eng., Vol 95, CO1, July 1969, p.49-54.
15. WINNEY, M. Mattress to control Nigerian meanders. New Civil Engineer, 18 March 1982, p.32.
16. ANWAR H.O. Inflatable dams. Proc. Am Soc Civil Eng, Vol 93, HY3, May 1967, p.99.119.
17. ANDERSON A.G. and DEVENPORT J.T. The use of submerged goins for the regulations of alluvial streams. Symposium on current problems in river training and sediment movement, Budapest, 1968.
18. VOLOTSKOI D.V., GUREVICH I.L. and NETFULLOV S.K. A study of the strengthening of earth slopes by chemical means. Translation from Gidrotekhnicheskoe Stroitel No 1, 1978, p.19-20.
19. BOYES R. Protective soil stabilisation for river slopes. Civil Eng (London), Sept 1982, p1.9.

11 Experience with a flexible interlocking revetment system at the Mittellandkanal in Germany since 1973

Dr Ing. G. HEERTEN, Naue Fasertechnik, Espelkamp, Dipl.Ing. H. MEYER, Wasser und Schiffahrtsdirektion Mitte, and Dipl.Ing. W. MÜHRING, Neubauamt Mittellandkanal, Osnabrück, West Germany

SYNOPSIS. The experience with a flexible interlocking revetment system which has been used at the Mittellandkanal in Germany since 1973 is reported. Technical informations about the special designed needlepunched nonwoven geotextile being the revetment filter layer and the interlocking concrete blocks with pegs, peg holes and anchoring wires being the revetment armour layer are given. Results of an official investigation and controlling program, measuring the pull in the anchor wires and six years of profile soundings, verify the very good experience with the terrafix-revetment system which shows the additional advantage of excellent greening.

INTRODUCTION

1. The design of bank protection structures has always played a major role in constructing and maintaining waterways. In the early 1960's there was a demand for new technical solutions for revetment design after significant damage to river and canal embankments became apparent as a result of the change from towed trains of barges to self-propelled motor vessels. In building and improving ship canals investment in bank protection accounts for a considerable percentage of total cost.

2. Whilst in the past it was mainly the search for low-cost, durable materials which stimulated progress, in the last 10 years it is the expenditure on manpower and equipment which has been the deciding factor. Nowadays in addition we have to consider a most favourable revetment design which will allow the system to blend into the natural surroundings in promoting the development of vegetation and providing a habitat for small aquatic life.

3. Taking these aspects into account, the design of a bank protection structure has to fulfil the following requirements:

- optimum technical layout for long-term use and for minimizing the manufacturing and maintenance cost
- approved installation technique for quick and safe installation in the dry and underwater
- most favourable layout considering environmental aspects.

4. In the light of this development the application of interlocking concrete blocks has gained special significance in the

field of bank protection. In 1973 the terrafix revetment system was used for the first time on a large scale for developing the Mittellandkanal in Germany, thus providing hydraulic engineering with a new economical method of construction. In the meantime about 500,000 m² of this type of bank protection have been placed in several european and overseas countries.

5. This report will be dealing with the results obtained from field measurements which have been carried out on this method of construction on the Mittellandkanal, giving a 10 years experience report.

THE MITTELLANDKANAL DEVELOPMENT SCHEME

6. The Mittelland Canal is being developed according to a skeleton draft drawn up in 1965 based on know-how available at that time. Set criteria for development were as follows:

Standard vessel:
    1350 t motor vessel (Waterway Class IV)
Permissible speed:
    10 to 12 km/h
Cross-sectional ratio (canal/ship):
    n = 7

7. Owing to technical and economic considerations, three standard cross-sections were developed:

- trapezoid cross-section with 53 m width at water level 1:3 slopes on both sides and 4,0 m water depth
- rectangular cross-section with 42 m width at water level and 4,0 m water depth (sheet piling cross-section)
- rectangular/trapezoid cross-section with 47 m width at water level with 1:3 slope on one side and vertical wall (sheet piling) at the other side, water depth 4,0 m.

FUNDAMENTALS FOR REVETMENT DESIGN

8. To secure sloping embankments one endeavoured to design revetments which could be regarded as having sufficient strength to withstand the forces exerted by shipping traffic. This resulted in the formulation of two basic requirements:

- Construction of an effective, erosion-proof high permeable filter layer below and above water with traffic passing constantly. This could only be achieved by placing geotextile filters, the development of which was considerably influenced by the Mittellandkanal project.
- The armour layer protecting the filter layer against the attack of waves, currents, ice and damage by ships should be made of bonded but sufficiently permeable structures, as flexible as possible. On the one hand there was the possibility of riprap layers with a partial grouting of bituminously or hydraulically bonded grouting material and concrete interlocking blocks.

9. It was particularly the last method which offered a series of advantages over conventional methods which had

essentially originated in the use of precast construction elements to ensure maximum adaptation to local conditions, not only during placement but also in subsequent operation.

10. The filter layer of a revetment structure has to stabilize the subsoil with a sufficient soil tightness and permeability to water. Layers of sand, gravel or bushy twigs are traditionally used but normally failed after a short service time caused by unsufficient installation technique or filtration properties, especially if underwater installation is necessary.

11. An important improvement in revetment construction could be observed in using synthetic filter fabrics (geotextiles) for underwater installation. Nowadays, numerous synthetic filter fabrics are available which meet current design criteria in regard to filtering capability and permeability both normal and parallel to the filter plane for a wide range of soil types. These filter fabrics offer the advantage of continuous underwater installation without interrupting the shipping traffic.

12. We have to distinguish woven and nonwoven geotextiles. The properties of fabrics are very different, influenced by the polymer properties and by the manufacturing process. For woven fabrics we have to distinguish e.g. the kinds of the threads, the kind of weaving, the used polymer and the fabric finish. Nonwoven fabrics also are produced by different polymers and we have to distinguish the method to obtain the cohesion of the fibres or filaments.

13. Because of its high resistance against ultra-violet irradiation, high specific strength and specific gravity the use of geotextiles produced from polyester fibres is advantageous especially for under water installations.

14. The long-term behaviour of a revetment structure mainly depends on the filtration properties of the filter fabric after geotextile and armour layer being carefully installed.

15. The traditionally used filter materials like sand and gravel are dimensioned after the well known filtration rules e.g. from Terzaghi or the U.S. Corps of Engineers. By this a coordination between the diameters of the soil particles of the subsoil and the filter layer is given. In many cases the filter on fine soils has to be built up from two or more separate filter layers. Limited by the accuracy of installation technique the thickness of these filter layers has to be 0.2m minimum. This minimum thickness is not given by the filtration rules mentioned above but is given by experience of construction work. The literature showed that the filter layer thickness also is very important for its working. Many of the filter layers designed by the given filtration rules would fail having not the thickness or "filtration length" of about 20 cm minimum. A filter layer of soil particles is not working as a thin sieve but is working as a filtration body with a given pore size distribution built up from all the soil particles of the filter layer and the incorporated sub soil particles. The interaction of the original sub soil and the soil particles of

the filter layer is very important for forming a stable, long-term working filter layer. With an increasing filtration length an increasing probability is given for a subsoil partical migrating through the filter layer being stopped by a smaller pore.

16. Discussing the filtration properties of geotextiles we have to distinguish the properties of woven and nonwoven fabrics. The filtration properties of woven fabrics are given by the mesh size or the fabric openings. The woven geotextile is acting as a thin sieve. The filter conditions could be stable with nearly all soil particles being larger than the mesh size or unstable with nearly all soil particles being smaller than the mesh size. This unstable conditions often are given on sub soils in the range from silty sands to clay.

17. The filtration properties of nonwoven geotextiles are influenced by the fibre size, the fabric weight and thickness. Thermal bonded nonwoven fabrics are relatively thin and they would act nearly as a woven fabric with irragular openings. Needle-punched nonwoven fabrics are considerable thicker than all other types of geotextiles. Caused by the needle-punching process the voids volume of needle-punched geotextiles is about 85 % or more. The filter conditions are comparable to soil-filter conditions. The interaction of fibres and soil particles is forming a stable, long-term working filter layer. Investigations on dug up fabrics have confirmed these conditions. In Fig. 1 some data of virgin nonwoven fabrics (porosity n, permeability k_n) and of the dug up fabrics (pore space clogged by soil, remaining porosity n', remaining permeability k_n') are given. The estimated permeability of the clogged geotextiles is 5 to 12 times higher as the measured soil permeability, which is in the range of $k \sim 1{,}0$ to $5{,}0 \quad 10^{-5}$ m/s. The remaining porosity of n' = 0,32 to 0,74 guarantees a sufficient long-term permeability. In contrast to these results for most of the investigated woven fabrics a lower permeability as given by the soil was estimated. The relation of the permeability of the woven geotextiles and the permeability of the soils was in the range of 0,16 to 1,8.

18. Based on the given difference in filtering and on bad experience in using grain filters and woven filters the application of heavy needlepunched nonwoven geotextiles really is a standard in revetment construction on waterways in Germany for about 15 years. Special guiding rules of the Bundesanstalt für Wasserbau (BAW, Federal Institute for Waterways Engineering), Karlsruhe, have to be considered for geotextile application (1).

THE TERRAFIX REVETMENT SYSTEM

19. The terrafix revetment system consists of two components complementing each other functionally: heavy needlepunched nonwoven geotextiles being the revetment filter layer and interlocking concrete blocks being the revetment armour layer.

20. Relating to the subsoil requirements or given guiding rules differenttypes of fabrics are available. But all terrafix

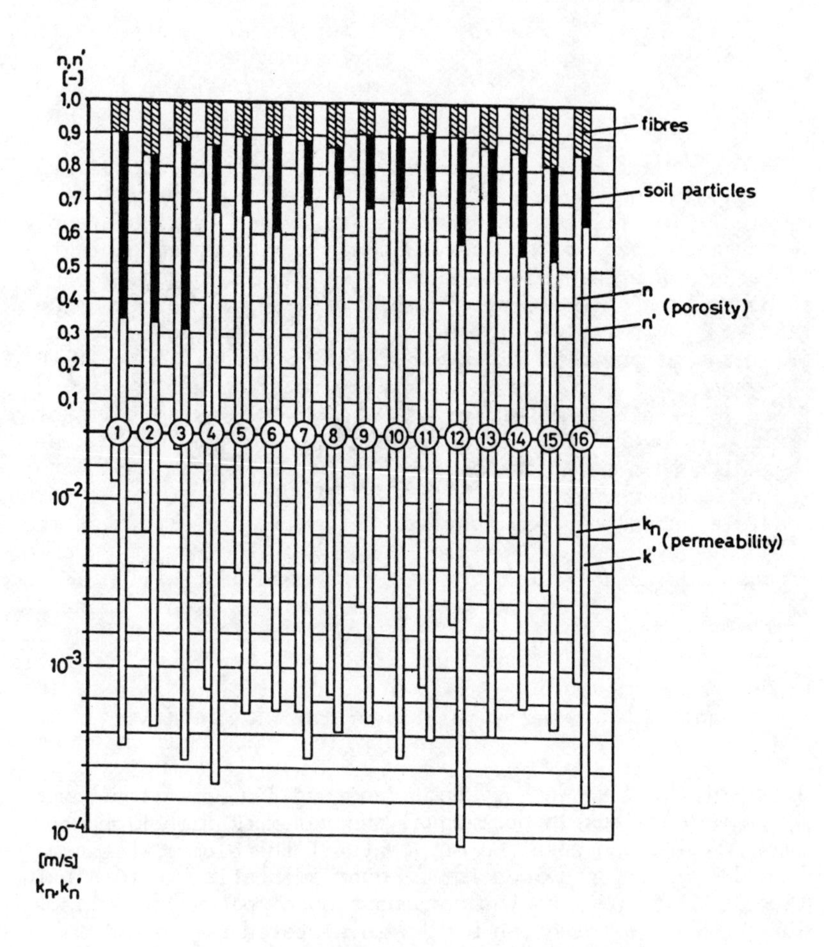

Fig. 1
Clogging of voids volume and permeability decrease of dug up needlepunched nonwovens

geotextiles are made from synthetic fibres building single-or multilayer needlepunched nonwoven geotextiles. To meet the strong guiding rules of the BAW for revetments on class IV waterways for 1350 t motor vessels on steep slopes (steeper than 1:4) of silty sands or finer soils the fabrics e.g. are composed of a fine and coarser filterlayer with a minimum thickness of 4.5 or 6.0 mm and a very coarse roughness layer for stabilizing the boundary layer between geotextile and subsoil with a minimum thickness of 10 mm (Fig. 2). These heavy multilayer fabrics with overall thicknesses of more than 15 mm and a weight up to 1800 g/m² are giving best properties for a safe installation without damage and for long-term filtering.

21. The interlocking concrete blocks are trapezoid-shaped with moulded-on conical pegs at the front and matching holes at the rear. The blocks protect and fix the geotextile filter

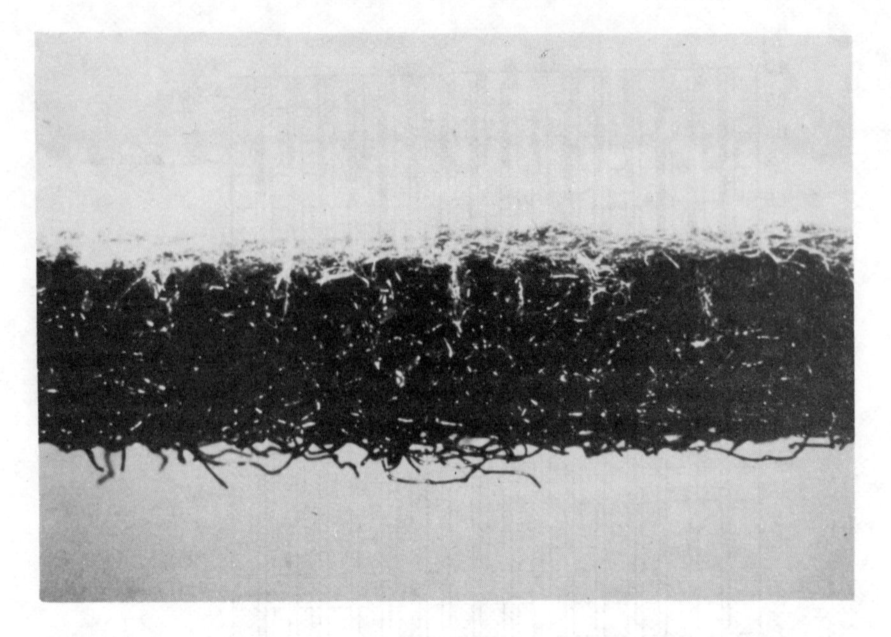

Fig. 2
Cross section of a heavy multilayer terrafix geotextile

layer of the revetment system. Pegs and peg holes ensure an optimum interlock in horizontal and vertical direction permitting tilt and rotation movements of the blocks (Fig. 3). The flexibility provides for a good adaptation to the installation level as well as the compensation of possible settlements. The distance between the blocks as dictated by the interlocking elements and the special shape of the blocks guarantee the necessary permeability to water, greenability and wave dampening effect. On account of the interlock the weight per unit area of the revetment system can be reduced considerably as compared with e.g. riprap revetments. Weights of approx. 1300 to 2500 N/m² have proved very successful even in the case of highly stressed waterways. At high and steep slopes, with danger of toe scouring or with bad subsoil conditions the hanging terrafix revetment constitutes a safe and proven solution. Wires passed through special holes in the interlocking concrete blocks permit a transmission of longitudinal forces from the toe of the embankment to its upper edge and thus a safe force distribution within the hanging revetment.

THE MITTELLANDKANAL EXAMPLE

Construction

22. In the years 1974 and 1975 about 75.000 m² terrafix revetment system have been installed at the Mittellandkanal from km 79,6 to km 85,7. In this area the canal embankments mainly

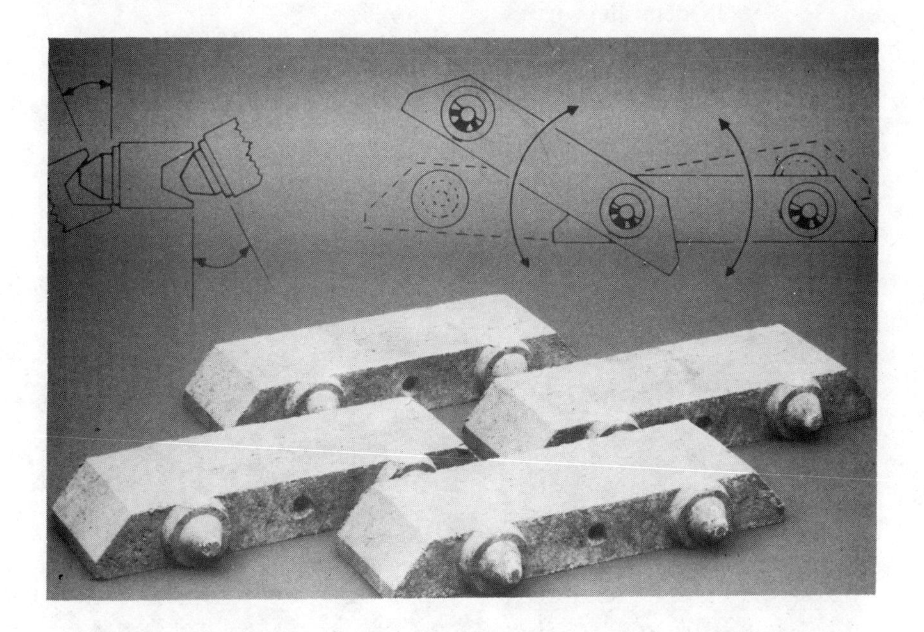

Fig. 3
Interlocking concrete blocks of the terrafix revetment system

consist of silty fine to medium sand for which the following average characteristic values were assumed:

Permeability	k	$= 6 \cdot 10^{-5}$ m/s
Angle of friction		$= 32{,}5\ ^{o}$
Cohesion	c'	$= 0$
Specific gravity	γ	$= 2{,}0$ kN/m³ , $\gamma' = 1{,}0$ kN/m³

23. Below the concrete blocks a terrafix 800 NSK geotextile was placed whose properties had been previously matched to the in-situ embankment soil. The characteristic data of the terrafix 800 NSK is as follows:

Thickness of filter layers	6.0 mm (4,5 + 1,5 mm)
Thickness of roughness layer	31 mm
Weight	1005 g/m²
Tensile strength (DIN 53858) length	2710 N/width 2590 N
Elongation (DIN 53858) length	90 % /width 80 %
Permeability to water	$9.2 \cdot 10^{-3}$ m/s
BAW-test soil type 3 (silty sand)	
a) soil tightness	2.0 g/34 h
b) permeability to soil type 3	$4.3 \cdot 10^{-3}$ m/s

24. The interlocking blocks were of the type NV 12 with the following characteristic values:

length/width/height	660/140/120 mm
block weight	G = 230 N, G' = 140 N

blocks per area $9{,}1\ m^{-2}$
area weight $g=2130\ N/m^2$, $g'=1230\ N/m^2$

G' and g' are giving the weight under water considering buoyancy.

25. For terrafix revetment construction at the Mittellandkanal the NV 12 interlocking blocks were assembled to revetment sections of 14.0 m in length and 5,60 m in width on a moveable pallet of a special designed floating barge (Fig. 4). Anchor

Fig. 4
Floating barge for underwater installation of the terrafix revetment system

wires 5 mm in diameter, hot-dip galvanized and plastic jacketed were inserted at the lower edge. The anchor wires were attached to a wooden toe beam and the wooden anchor stakes (bongossi) at the top of the embankment without preloading. By retracting and tipping the pallet of the floating barge the section hanging from the anchor wires was gradually lowered onto the slope upon which the heavy needlepunched nonwoven geotextile type terrafix 800 NSK had just been laid. Fig. 5 is showing the construction sequence. When the concrete block sections had been placed, the upper part of the embankment was fixed. The joints between the revetment sections were filled with hydraulically bonded grouting material. As a result of development the joints are not grouted today but half blocks are used at the edges of the revetment sections forming a flexible close section joint.

Fig. 5
Construction sequence installing the terrafix revetment system at the Mittellandkanal

Field measurements

26. The amount of load placed on the anchor wires of the terrafix revetment system was to be subjected to both individual and continuous measurements in the field, first of all upon placement and then during subsequent operation, together with the relevant loads exerted by the passage of shipping traffic. To measure the tensile forces in the wires special tensile force probes were installed above the revetment sections having a measuring range up to 8000 N.

27. In order to measure the water level changes when vessels pass, thus serving as a representative characteristic quantity of load being exerted on the revetment, special pressure transducers were installed 2 m below water level.They were used for both, continuous and individual measurements and were designed to cover a measuring range of 7,5 m water column.

28. At the Mittellandkanal measurements began during installation of the pre-assembled revetment sections in 1975 and ended in November 1980.

Results during revetment construction

29. The tensile force probes were fitted shortly before the sections were submerged. As a rough check on the behaviour of the anchor stakes, unstressed control stakes were placed at

intervals of 50 cm behind the stakes for the measuring wires. Whilst the sections were being laid the tensile forces in the wires fitted with tensile force transducers were measured and recorded with analogue results.

30. The anchor wires are preloaded during laying and are evidently subjected to their maximum load when the laying platform below the revetment section suspended from eight anchor wires is drawn away. At the end of laying the tensile forces in the anchor wires are reduced again. The maximum tensile forces measured when laying the sections gave a safety factor of 2.0 as the minimum section safety with regard to the wire breaking load. Based on the measurements the relative coefficient of friction between the concrete blocks and the barge assembling platform could be calculated to tan φ = 0,17. Fig. 6 is showing an example of measured tensile forces in three anchor wires during laying operation.

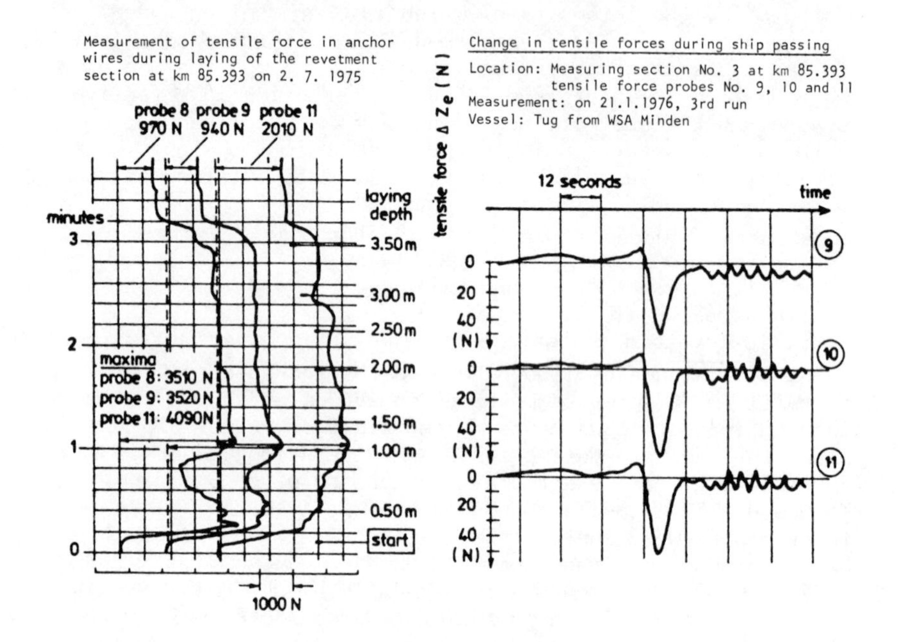

Fig. 6
Tensile forces in the revetment anchor wires during installation and ship passing

Results after revetment completion

31. After completion in additional measurments the behaviour of the anchor wires fitted with tensile force probes was registered from time to time and in addition soundings were taken on the profiles of the slopes.

32. However, the measurements were affected by disturbances which were not completely clarified. Most of the tensile force

probes experienced a zero point movement into the pressure zone so the values measured can merely be assessed according to their tendency.

33. The tensile forces were constantly reduced in all measuring zones, which would seem to indicate a sufficient shear strength between the subsoil and the geotextile and between the geotextile and the concrete blocks. These results are confirmed by soundings on the slope profiles which have, all in all, shown no significant changes since completion of the revetment during the 6 years of measurement.

Behaviour during passing of ships

34. The effect of the rapid lowering of the water level, caused by the passing of motor ships, on the tensile forces acting in the anchor wires was measured in 1976 and 1980.

35. The passage of ships produces a strong current from bow to stern with a lowering of the water level at the side of the ship, the intensity of which is dependent on speed, the shape of the vessel, the cross-sectional ratio canal/ship and the distance of the vessel from the bank. On the one hand, the reduction in water level causes a reduction in effective buoyancy in the bank protection and on the other hand a current in the ground water running perpendicular to the slopes with a relativ high hydraulic head. From the data obtained on the passage of vessels, it can be concluded that the forces acting on the revetment structure can be absorbed under safe conditions without any problems. As also shown in Fig. 6 the changes in tensile force in the anchor wires are minimal in relation to wire breaking loads.

36. To investigate the actual load on the revetment structure caused by normal ship traffic conditions continuous measurements were carried out at the Mittellandkanal in a developed cross-section with a cross-sectional ratio of n = 7. Over a period of seven weeks the lowering of the water level caused by passing ships was recorded. The evaluation included all the lowering values which were equal to or greater than 10 cm. For smaller water level changes there is no clear separation from wind-generated waves. Based on these investigations with an average traffic density of 75 ships per day it can be shown that only 40 % of all ships produce any significant loads and only about 10 % produce loads which have any substantial effect on revetment design with lowering values of about 30 cm and more (Fig. 7).

THE TERRAFIX REVETMENT DEVELOPMENT

37. Since the first test sections at the Mittellandkanal in 1973 and the first big job in 1974/75 described above the terrafix revetment has been used on different hydraulic structures in several european and overseas countries. It has been used mainly for bank protection purposes on rivers and canals but also as revetment and bottom protection on big culverts and spillwaqs with design current velocities of about 5 m/s.

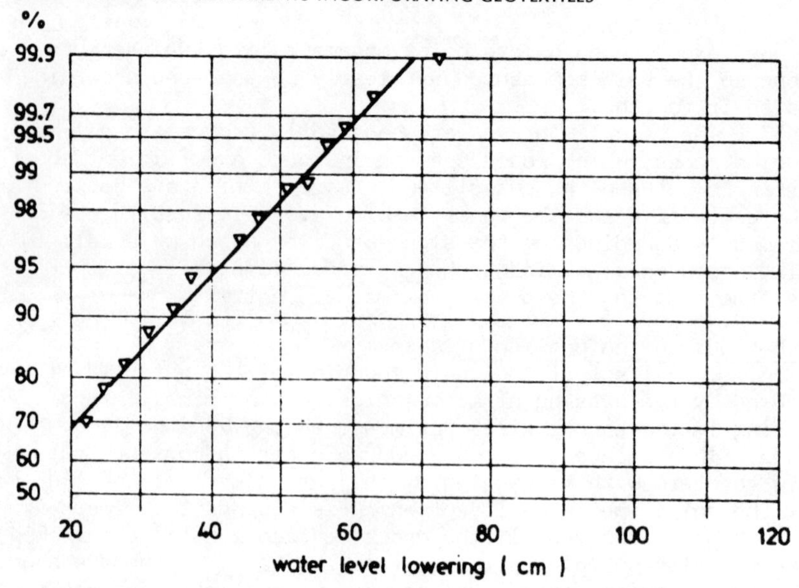

Fig. 7
Relative total frequency of water level lowering as a percentage of all vessels passing.

38. Besides the underwater installation method using a special floating barge the installation of terrafix revetment sections by crane using special traverses has been developed. Traverses for vertical hanging concrete block sections and horizontal hanging block sections are available for different installation purposes. (Fig. 8)

39. Additional investigations and model studies at the National Water Research Institute, Hydraulics Research Division Toronto, Canada and the Leichtweiß-Institut für Wasserbau of the Technical University of Braunschweig, Germany were carried out to investigate the resistance of the revetment system against high current velocities and wave attack. The model studies showed that the interlocking blocks type NV12 could withstand current velocities up to v = 8 m/s and that rectangular shaped blocks 15 cm thick showed no damage under 20 hour wave attack with a significant wave hight of $H_S = 1,6$ m .

40. In the last years the requirement for a most favourable layout of revetment structures considering environmental aspects became more and more important. Therefore it is a big advantage of the terrafix revetment system that only a few vegetation periods are necessary to green over the revetment (Fig. 9). Caused by the large pore volume and the movability of the fibres in the fibre labyrinth of the needlepunched nonwoven fabric and the special joints between the concrete blocks aquatic growth generally developes above and below the water level, also encouraging the development of small marine life.

Fig. 8
Horizontal traverse for terrafix revetment installation

Fig. 9
Terrafix revetment greened over

CONCLUSION

41. The measurements and profile soundings carried out on the Mittellandkanal confirm the overall positive experience to date involved in using bank protection structures made of the terrafix revetment system for about 10 years. The terrafix revetment system is an important example for modern revetment design performing all actual requirements - minimizing installation and maintenance costs, quick and safe installation in the dry and underwater, environmental friendly. The actual experience is based on more than 500.000 m² terrafix revetment being installed in european and overseas countries.

References:

U. Abromeit: Biding Procedures and Placing Operation of Geotextile Filter Layers. Proceedings Int. Conf. on Flexible Armoured Revetments, London 1984.

G. Heerten : Dimensioning the Filtration propertiesof Geotextiles Considering Long-Term Conditions. Proceedings 2. Int. Conf. on Geotextiles, Las Vegas, 1982.

G. Heerten : Modern Technique in Bank Protection. Int. Symposium Polders of the World, Lelystad, The Netherlands, 1982.

H.-G.Knieß,
H. Meyer : Naturmessungen an Uferdeckwerken aus Betonverbundsteinen. Zeitschrift für Binnenschiffahrt und Wasserstraßen, Nr. 8, 19 3.

L.Wittmann : Soil Filtration Phenomena of Geotextiles. Proceedings 2. Int. Conf. on Geotextiles, Las Vegas, 1982.

T2 The ACZ-DELTA mat

Ing. J. C. DORR, Ir. A. H. M. BARTELS and Ing. P. SCHUIT, ACZ Marine Contractors, Gorinchem, The Netherlands

SYNOPSIS. ACZ Marine Contractors B.V. has developed a concrete block mattress especially for use in slope protection works: the ACZ-DELTA mat. The prefabricated ACZ- DELTA mat comprises a polypropylene filter fabric to which concrete blocks are firmly connected. Thorough research of the relevant parameters based upon many years of experience and extensive studies carried out by the Delft Hydraulics Laboratory and the Delft Soil Mechanics Laboratory during the last decade have led to the design of the ACZ-DELTA mat. This paper presents a description of the mattress and its properties in relation to its unique blockform and pattern.

INTRODUCTION

Over the years the Dutch have made many in-depth studies of river- and canalbank protection and with an ever increasing knowledge of the mechanics of the forces acting on banks it was found that the constructions made in the past either required high maintenance or were not dimensioned to handle modern ship traffic.

During the last two decades much research has been carried out into the different types of shore protection. This has led to a tendency to choose slope protection rather than vertical sheetpiling. ACZ has, therefore, developed its existing concrete blockmattress which is used over very large areas (4.4 million m^2) as bottom protection against scouring in the Eastern Scheldt Storm Surge Barrier Project, a part of the Deltaworks, to form the ACZ-DELTA mat.

DESCRIPTION OF THE MATTRESS

The prefabricated ACZ-DELTA mat comprises a filter fabric to which concrete blocks are firmly connected. The mat is made on a production plant, transported to the site and placed on the slopes with the aid of a crane and spreader.

The construction of the mat can be divided into three items: the fabric, the concrete blocks and the connection between concrete blocks and fabric.

The fabric

The fabric which is used serves both as a filter for the subsoil and as a connector between the concrete blocks. Therefore, the choice of the fabric depends on the local circumstances and the required dimensions. The properties of the slope that has to be protected determine the filter-qualities of the fabric: sand tightness must be guaranteed, but the waterpermeability must be as high as possible to prevent overpressures underneath the revetment. The required tensile strength and elongation properties depend on the unit weight of the mattress and the dimensions in connection with the transport and placing method.

In any case a fabric can be chosen tailormade for the purpose. Normally a polypropylene fabric in the range of 400 to 800 gr./m^2 is used.

The concrete blocks

To meet all the requirements concrete blocks have been developed with roughly a sloping parallelepiped form.

Fig. 1. Concrete block

The bottom dimensions are 263 x 371 mm. The height can be varied between 80 mm and 200 mm, permitting the construction of the ideal mattress for every application.(Fig. 2).
The spacing between the blocks at the bottom is 20 mm, which results in 9 blocks per square metre, corresponding with a degree of occupation of 88%. The 12% open space has proved to be enough for permeability.
The wedge-shape of the gaps between the blocks is such that

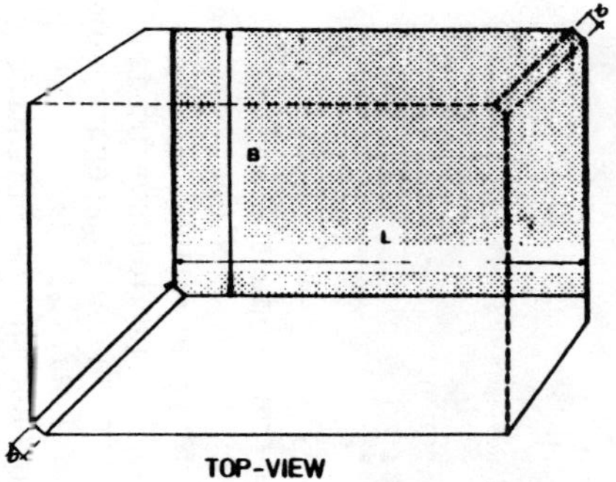

B
L
TOP-VIEW

H (mm)	L (mm)	B (mm)	Mattress weight (N/m²) ρ = 2400	2600	2800 kg/m²
80	338	230	1537	1665	1793
100	333	225	1898	2056	2214
120	328	220	2247	2435	2622
140	324	216	2585	2801	3016
160	319	211	2912	3154	3397
180	314	206	3227	3496	3765
200	310	202	3531	3826	4120

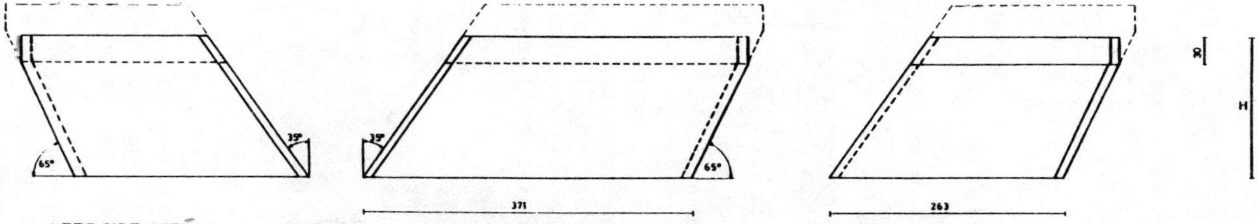

65°
35°
35°
65°
371
263
H
LEFT SIDE-VIEW
FRONT-VIEW
RIGHT SIDE-VIEW

Fig. 2. ACZ-DELTA mat block

once the mattress has been laid the gaps may be filled with crushed gravel. By the clenching action of the fill material a substantial increase of the overall stability is achieved. The quality of the concrete meets all the requirements for concrete applied in marine environments.

The connection between blocks and fabric.

The connection is not only of great value during transport and placing operations, but is also important to improve the stability of the blocks under wave-attack.
Therefore, a strong and abrasion-resistant connection is required. The connection consists of polypropylene cords woven in in the warp of the fabric and forming loops on the spots where the concrete blocks are situated. The loops are raised before the concrete is poured, so after curing a strong and slipproof connection is guaranteed.

PROPERTIES OF THE ACZ-DELTA MAT

An ideal revetment mat should fulfil the following more or less important design criteria.

1. The mat has to be strong and stable.
2. Energy absorbtion should be as high as possible to reduce wave run-up.
3. Preferably no concentration of waterflows due to the presence of vertical or horizontal successive openings between the blocks.
4. Flexibility is required.
5. The filter fabric should be protected against direct ultra violet radiation.
6. Possible vandalism should be reduced to a minimum.
7. The materials used in the mat must be ecologically harmless. Development of vegetation must be possible if required.
8. The mat should be accessable for human beings and animals.
9. Aesthetically pleasing.
10. Possibility to replace and repair.
11. Useful in all types of slope and bottomprotection works.
12. Economic.

Ref. 1

The stability of a structure generally depends on the loads, where and when they occur, on one hand, and on the strength of the structure on the other hand. A slope revetment is subject to loads from the free water, waves and currents, and to loads from the subsoil. The loads from the subsoil can be quasi-static and dynamic and are generated by the wave-action on the slope in several ways.

1. Due to the fact that seepage through the revetment into the subsoil takes place over a larger area than seapage out of the revetment the result of cumulative waves is an elevation of the mean phreatic level and so an increase of

quasi-static pressures underneath the revetment.

2. When a wave approaches a slope an increase of pressures below the wave crest can be transmitted under the slope revetment, thus causing uplift pressures over a limited area just in front of the wave top only.

3. Depending on the surf similarity parameter

$$\zeta = \tan\alpha \,/\, \sqrt{H/L_o}$$

wave breaking may occur, where:

α = slope of the revetment
H = wave height
L_o = wave length (deep water)

Strong increases of pressure due to wave breaking may propagate under the slope revetment, resulting in short duration uplift pressures.

4. After breaking, a strong reduction of pressures above the revetment may occur, intensifying the pressure gradient over the revetment.

If an individual block of a slope revetment has been raised by uplift forces, wave run-up and run-down can cause dragforces, inertia forces and additional lift forces and so cause further collaps of the construction.

The Delft Hydraulics Laboratory (DHL) and the Delft Soil Mechanics Laboratory (DSML) in the Netherlands, have developed a numerical model called 'STEENZET' with which the uplift forces on blocks of a revetment can be determined as a function of the block dimensions, permeabilities of the subsoil and of the revetment and the wave properties. The model has been calibrated with measurements in large scale modeltests. An important number for the determination of the quasi-static pressures underneath the revetment is the leach length, defined as

$$\lambda = \sin\alpha \sqrt{bd\,\frac{k}{k'}}$$

where:

α = slope of the revetment
b = thickness of the filter-layer underneath the revetment
d = thickness of the revetment
k = permeability of the filter-layer
k' = overall permeability of the revetment

It proved that the pore pressures are smaller if the revetment is more permeable. Therefore, a rather open construction such as the ACZ-DELTA mat with its wide openings between the blocks is a favourable revetment.

With the knowledge of the phenomena which can cause damage to a revetment the design of the ACZ-DELTA mat has been optimised.

1. A permeable construction with wide openings between the blocks has been chosen, resulting in relatively low uplift pressures underneath the revetment.

2. Uplifting of individual blocks is prevented by the weight and by the special shape of the blocks resulting in an interlocking of the blocks and cooperation of adjacent blocks. Stability of the blocks is increased by the use of crushed stone in the openings between the blocks, where the wedge-shape of the openings and the dimensions of the fill material concur.

3. The connection with the underlaying filter fabric gives an additional safety against uplifting.

In the summer of 1983 prototype tests on a number of revetment constructions have been done in the Hartel Canal, the Netherlands, by the Public Works Department (Rijkswaterstaat). With help of the results further optimisation of the block form has taken place.

By order of ACZ the Delft Soil Mechanics Laboratory in connection with the Delft Hydraulics Laboratory in the Netherlands have made a study of the stability of the ACZ- DELTA mat under wave-attack. Using the mentioned DSML/DHL numerical model 'STEENZET' first the stability in terms of maximum uplift force on one block and on an area of three blocks was considered For the computation of these forces pressure peaks with a duration shorter than 0.1 seconds were neglected. Comparison of the maximum force on one block and on three blocks gives an idea of the effect of a mattress-structure on the stability of the slope protection: the blockweight, necessary to counteract the uplift forces, can be lowered. Additionally, when using an engineered fill material (crushed stone) in the spaces between the blocks, the stability of the ACZ-DELTA mat can be heightened with at least 50%. The construction of the ACZ-DELTA mat and the orientation of the blocks is such that it is not likely that washing out of material will occur. The ultimate result of this study is a, rather conservative, relation between wave properties, wave height and wave period and the required mattress weight, presented in fig. 3 and 4.

ACZ is currently preparing large scale modeltests (1:2 or 1:1) in the DHL Delta-flume in order to test the influence of the connection filtermat - blocks and to obtain the ultimate damage wave height of the complete system and to improve the conservative approach used now.

Ref. 2

The herring-bone pattern of blocks and the filling of the openings between the blocks with crushed stones give the revetment a relative rough surface. This results in a relatively low wave run-up, which can be compared with an at random placed stone revetment.

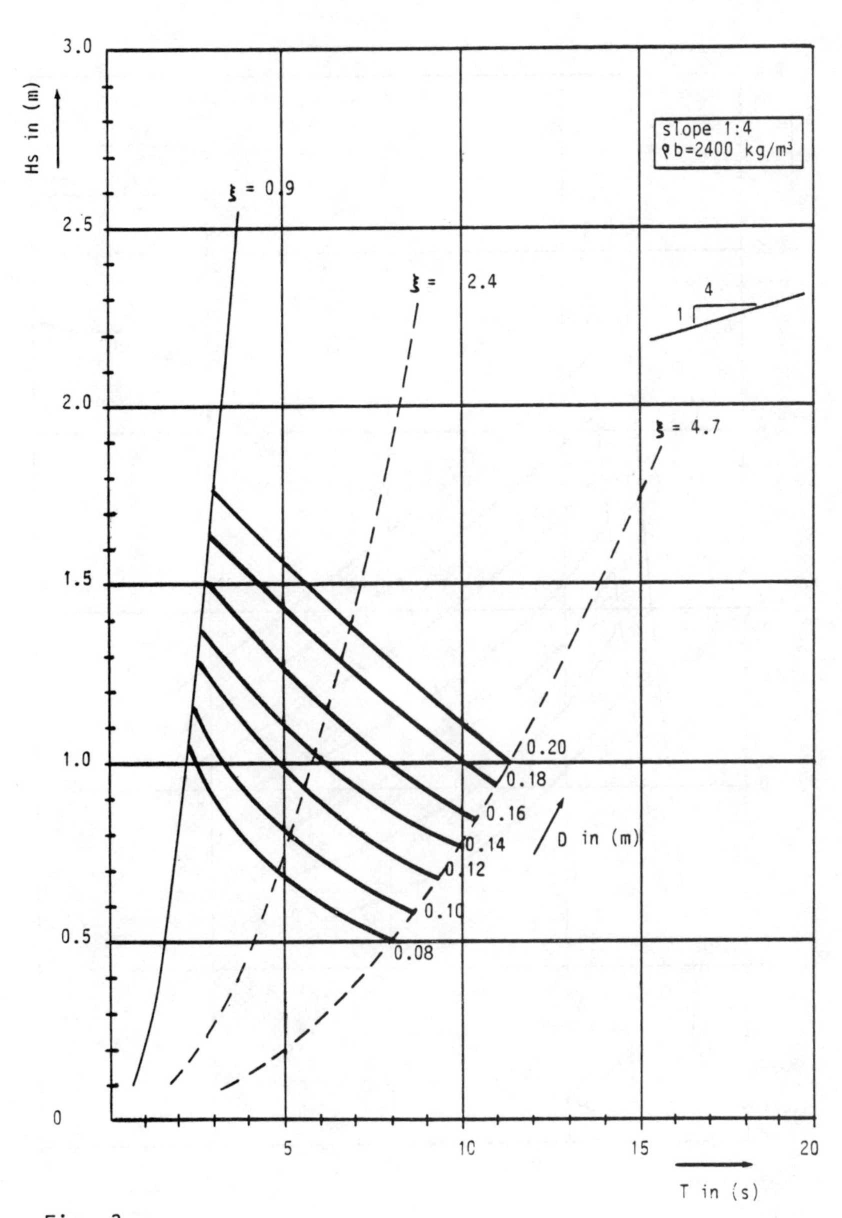
3.0
2.5
2.0
1.5
1.0
0.5
0
Hs in (m)
slope 1:4
ρb=2400 kg/m³
ξ = 0.9
ξ = 2.4
ξ = 4.7
4
1
0.20
0.18
0.16
0.14
0.12
0.10
0.08
D in (m)
5
10
15
20
T in (s)

Fig. 3.

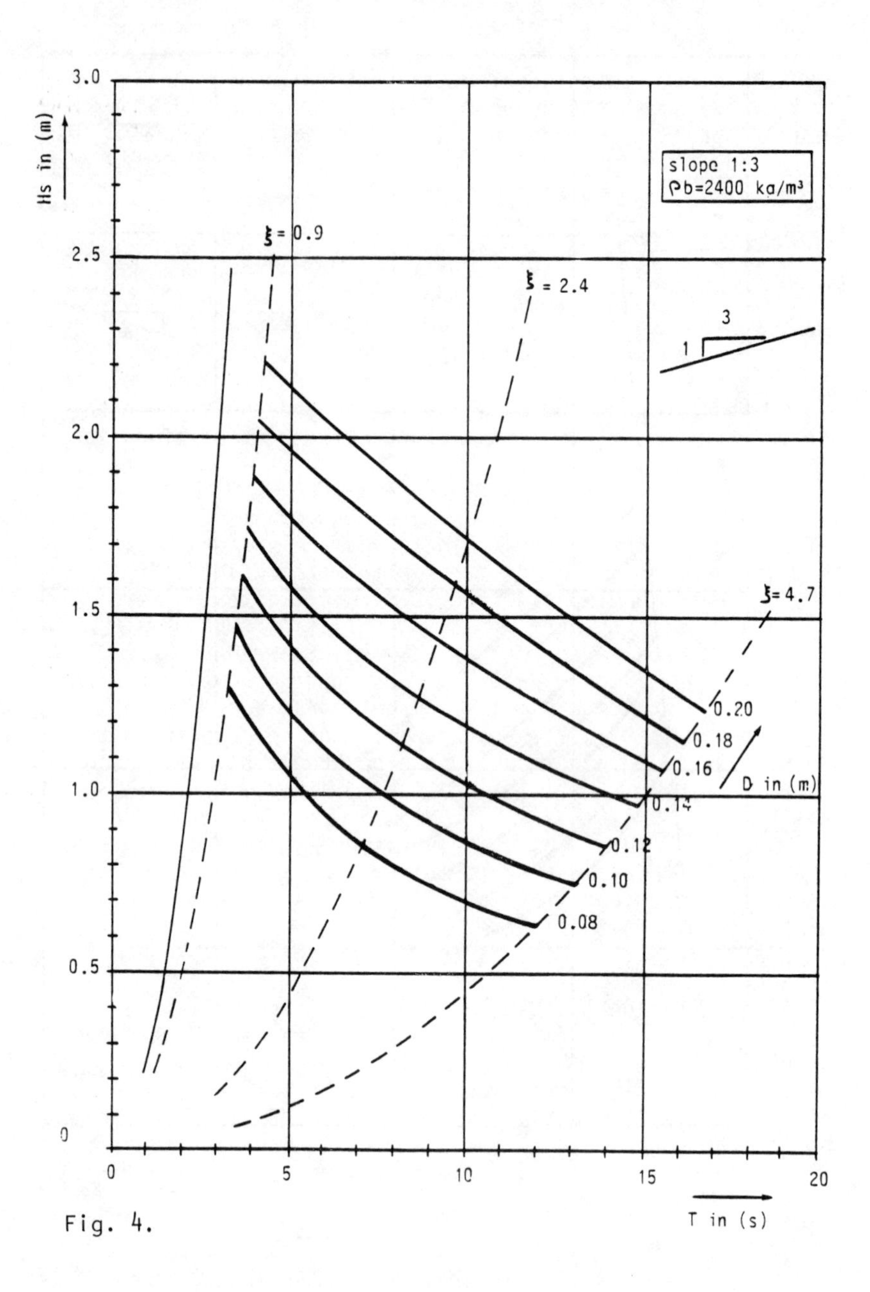
3.0
2.5
2.0
1.5
1.0
0.5
0
Hs in (m)
slope 1:3
ρb=2400 kg/m³
ξ = 0.9
ξ = 2.4
ξ = 4.7
3
1
0.20
0.18
0.16
0.14
0.12
0.10
0.08
D in (m)
0
5
10
15
20
T in (s)

Fig. 4.

Ref. 3.

When the strength of a slope protection increases, other parts of the structure may become the weakest link.
A concentration of waterflow during run-down of breaking waves can result in sandtransport under the filter fabric and can so cause erosion or settlements. The herring-bone pattern overcomes the problem of waterflow concentration so erosion under the mat and resulting settlements are minimized.

Ref. 4

The herring-bone pattern and the special block shape used in the ACZ-DELTA mat make an extremely flexible construction. This means that the mattress will follow settlements exactly. Thus contact with the underlaying soil is always guaranteed and the overall filter construction remains intact despite the possibility of localised breakdowns. It is possible to continue the mattress beyond the slope onto a berm, since the mat permits a change of direction of up to 70 degrees.

Ref. 5

The shape of the concrete blocks and the pattern in which they are laid form a major part in the protection of the fabric against U.V.-radiation. It is well-known that even stabilised polypropylene loses some of its strength with prolongate exposure to the U.V.-component of sunlight.
The blocks have, therefore, been designed to give maximum shade to the whole mat including the joints and thus to minimize damage by U.V.-radiation. If the voids between the blocks are filled with gravel no exposure at all will occur.

Ref. 6

Human damage to the existing revetment systems usually takes one of the following forms:

- removal of elements (blocks or stones)
- knife cuts in the fabric
- pushing objects such as fishing rods between the blocks and thereby damaging the fabric.

Pulling out blocks of the ACZ-DELTA mat is impossible since the blocks are firmly fixed to the fabric. Damage due to cuts is minimized due to the block shape and the pattern. Similarly the angle that the block faces make with the slope make it unattractive for anglers to place their rods in the gaps. In both cases damage is further minimized by placing crushed gravel between the blocks.

Ref. 7

Providing no toxins are used in the manufacture of the concrete the mattress is harmless to the environment.
If for environmental or aesthetical reasons it is required that the slope is able to support plant life it is possible

to fill the inter-block voids with soil instead of gravel. In due course the indigenous plant life will return and the slope protection will blend in entirely with the surroundings. If necessary extra holes can be formed in the blocks to increase the soil area.

Ref. 8

Even on steep slopes the mat is easy to walk on (by human beings as well as animals).

Ref. 9

Although peoples' thoughts about aesthetical aspects of concrete mattresses may vary the ACZ-DELTA mat has great advantages. The herring-bone lay-out makes a pleasant surface pattern without the need for horizontal and vertical lines.

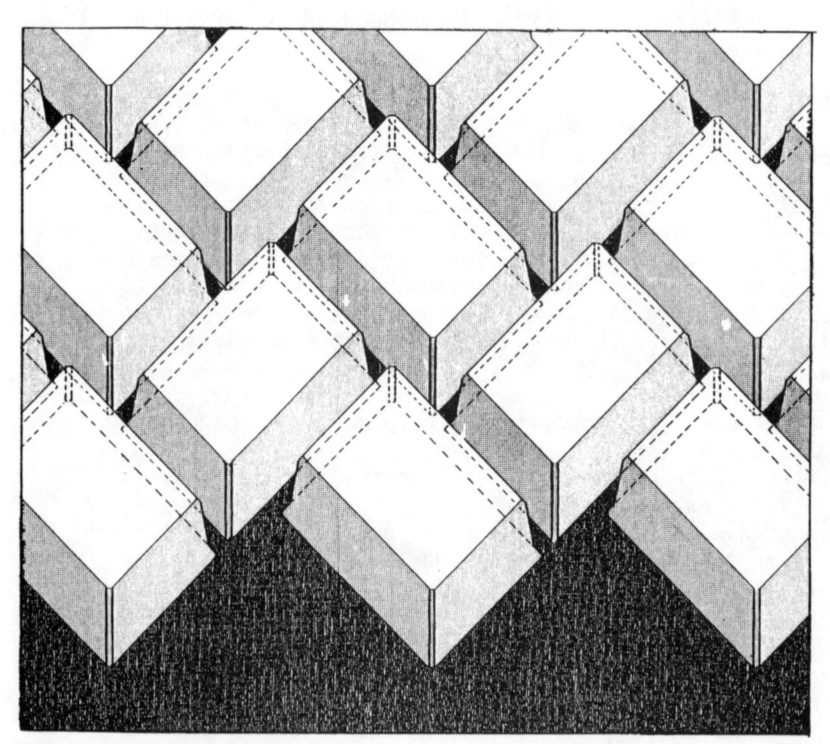

Fig. 5. Herring-bone pattern

Ref. 10

If necessary, in the case of a collision for instance, the mat can be repaired by placing a new one or pouring new concrete blocks onto the mat.

Ref. 11
The ACZ-DELTA mat can be used for the protection of the banks of rivers and canals, the covering of pipelines, shore protection and the protection of artificial islands. By increasing the height of the blocks and thus increasing the total weight per square metre it is possible to tailor the mat to fit the hydraulic circumstances of the different applications.

PRODUCTION AND PLACING METHOD

Although block form and pattern do look complicated a singular mould is sufficient for production. So production is easy by stretching out the fabric, positioning the mould and casting concrete. After one day the concrete is strong enough to pick up the mat by crane and spreader for stock.

Placing the mat on a slope takes place by crane and spreader. Overlaps are created by pieces of fabric without concrete blocks. The mats fit to each other above the waterline within the herring-bone bond, so seams are not visible when the mat is placed accurately.

CONCLUSION

The design of the ACZ-DELTA mat is an important step forward in low cost revetments and is an example how, by the results of independent research combined with decades of contractors-experience, it is possible to develop a new product which is universal in its applications and fits all the requirements principals can aim.

REFERENCES

1. BOER K. DEN, KENTER C.J., PILARCZYCK K.W. Large scale model tests on placed blocks revetment. Delft Hydraulics Laboratory, Delft, January 1983, publication no. 288.
2. ANONYMUS. Behaviour ACZ-mattress under influence of currents. Report model-tests. Delft Hydraulics Laboratory, Delft, August 1973, R 460 part X. (Original Dutch title: Gedrag ACZ-mat onder invloed van stroom. Verslag model-onderzoek).
3. BEZUIJEN A. 'STEENZET', model for the computation of waterpressures underneath a revetment. Delft Soil Mechanics Laboratory, Delft, 1983 CO-258901-91. (Original Dutch title: 'STEENZET', model ter berekening van waterdrukken onder een steenzetting).
4.ANONYMUS. Damage to cross-sections in waterways. Delft Hydraulics Laboratory, Delt, 1971-1986, M1115. (Original Dutch title: Aantasting dwarsprofielen in vaarwegen).

Discussion on Session 3: Revetment design

Mr L. Summers, Sir William Halcrow & Partners, Swindon

In Fig. 4 of Paper 8, dimensionless run-up is plotted against a function of breaker parameter and is compared to the wave run-up relation according to Hunt.

It is noted that for the low waves in the tests the run-up lies below the Hunt line. This of course is not a feature of the type of revetment but is to be expected, as the Hunt data ignores the tendency towards a limit as wave steepness gets smaller. Work done by Issacson and Miche shows how the Hunt data should be modified, and a paper by Chue attempts to develop universal curves. Fig 1. shows the Issacson and Miche modification calculated from the data provided in Table 1 in the paper.

It has been indicated that it is possible to include cable strength in assessing the stability of the flexible revetment. This is based on a possible safety factor of 5 in the tensile strength of the cables. However, the blocks in the mat would inevitably rock when subjected to cyclic wave loading, despite the assumed perfect meeting together of the blocks. This rocking will subject the cables to shear forces and any cable subjected to tension and cyclic shear stress will soon fail.

A further problem in incorporating cable strength in design of such a mat appears to arise when damage occurs and must be repaired. Would it be necessary to take up the whole mat and replace it?

References

1 Issacson E. Water waves on a sloping bottom. Common Pure Appl. Math., 1950, 3, 11-31.

2 Miche R. Le pouvoir reflechissant des ouvrages maritimes exposes a l'action de la houle. Annl Ponts Chaus, 1951, 121, 285-319.

3 Chue S. H. Wave run-up formula of universal applicability. Proc. Institution of Civil Engineers, Part 2, 1980, 69, 1035-41.

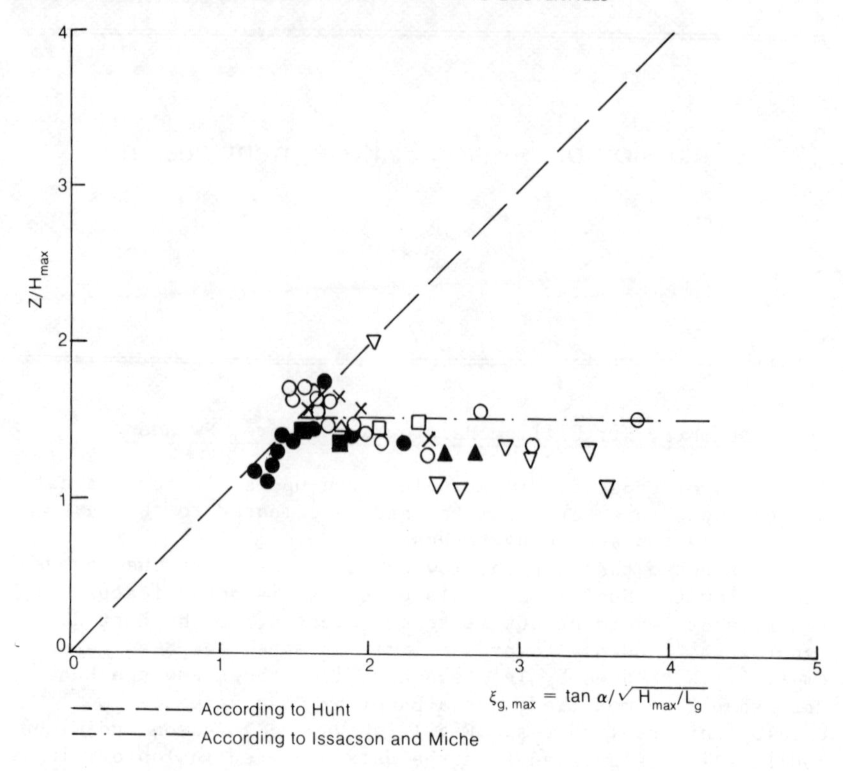

Fig. 1. Modifications to Fig. 4 of Paper 8 to take account of run-up limits according to Issacson and Miche

Major E. G. Wise

In the historical timescale of generally accepted new civil engineering components, parallel-laid polyester filament cables are a comparatively recent innovation; therefore, the long-term proving of their superior characteristics when adapted as the orthogonal integrating meshes of flexible single-layer revetment structures is necessarily an on-going operation. Because of this continuing monitoring, and despite the fact that records of variable hostile site conditions and the conclusions drawn therefrom, have, to date, been consistently encouraging I have opted for the restrained propositions put forward in my paper.

However, a quarter-century of already well established synthetic cable performance undoubtedly provides a most cogent argument for its use to orthogonally integrate flexible armour, and it is surely one which should be of particular relevance in the prevailing economic pressures adversely affecting major revetment requirements.

Beyond the general benefits of cabling, the possibility of a rocking of armour units, raised by Mr Summers, is an interesting one; whether or not such movement in fact were to occur under design parameters or only under extreme and

Fig. 2.

unprecedented attack; and whether such armour rocking units be structurally interconnected or not. If not structurally interconnected, then such rocking would seem to largely invalidate design stability assumptions applicable to discrete units. On the other hand, accepting that a unit rocking may always be induced, or perhaps more likely a three dimensional random rhythmical movement of a module of interconnected units, then the added restraint to such displacement provided by a two-dimensional cabling interconnection of such a module would tend, not to place the cables in immediate shear, but rather to cause a mutual longitudinal sliding of their parallel fibres to accommodate small cable deflection angles accompanied by a cyclic variation in their tensions.

To this end it should be added that the cable conduits, through armour units, must invariably be provided with anti-abrasion orifices, and be over-sized relative to design cable diameters.

Regarding the possible damage to integrated structures raised by Mr Summers: where such damage is specifically confined to its cabling mesh (perhaps through vandalism or from faulty installation) affected panels, or parts thereof, may be removed either by their remaining longitudinal or their lateral cables. Following the replacement of the damaged cables, they may thereafter be reinstated and reintegrated into the structure.

Extensive fracturing of the actual concrete armour units, forming an integrated structure, will cause little reduction in their collective effectiveness, since mutual integration of the units is maintained through their interconnecting cables.

As an example of the benefits of orthogonal cabling, in damage situations and in amplification of my paragraph 4b, Fig. 2 shows the removal (and subsequent reinstatement and reintegration) of one of several orthogonally interconnected panels, following the collision of a manoevring bulk cargo vessel with the revetted embankment of an American ship canal. Some 100 linear yards of the canal earth embankment were totally destroyed, but less than 10% of its original cabled revetment had to be replaced, while the remaining 90% required only repair and reinstatement.

Mr L. Summers

A common method of failure of gabions is bursting of the cages, especially when the cages are piled one on another to a substantial, fairly vertical height.

It would be interesting to know the basis of design of the very long tubular sacks in order to resist bursting. In view of the efficient compaction of the contained rock by cyclic wave loading one would expect the bursting loads at the bottom of the rock to be very high.

Mr C. D. Hall

Mr Summers refers to his experiences with more conventional gabions in which a finite volume of rock is contained in a rectangular cell. Any 'bursting pressure' in the toe of the tubular gabion is controlled by the internal stability that any frictional material possesses and also, on further compaction by waves and currents, by internal arching of the rock fill across the walls of the tube.

As for stress in the tubular gabion, the top surface acts as a membrane subjected to an internal pressure. An upper bound estimate of the membrane tension can be made by adopting the active soil pressure as the 'bursting pressure', giving:

$$T = \frac{K_a h \gamma D}{2 \sin\theta}$$

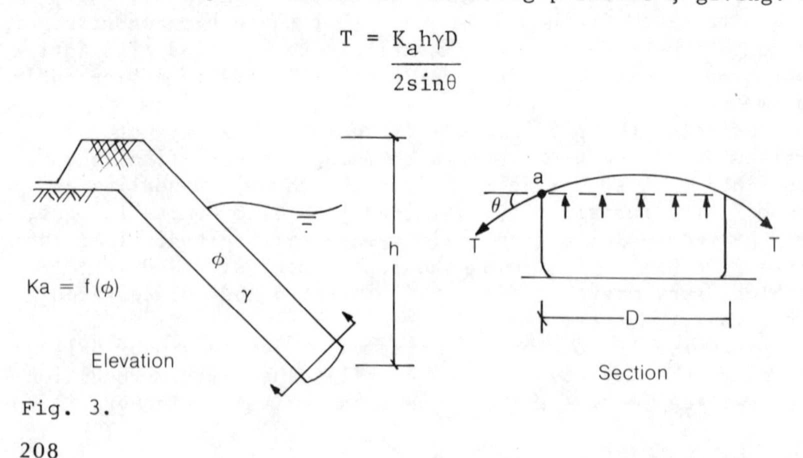

Fig. 3.

The terms are defined in Fig. 3. In such an evaluation, the curvature of the top surface (and hence the value of θ) is theoretically contentious but can be measured on site installations. A value of about 40° at point a appears to be typical.

Mr K. Pilarczyk, Delta Department, The Netherlands

Many different kinds of revetment blocks and/or systems are used for coastal protection and protection of navigation channel banks under a variety of conditions. These blocks are often provided with various kinds of joint arrangements, which include ship-laps, wedge-formed laps, single or double wedges for asphalt filling, etc., or flexible joints by means of cables of pins (to the geotextile), or by means of another (finer) material (grouting). All these arrangements have their advantages and disadvantages. In all these cases, the type of sublayer (permeable/impermeable) and the grade of permeability of the blocks are very important factors for the stability of the protective system. All these blocks may also be with or without roughness elements of different shape to reduce the run-up.

In general, no reliable design criteria are yet available for this kind of construction. On the other hand, the available data from investigations in the USA (mainly regarding interlocking blocks) and recent systematic research in Holland (on free blocks and flexible interlocked blocks), and some data from other countries permit a preliminary attempt at defining the stability criteria. I have summarized most of these results elsewhere ('Stability of revetments' in 'Closing tidal basins', Delft University Press, 1984). Some of these results (for wind-wave attack) are presented in Fig. 4.

In this design process the following aspects have to be taken under consideration:

(a) stability (toplayer, sublayer, subsoil)
(b) flexibility
(c) durability (toplayer/concrete, geotextile, cables, etc)
(d) possibility of inspection of failure (monitoring damage)
(e) easy repair
(f) low cost (construction/maintenance).

The best revetment is one which combines all these functions.

Stability means that the single block stays in place and can be considered of primary importance in revetment design. One may distinguish between external and internal stability. The former refers to the forces that act on the blocks themselves; the latter deals with the conditions of stability of the filter construction and subsoil supporting the blocks. Because of the complexity of the problem, external and

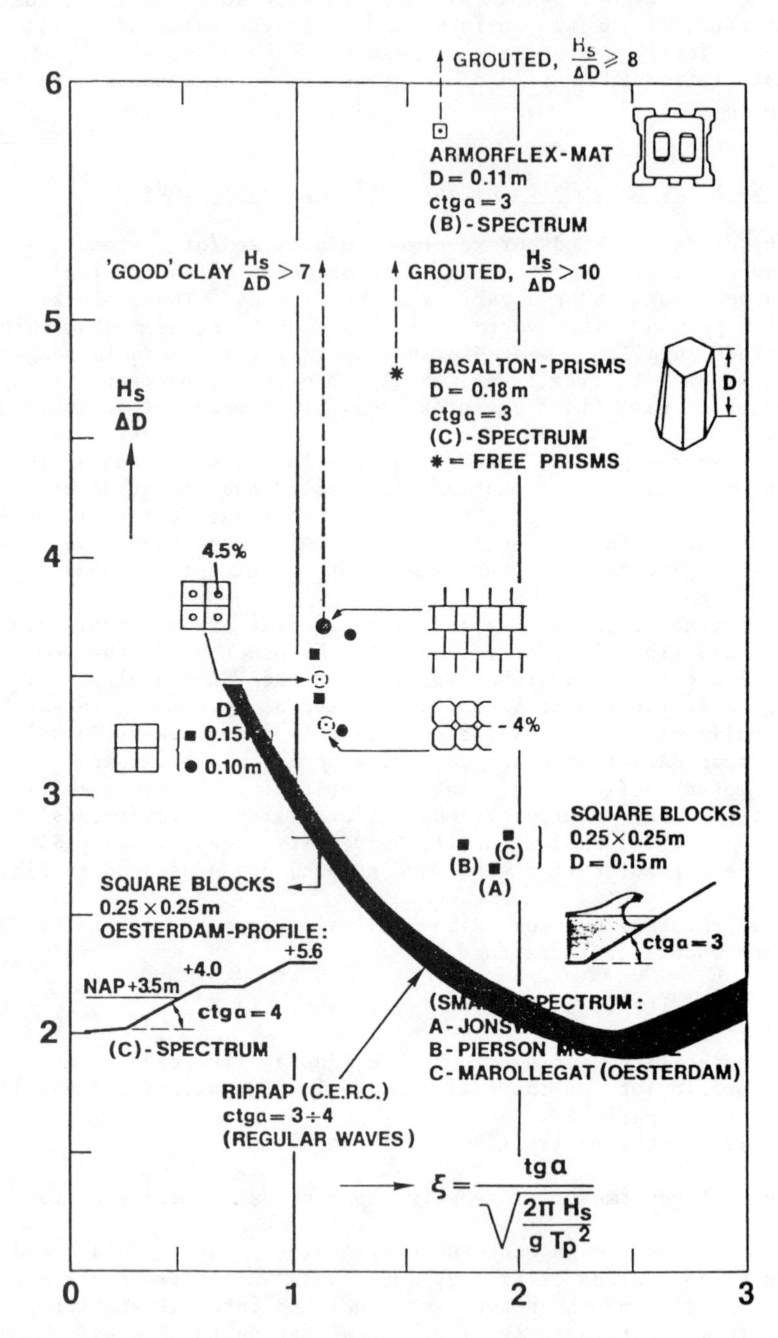

Fig. 4. Stability numbers for block revetments (relative density of blocks Δ = 1.24 ÷ 1.38)

internal stability have to be evaluated mostly by hydraulic model tests.

Stability of armour units is mostly related to Hudson's formula

$$G = \frac{r_s H^3}{K_D \Delta^3 \cot\alpha}$$

where G = weight of unit, r_s = unit weight, Δ = relative density, H= wave height, α = angle of slope and K_D = stability coefficient.
However, in the case of block-revetments it is more proper to express the stability as a function of the block thickness,

i.e.

$$\frac{H}{\Delta D} = (K_D \cot\alpha)^{\frac{1}{3}} \quad \text{or } K_D = \left(\frac{H}{D}\right)^3 \tan\alpha$$

The example given below will show that K_D should not be used for comparison of stability of revetments.

$\frac{H}{\Delta D}$	Relative stability (w.r.t. D)	K_D (i.e. α = 3)	Relative stability (w.r.t. G)
2	1	2.67	1
10	5	243	91

The stability of revetments with H/ΔD = 10 (i.e. grouted/cabled systems) is 5 times stronger than revetments with H/ΔD=2 (i.e. riprap, free blocks), while the K_D value indicates an increase in stability by a factor of 91.

The external stability has to be complemented by internal stability. To fulfill this requirement it is advisable to use a highly permeable toplayer in the case of highly permeable sublayer and less permeable toplayer on less permeable sublayer/subsoil (i.e. clay) (see Fig. 5).

In the case of good quality clay it is preferable to place full-blocks (with rather low permeability) directly on clay. It increases the stability of the toplayer because the build-up of the pressure underneath the blocks will be avoided or diminished. In the case of poor clay (sandy clay) or sand (properly compacted), it is preferable to use full blocks with geotextile inbetween. The use of highly permeable toplayer in the cases mentioned above will lead to increased erosion of subsoil, and even to a spontaneous liquefaction of sand. It will result in higher uplift and deformation of revetments.

For total stability it is also important that the revetment is properly supported at the toe and provided with proper drainage unless it is laid on clay. A toe failure may prove

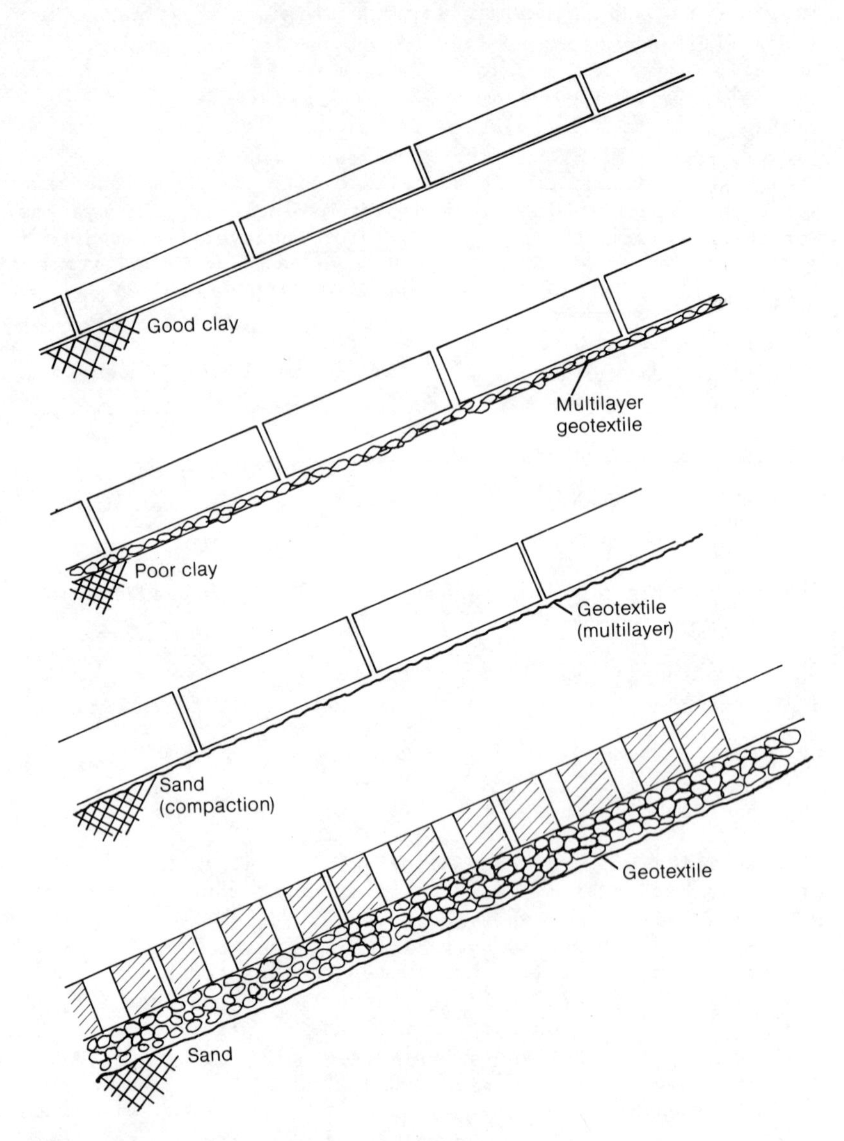

Fig. 5. Principles of revetment design

to be fatal for the whole revetment. An important aspect for overall stability is the relation between the revetment's armour layer and the condition of the core fill upon which it is placed. A new fill may be loose, and must be well compacted to serve as a foundation for a new revetment. In addition, blocks have to be placed accurately which, in turn, means that they must be cast within narrow limits of tolerance. Even the best design may fail as a result of poor workmanship.

Flexibilty means that a block is able to move independently (to a limited extent) of its neighbours without jamming. Normally, some settling (or erosion) may be expected, and it is therefore important that the revetment be built flexibly to prevent it from being damaged by movements that open up the joints, thereby leaving part of the slope open to direct attack by waves and currents. Flexibility is the main weak point in respect to interlocked systems including mat systems. For these systems the proper design of sublayer (related to eventual erosion of subsoil) is of major importance.

The main advantage of applying mats and/or interlocked systems are:

(a) in general, high stability of toplayer
(b) mechanical placement (also from water side)

and the main disadvantages (or uncertainties)

(c) damage/durability of binders (cables, pins, etc)
(d) inspection problems in respect to failure sublayer/subsoil
(e) connection of adjoining mats (especially under water)
(f) repair problems

Many of these problems will be omitted if the sublayer is designed properly. Although the grouting of these systems (i.e. grouting by fine broken stone) decreases the flexibility, it can be helpful for some reasons.

For non-grouted systems:

(a) individual blocks can move (less stability of toplayer)
(b) cables/pins loaded more frequently
(c) abrasion of geotextile possible
(d) erosion sublayer/subsoil can take place more easily

Grouted systems:

(a) higher stability of toplayer (no movement of individual blocks)
(b) if (at heavy attack) the mat will be lifted-up, then there is less loading of cables and pins and less abrasion of the geotextile, because such mats will move as an integral unit.

However, not every system can be grouted; if the interspaces between the blocks are too large, the grouting material will be easily washed out. In this respect the basalton-system (grouted but not cabled - see paper 13) has more advantages than other systems, because of the tapered vertical form and lack of cabling, the blocks can still follow a limited settlement (also, the settlement of the solid-body or erosion of the sublayer are immediately evident) and a

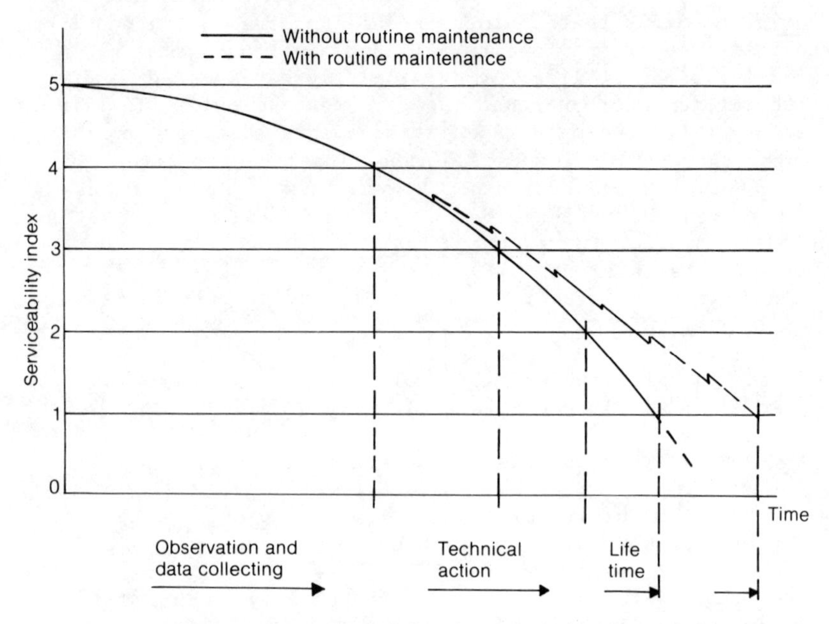

Fig. 6. Performance model of long term behaviour

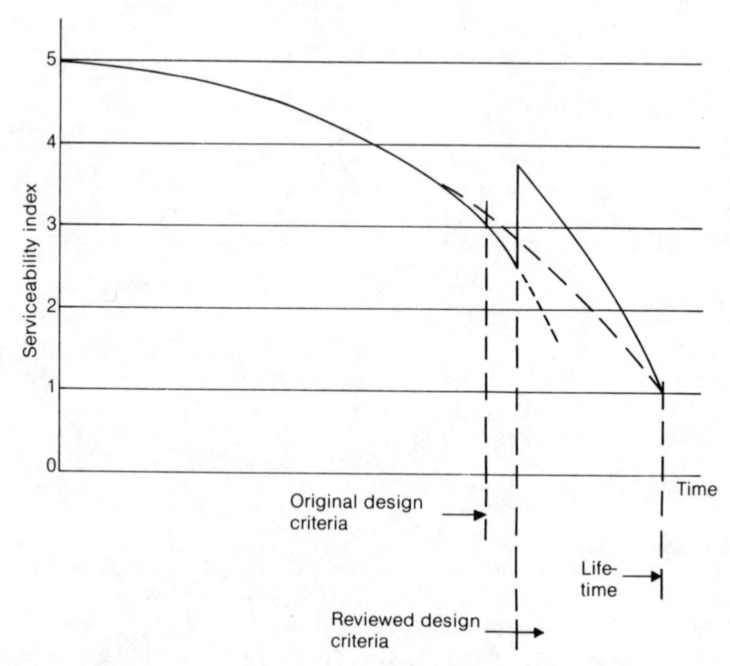

Fig. 7. Performance model after review and structural design correction

washing out of grouting material is limited.

The general conclusion is that in the case of interlocked and mat systems the proper design of sublayer is of major importance (a higher safety reserve has to be included in the design) because of problems regarding monitoring of sublayer-failure and repair of revetment.

It has to be pointed out that the present evaluation of the available data on design of revetments still has many limitations regarding their practical use. This is due to: insufficient knowledge about failure mechanisms, restricted variety of boundary conditions, possible scale effects in model results, insufficient controlled examination under in situ conditions, and non-uniformity in definitions and presentations.

Considerably more study is required before definitive answers on the selection of the best types of revetment for a variety of conditions can be given.

On other hand, in many cases these data could be sufficient for design purposes, especially if some reserve stability were taken into account.

Professor H. J. Span, Technical University, Delft, The Netherlands

A qualified design is a product of decision making in: (a) technical solutions with acceptable risks in life time; and (b) cost-benefit strategy and financial budget planning, because design costs are 2-4% but construction costs about 50-60% and maintenance costs about 40-50% afterwards!

Construction performance can be expressed in a behaviour model related to service ability (fitness for purpose) during design lifetime (Fig. 6). By means of routine technical maintenance, this design lifetime measuring is included and is necessary in this approach for preparing technical action for structural correction.

Technical failure of an essential part of the construction, or essential changing of loading criteria, strongly influences the performance curve (Fig. 7). This has to be translated into structural action for restrengthening or reconstruction of the basic design.

It is now essential in this field to have a basic design philosophy and that the revetment can be restrengthened if required by changed design criteria.

In my opinion this is a marginal problem depending on the structural system chosen for the bank revetment. A rip-rap material can simply be strengthened by sealing compound grouting, but a armoured block revetment must be overlaid!

This approach is essential for a qualified and economic design.

I would like to ask the authors of Papers 7 and 11 if they agree that a rational revetment mangement system must be a part of the design philosophy and strategy. My suggestion to the working group is that they make this a focal point of their study.

Dipl. Ing. H.-U. Abromeit

While we agree with these observations, the question arises whether the conditions mentioned can be numerically fixed. In any case it would be important and desirable.

Mr C. T. Brown

Gabions have failed by stones being sucked towards the toe. Use of diaphragms is made to control this. How does the material respond to fire, sun and high temperature?

Mr J. G. Berry, Bertlin & Partners, Surrey

Revetments for sea walls rely on having good toe protection for their stability. The usual solutions are sheet piling or surface (or near-surface) mats. Cable connected precast concrete mats may be a suitable solution, and presumably react to formation of scour depressions more rapidly than mastic asphalt mats.

In model hydraulic tests the design of the toe is not tested because of scale effects. However, the new Delft flume can test at prototype size. Has testing of cable connected precast concrete mats when used as a toe protection, rather than a slope protection, been carried out?

In view of their relatively slender dimensions, do hollow precast units used in cable-connected mats withstand attack by 50 mm shingle under wave attack?

The use of exposed geotextiles is undesirable because of ultra violet light attacks, and most examples of use keep them under the surface. Mr Hall has referred to a 15 year life time. Clients would not accept 15 years as the life of a revetment. Could Mr Hall comment on this?

Mr C. D. Hall

Mr Brown and Mr Berry's questions relate to the suitability of polymer grids in tubular gabions. Polymers respond to fire by burning, and to elevated temperature by softening (1).

The life of any engineering material is the sum of all the deleterious effects of its working environment. This has always been recognized in assessing, say, corrosion of steel, chemical attack on concrete or the rotting of timber. Ultra-violet irradiation is the main cause of the decline in the physical properties of polymers, yet a simple prediction of service life proves to be elusive. Some field data is currently being acquired in the Middle and Far East, USA and South Africa and I look forward to the reporting of these findings. I hope that they will support the notion of a half-life in quantifying the UV effect. In the case of carbon impregnated high density polyethylene, for example, the 15 years that has been mentioned indicates the half-life of the strain to failure in a UK exposure (2). While the tensile strength is upheld, irradiation of the polymer causes embrittlement. In common with the methods of civil engineering design with polymers (3), the value of permissible stress is based on long term strain behaviour. Nothing terminal suddenly happens at 15 years therefore, and this is not a projected life of a tubular gabion revetment. Clients do find tubular gabion revetments acceptable in the circumstances described in the paper.

References

1. McGown A. et al. The load-strain-time behaviour of 'tensar' geogrids. Symposium on Polymer grid reinforcement in civil engineering, London, 1984.
2. Landsberg H. E. et al. Weltkarten & Klimakunde. Springer Verlag, Berlin, 1966.
3. Netlon Ltd. Test methods and physical properties of 'tensar' geogrids.

Mr M. T. de Groot, Delft Soil Mechanics Laboratory, The Netherlands

In Fig. 5 of Paper 7, Mr Abromeit presents registrations of pore pressures at five points in the subsoil perpendicular to the slope. The figure suggests that the greater the depth in the subsoil, the more the response of the pore pressure is damped. Now, for a completely homogeneous soil this will indeed be the case.

In general, ground-water flow can be described by equilibrium equations and stress-strain relations for the grain skeleton and by the Biot storage equation.

$$\frac{k}{\gamma_w}\nabla^2 w = \frac{n}{k_w}\frac{\partial w}{\partial t} + \frac{\partial e}{\partial t}$$

k = permeability
γ_w = mass density pore-water
w = pore pressure
n = void ratio
k_w = compressibility modulus pore-water
e = volume strain

It follows that wave induced pore pressures depend on these parameters, and in particular on the air content of the pore-water. For example, if the pore-water contains 2% (volume) of

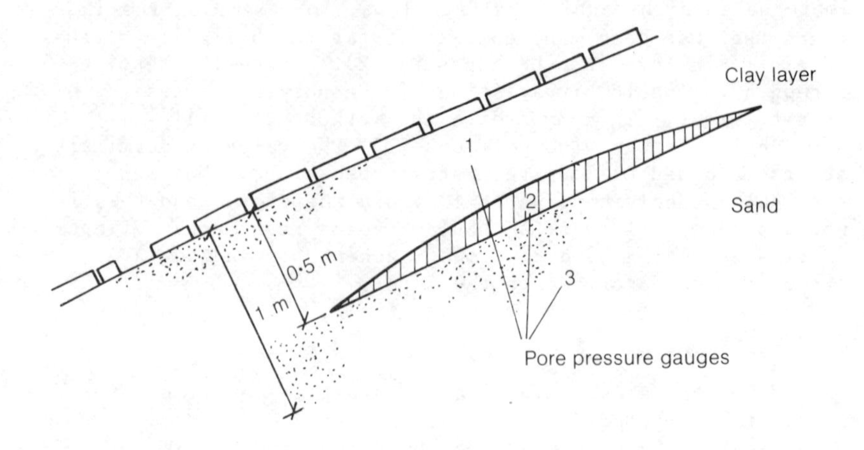

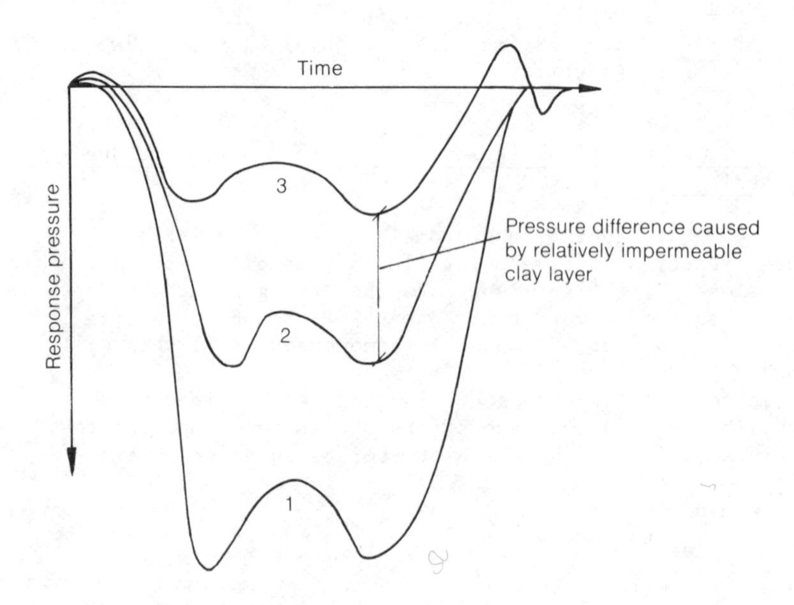

Fig. 8.

air its compressibility modulus reduces from 2.10 kN/m^2 to 10^4 kN/m^2. The influence of air content and soil permeability on the response is very significant.

When the subsoil of an embankment consists of sand with thin layers of clay the anisotropy of the soil may very well lead to a response of pore pressures which differs from Mr Abromeit's figure.

From prototype measurements in the Hartel Canal in the harbour area of Rotterdam we found at some places the response of pore pressures in the subsoil shown schematically in Fig. 8.

On the other hand, for homogeneous subsoils Mr Abromeit's figure may well apply.

Dipl. Ing. H.-U. Abromeit

The distribution of pore-water pressure presented in Fig. 5 of our paper is only valid in case of rapid water level depression on banks consisting of a homogeneous silty fine sand with a permeability of $k = 6 \times 10^{-5}$ m/s (Darcy's coefficient).

Mr F. C. M. van der Knaap, Delft Hydraulics Laboratory, The Netherlands

In Mr Hall's paper a very attractive bank revetment system of tubular gabions is treated.

The computation of the hydraulic attack forms the background of my question in particular. The hydraulic attack is computed in this paper by applying the method of Schijf. This method is based on a one-dimensional potential theory by which it is only possible to compute average values for the return current velocity and water level depression. This is also true of the theory of Balauin and Bykov, which is shown in the DHL publication No. 302, 'Prediction of ships sailing in restricted water'. Therefore these methods do not take into account the influence of the ships position in the cross section of the canal. This influence can be great importance in determining the attack on the revetment.

Does Mr Hall include the influence of the ship's position in the prediction of the attack on the bank revetment?

Mr C. D. Hall

PIANC literature was reviewed to obtain predictions of both the return current and bow wave height at the canal bank for use in the stability equations 3, 4 and 8 of my Paper. As Mr van der Knaap rightly points out, Balanin & Bykov (Selection of leading dimensions of navigation canal sections, Int. Navigation Congress, PIANC, Stockholm, 1965) do not deal explicitly with the case of a vessel's passage being eccentric to the canal centre line:

$$V_b = 0.7u\left(\frac{1}{1-CB}\right)$$

where V_b = current velocity at bank, u = Schijf limiting speed and CB = blockage ratio.

For 'small' eccentricities measured towards the bank in question, the return current may well increase to a value of that created by a concentric vessel whose beam is broadened by 2e. The blockage ratio in the above expression can be factored by (1 + 2e/B) to represent the apparent blockage of an eccentric vessel:

$$V_b = 0.7u\left(\frac{1}{1 - CB(1+2e/B)}\right)$$

where e = eccentricity, B = beam

The effect of small eccentricities on wave height at the canal bank appears to be slight. The notion of apparent blockage can also be applied to the expressions of Balanin & Bykov.

For 'large' eccentricities, where the flow field becomes more complex as bank effects become significant, I look to establishments such as Mr van der Knaap's for guidance.

Mr C. T. Brown, Seabee Developments, Australia

Cabling existed prior to 1981, the date suggested by Major Wise: examples are gabions (1890), name blocks (1972), Palmers Namiring (1975), Seabees (1975).

Major E. G. Wise

I am indebted to Mr Brown for drawing attention to the inter-cabling of numerous revetment structures within the last century, and would add to his list the cabling of the original Atlantic Beach Coast Defences, east of Banjel in The Gambia. My remarks were certainly not intended to imply that no cabling of armour units had taken place before 1981; but rather that I personally have been unable to trace any monitoring records, or conclusions concerning the contribution such cable interconnections may have made to the long term stability of their structures.

Mr C. T. Brown

What is the term BL referred to in Mr Abromeit's paper? Why did he use $\hat{H}$ (maximum water level excursion), instead of the more general H_s?

Dipl. Ing. H.-U. Abromeit

BL is the abbreviation of 'Belastungsgrad'. This is the

relative frequency of loads (waves, return flow, water level depression) important for design of revetments, given as a percentage. These statements can be gained either by spot-check measuring of size and frequency of the single loads on the waterway concerned, or if these data are already available for comparable waterways they can be taken from there.

$\hat{H}$ was chosen because the height of waves is composed in this case of two different components (water level depression and stern transverse wave).

Mr C. T. Brown

I agree with Mr Pilarczyk that K_D is not an economic comparator. Tightly jointed blocks can fail by jet erosion through the joints. Are we after maximum stability or maximum economy?

Incorporation of porosity can materially reduce run-up and reflection.

Dr G. Heerten, Naue-Fasertechnik, Germany

Concern has been expressed about the long-term behaviour of cables in revetment structures. I would like to give some detailed information about the construction of the Port Kembla Seawall in Australia, where polyester ropes were used for the toe construction.

The stability of the seawall depended on the integrity of the toe of the structure. The difficult construction requirement of excavating a tow within the surf zone where there were large quantities of mobile sand, and the impossibility of satisfactorily placing conventional filter material, led to the adoption of a synthetic filter cloth material. Materials were subject to extensive field tests to ensure that a suitably robust cloth was available to withstand placement and movement of sub-armour and primary armour without damage to the cloth. Heavy needle-punched nonwoven fabrics with a weight of 1200 g/m^2 and 2000 g/m^2 had been selected.

Potential problems associated with the placement of the filter cloth, and securement before overlying material was placed, were recognized. A concrete block filter mat was designed as a method of holding the filter cloth in position until rock and armour units were placed. The prefabricated block mat consisted of twenty five 670 kg concrete blocks laced together with polyester rope. The concrete blocks were cast directly around the polyester ropes and PVC sheaths which covered and protected the ropes between the blocks.

The main functions of the ropes occur during installation and construction. The ropes are of little importance for the stability of the structure because after completion the stability of the structure has to be given by the stability of

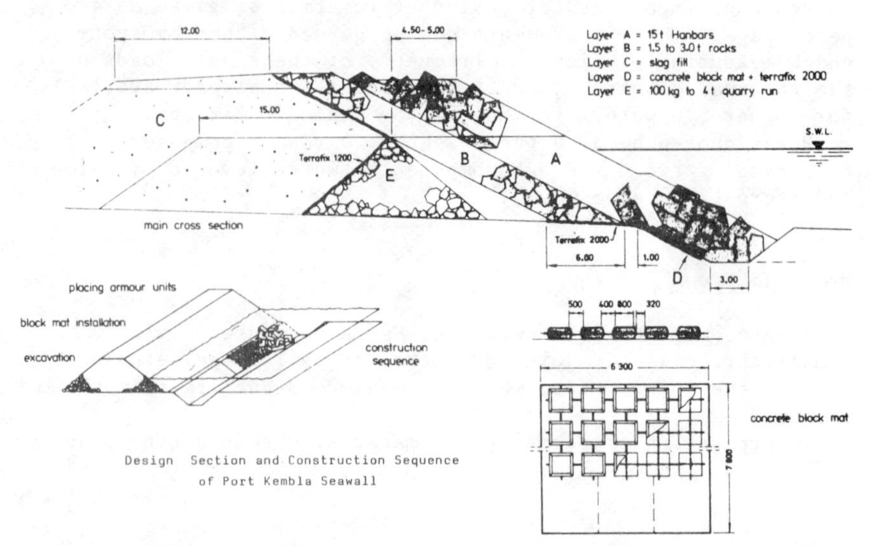

Fig. 9. Design section and construction sequence of the Port Kembla seawall

Fig. 10.

the armour layer itself (15 t Hanbars). Therefore the long-term behaviour of the ropes is not important in this special case.

Fig. 9 shows the design section and construction sequence of the Port Kembla Seawall, and Fig. 10 shows the installation of the block mat.

Major E. G. Wise

The introduction to my paper included the suggestion that a fundamental distinction be drawn between a revetment whose armour units or components are simply inter-meshed or fitted together to form a plane non-rigid surface, and a revetment whose armour is structurally interconnected in two mutually opposed, or orthogonal, directions, thereby forming an integrated flexible structure of any variable configuration.

The suggestion to differentiate betweeen non-rigid surfaces and flexible structures is no mere exercise in semantics, since for any surface assembled from fitted or inter-meshing parts there will always remain at least one direction along which those parts may be separated or unmeshed, and the surface thereby dismantled. The suggestion was intended as a prelude to my general theme that a finite two dimensional integration of an armoured structure would both simplify its design and add significantly to its stability as a whole; especially if that structure were called upon to withstand wave or current attack beyond the limits imposed by only the static masses of its parts and an indeterminate degree of mutual interlock.

Until quite recently outside of the USA there had been no opportunity to monitor the actual effects of a two-dimensional flexible interconnection of armour units when under real and severe wave attack. No such structure then existed in this country. Furthermore, up to a year or so ago no attention appears to have been paid to the possible role or likely potential of the orthogonal cabling of armour when laid on surfaces at varying inclinations, when subjected to variable wave attack, or even in the event of unforeseen embankment movement. Again, the anchoring of systems seems generally to have been undertaken as a rather ad hoc added insurance against undefined disaster, and to have been based largely upon strongly held opinions and much subsequent faith.

However, recently there have been several significant changes in the integrated flexible revetment scene; brought about, in part, by the careful observation of an increasing number of installed structures at coastal, inland water and reservoir sites in Britain. Also of great importance has been the large-scale model wave testing of integrated and anchored flexible armour, undertaken during 1983 at the Oregon State University Wave Research Faculty in the USA (to be published later in 1984).

The initial results of these tests have not only confirmed early beliefs in a general improved stability of flexibly

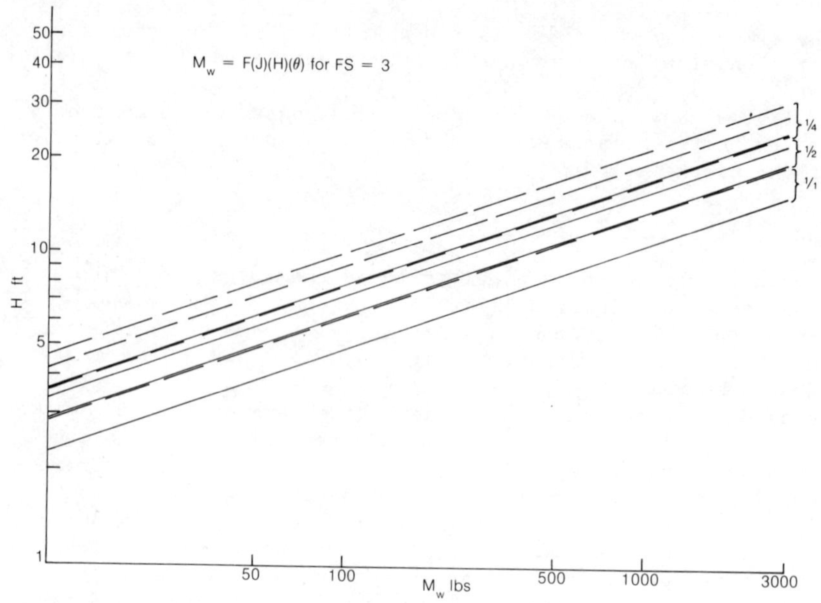

Fig. 11.

integrated armour, but have indicated that predictions of the likely degree of such improvements were at least 50% too conservative.

Nevertheless, because the orthogonal total interconnection of armour units will appear as an important factor of revetment stability under conditions of extreme wave or current attack it is interesting to now pursue further the stability of a single finite area of such an integrated revetment. This has led to the concept of a flexible integrated armour module (FIAM) which I define as being the maximum number of integrated armour units of any given mass or masses which could be supported by an orthogonal system of any number of interconnecting cables, without any cable of the supporting system being loaded beyond an assigned safe working load.

The periphery of a particular FIAM will vary with the mode and strengths of its cable interconnections and with the mass or masses of its individual armour units. Such modes, cable strengths and unit masses will all be known as design parameters, and all may be varied at will, since the guaranteed minimum breaking strengths of the cables are known and armour unit masses as well as their shapes and the arranging of their cabling conduits are conditioned primarily by production and economic considerations.

Based upon the Oregon Tests and subsequent work, it will become possible to develop FIAM coefficients, combining integration and mass factors, applicable to the several already well established stability expressions such as those of Hudson or Shields or Irrabarron whose forms, as pointed out

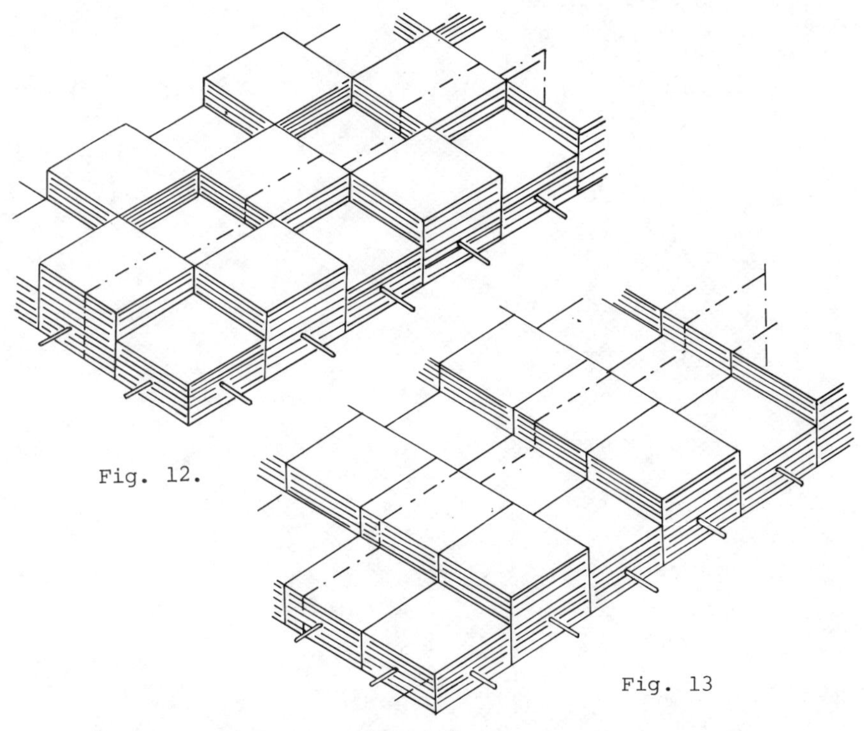

Fig. 12.

Fig. 13

earlier by Mr Brown, are all closely related.

For example, Fig. 11 shows the general relationship of wave energy and embankment inclinations to module armour unit or block weights (Mw) for a single orthogonal cable interconnecting system to which a safety factor of 3 has been applied.

Having determined a modular weight in relation to wave or current attack, there remains the need to ensure that each such module, and therefore the flexible structure as a whole, is adequately anchored to the protective surface in a manner which will limit all cable loadings within each module to their safe designated values.

Because the strengths of cables can be related to any weight of armour unit it may be that a future dominant factor in the use of heavy duty integrated flexible revetments could become their economic handling and cranage. If this were the case then an alternative option, to provide a lighter and more manageable (and therefore cheaper) revetment, would be to depart from the generally accepted armour surface, which lies in a single plane at any point, and substitute instead surfaces which lie in alternating parallel planes, whose effect would be to increase the extraction of wave energy from any given area of the revetment.

Examples of two such arrangements, having castellated profiles and therefore described in their patent application as castellated flexible revetments, are illustrated in Figs 12 and 13.

12 US Army Corps of Engineers experience with filter fabric for streambank protection applications

M. P. KEOWN and N. R. OSWALT, US Army Engineer Waterways Experiment Station, Vicksburg, Mississippi

SYNOPSIS. The major use of filter fabric by the U. S. Army Corps of Engineers (CE) for streambank protection applications has been as a filter for riprap blankets; however, fabric has also been used for placement under articulated concrete mattresses, gabions, and precast cellular blocks. Filter fabric was initially used by the CE in Memphis District as part of a riprap revetment repair project (1962). Further field experience gained by Memphis District and fabric testing by the Waterways Experiment Station resulted in issuance of the first Corps-wide guidelines for filter fabric placement (1973). A recent survey of the Corps Districts (1983) indicated that only 10 Districts had reported unsuccessful use of filter fabric and, in most cases, these reports involved one or two projects in a District.

INTRODUCTION

1. As part of early efforts to stabilize streambanks with a layer of stone (riprap), U. S. Army Corps of Engineers (CE) observations indicated that placement of a granular filter (sand, gravel, crushed rock) between a riprap blanket and the prepared bank surface resulted in a measurable improvement of revetment stability at sites where the soil material was erodible or in a high energy environment (wave action, eddy currents, prop wash, etc.). A properly designed granular filter effectively reduced the amount of soil being eroded through the riprap blanket, provided a bedding layer for the riprap, and still allowed for natural drainage from the streambank (ref.1). Without the filter, the integrity of the structure could have been seriously compromised as more and more material was removed from the bank slope through the riprap. During the 1960's fabric materials were introduced as a revetment filter (Fig.1) for projects where suitable granular materials were not readily available or were not cost-effective due to transportation, quality control, or manpower constraints. Although use of granular filters is still considered as part of the "traditional" approach for revetment design and construction, filter fabric is being used for many projects.

Fig. 1. Placement of riprap on filter fabric

2. The initial use of filter fabric for hydraulic applications can be traced to projects placed by Dutch engineers in 1956. However, in the ensuing years, filter fabric did not find widespread acceptance in the American engineering community. As late as 1967, there were only two domestic sources of fabric, although the use of fabric as a filter under an interlocking block revetment had been reported as early as 1958 in Florida.

3. Prior to 1970, no site-specific cost comparisons for using filter fabric as a substitute for granular filters were readily available. The initial economic case study of record was conducted by the U. S. Army Engineer District, Memphis, in 1966. Results of this study indicated that filter fabric could be placed under articulated concrete mattresses (ACM) for \$9.71/square (100 ft^2) as opposed to \$8.03/square for a 4-in.-thick granular filter (ref.2). However, a factor not considered in this comparison was the cost to repair undermined ACM, which comprises a large percentage of upper bank repairs along the Lower Mississippi River. The repairs through this reach were reviewed to determine what cost reductions could have been realized by the use of adequate filter material. Although it was not possible to identify all repairs attributable to the loss of granular filter and subgrade material through the mattresses, costs compiled from many construction sites through fiscal years 1967, 1968, and 1969 indicated that the repair cost for undermined ACM was \$1.73/square. During the period 1967 to 1969, no repairs were needed where filter fabric had been placed under the mattresses by Memphis District. Thus, on a short-term basis, fabric was directly competitive with granular filters through this reach of the Lower Mississippi River.

4. As the utility of filter fabric became apparent, the Office, Chief of Engineers (OCE), directed the U. S. Army Engineer Waterways Experiment Station (WES) to conduct a study to determine the extent and diversity of use of this material by CE Divisions and Districts. The findings of the study (ref.3) indicated that although there was wide and varied use of filter fabrics by the CE, a test program was needed to define the engineering properties of the fabrics when used for filter and drainage applications. This initial study became the first phase of a broader program conducted at WES (1967-1972). As part of this program, several filter fabrics (six woven and one non-woven) were evaluated by chemical, physical, and filtration testing. Additional work by the Memphis District provided needed information on the large-scale field application of filter fabric. From the results of the WES program (ref.4 and 5) and CE project experience, OCE guide specifications were developed for field use of filter fabric (ref.6).

5. When nonwoven or random fiber fabrics became available, additional examination of fabrics and methods of evaluating their engineering properties was considered necessary. Laboratory testing was conducted at WES during 1974-1976 to refine existing test methods for woven fabrics and to develop new methods for the evaluation of nonwoven fabrics. The results of this effort and further field experience provided the basis for new CE guidelines (ref.7) for the field use of woven and nonwoven filter fabrics.

6. In recognition of the serious economic losses occurring throughout the nation due to streambank erosion, the U. S. Congress passed the Streambank Erosion Control Evaluation and Demonstration Act of 1974, which became known as the Section 32 Program. Under this program (1975-1982), filter fabric performance at 25 streambank protection projects was monitored; in addition, hydraulic research was conducted at WES to evaluate the performance of riprap in combination with filter fabric under wave attack and under various seepage and rapid drawdown conditions. The findings of the WES research and the results of monitoring at the 25 projects were published as part of the program's Report to Congress (ref.8). WES also conducted a survey of Corps filter fabric usage as part of the Section 32 Program. This survey (ref.9) indicated that 29 of the 38 Districts had used fabric for streambank protection purposes. A later survey compilation by WES in 1983 (ref.10) indicated that only 10 Districts had reported unsuccessful use of filter fabric and, in most cases, these reports involved only one or two projects in a District. Specific details of this survey are discussed later in this paper.

FILTER FABRIC PLACEMENT

7. Under current CE design guidance (ref.7), three fac-

tors must be evaluated during the selection of filter fabric for a specific project application:

a. Filtration. The fabric must act as a filter; i.e., the flow path through the fabric mesh must be fine enough to prevent continuous infiltration and passing of soil, yet large enough to allow water to pass freely.
b. Chemical and physical properties. The fabric's chemical composition must be such that it will resist deterioration from climatic conditions and from chemicals found in the soil and water, and must possess sufficient strength so that it will not be torn, punctured, or otherwise damaged during placement and through continued use.
c. Acceptance of mill certificates and compliance testing. The fabric must meet Government standards for acceptance of mill certificates and compliance testing.

8. Due to possible damage resulting from ultraviolet radiation or improper handling, the fabric should be wrapped in a heavy-duty protective covering such as burlap during shipment and storage. In addition, the fabric must be protected from mud, dust, and debris and from temperatures in excess of 140°F. Prior to fabric placement, the streambank soil surface should be graded to a relatively smooth plane, free of obstructions, depressions, and soft pockets of material. Depressions or holes in the soil should be filled before the fabric is spread since the fabric could bridge such depressions and be torn when the revetment materials are placed.

9. After bank preparation is completed, the fabric can be removed from the protective covering used for shipment and spread on the bank (Fig.2). Overlapping fabric edges should be joined by sewing (Fig.3) or bonding with cement or heat. After the fabric is joined, securing pins should be inserted through both strips of fabric along a line through the midpoint of the overlap. The pin spacing specifications for pins placed through the seams and over the remainder of the filter (Fig.4) are 2 ft for slopes steeper than 1V on 3H; 3 ft for slopes of 1V on 3H to 1V on 4H; and 5 ft for slopes flatter than 1V on 4H (ref.7). As reported in ref.4, several U. S. Army Engineer Districts have experienced tearing of filter fabric at the seams and pins due to stone sliding down 1V-on-2H slopes (Fig.5). This problem was minimized in the Divide Cut Section of the Tennessee-Tombigbee Waterway by placing the fabric loosely over the prepared bank and using only enough pins to hold the fabric in position prior to placing the stone (Fig.6). After placement, the stone moved toward its permanent resting position; as this occurred, the loosely placed fabric was allowed to move with the stone, thus avoiding the pin and seam tears that often occur on

Fig. 2. Filter fabric being spread on prepared bank (photograph courtesy of E. I. Dupont de Nemours and Co., Inc.)

steep slopes where many pins are used.

10. Fabric strips should be placed on a prepared bank surface with the longer dimension parallel to the current, when used along streams where currents acting parallel to the bank are the principal means of attack (Fig.7a). The upper strip of fabric should overlap the lower strip (as roofing shingles are commonly placed), and the upstream strip should overlap the downstream strip. To avoid long sections of continuous overlap, the overlaps at the ends of the strips should be staggered at least 5 ft as shown in Fig.7a. The revetment and fabric should extend below mean low water to minimize erosion at the toe. When the revetment materials and fabric are subject to wave attack, the customary construction practice is to place the fabric strips vertically down the slope of the bank (Fig.7b). The upper vertical strip should overlap the lower strip. The fabric usually needs to be keyed at the toe to prevent uplift or undermining.

11. When filter fabric is selected for a project, the placement of revetment materials on the fabric must be conducted in such a manner that the fabric is not torn or punctured. The most common material placed on fabric for streambank protection applications is stone riprap. Heavy and angular stone dropped from heights of even less than 1 ft can damage filter fabric. Displacement and settling of stone after placement could also result in ultimate failure. Various precautions have been taken in previous applications to prevent damage of fabric, such as a cushioning layer between the fabric and riprap (Fig.8). How-

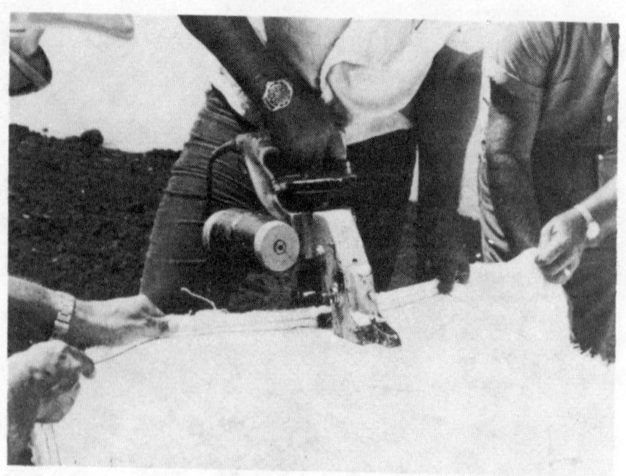

Fig. 3. Filter fabric sections being sewn together

ever, care should be taken to ensure that any cushioning layer does not form a low permeability layer between the stone and fabric.

RECENT CE EXPERIENCE WITH FILTER FABRIC

12. In March 1982 OCE sent a letter of inquiry to the CE Divisions and WES requesting descriptive information on those streambank protection projects where either failure or less than satisfactory performance had occurred that could be specifically attributed to the use of filter fabric. The survey responses (ref.10) not only identified problem areas, but also provided timely guidance, in the form of recommendations and cautions, for dealing with these problems. The problem areas that were identified, as discussed below, reflect the current spectrum of CE experience with filter fabric.

a. Erosion under the fabric. Small voids and loose fill areas are generally bridged by filter fabric, providing a site for potential erosion. As surface runoff moves downslope between the fabric and bank material, soil loss may occur. Silt, silty sand, and sand banks are particularly susceptible to this problem.

b. Slope failures. This was a widely reported problem. A typical sign that a failure has occurred is a bulge in the fabric near the bank toe and a depression upslope above the bulge. Although not a slope failure, a similar phenomenon may result when erosion occurs under the fabric and material is transported to the toe of the bank, which in turn clogs the fabric. Some survey responses indicated that failures could also occur on a saturated slope during rapid drawdown or due to the inability of the fabric to pass flow quickly enough to relieve pore pressure from groundwater flow.

Fig. 4. Filter fabric being pinned in place (photograph courtesy of Carthage Mills)

c. Tearing/puncture of the fabric. This problem may lead to entry of large volumes of water or exit of eroded soil.

d. Slippage of revetment material. This type of failure occurs primarily due to poor support at the bank toe or placement of the fabric on a steep slope (greater than 1V on 3H).

e. Ultraviolet light. Fabric exposed to sunlight for long periods during storage, construction, or maintenance can suffer a significant loss of strength.

f. Vandalism. Fire can destroy fabric. Designers should be aware of this problem, especially when placing revetments in recreation areas.

13. The following recommendations and cautions were offered through the survey comments:

a. Erosion under the fabric. During construction, strong emphasis should be placed on maintaining close contact between the filter fabric and bank slope. This can be accomplished by considering several design features:

(1) The bank slope should be smooth. Fill areas should be properly compacted so that settling does not occur after revetment placement.
(2) Fabric should not be placed in tension so that fabric/soil contact can be maintained.
(3) The fabric should be keyed in at top bank with an earth-fill trench.
(4) Overbank drainage should be minimized by routing flow parallel with top bank to a controlled discharge point.

Fig. 5. Tearing of filter fabric at securing pin (ballpoint pen for scale)

Fig. 6. Loosely placed filter fabric on the Divide Cut Section of the Tennessee-Tombigbee Waterway

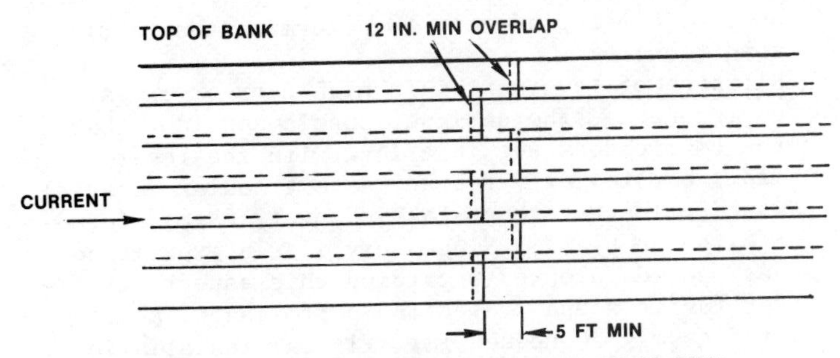

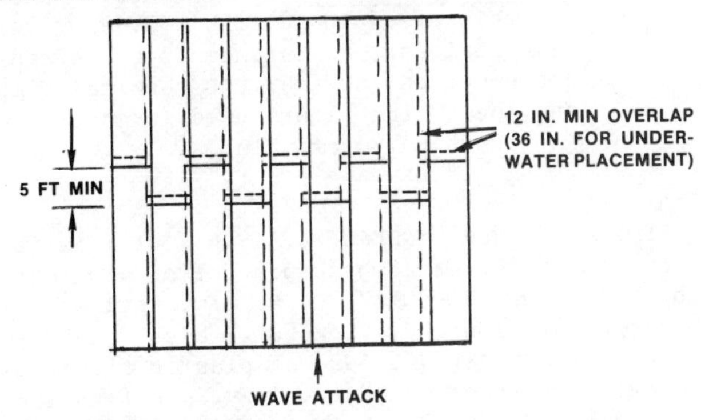

Fig. 7. Correct fabric placement for current acting parallel to bank or for wave attack on the bank

(5) If seepage from the bank slope is occurring or grading of the slope could result in surface erosion, then the filter fabric and revetment may need to be placed immediately after preparing the slope. Excessive seepage or piping during grading may require placement of granular filters before placement of the filter fabric.

b. Tearing/puncture of the fabric. After placement is completed, fabric tension should be only enough to hold the material gently in place and prevent bridging of any depressions on the slope. Excessive tension increases vulnerability to puncture and tears in the fabric. Because of the many variables influencing tearing/puncture, designers should consider preparing performance specifications that require placement of the fabric and revetment by the contractor on a test slope in order to qualify for a particular project. Project specifications should identify test sections where revetment could be removed shortly after placement to determine the fabric condition.

c. Slippage of revetment. In high current or wave environments, the designer should consider that the fabric does not interlock with the revetment, particularly riprap, as a granular material would. The revetment-fabric interface may form a slippage plane. Since very little experience has been documented regarding this aspect of filter fabric slope protection performance, model studies may be needed for critical installations.

d. Ultraviolet light. Filter fabric material should not be exposed to sunlight (UV) for very long time periods. Resistance to UV degradation is dependent on the fabric's chemical composition and the UV inhibitors used in manufacturing and is thus highly variable.

REFERENCES

1. KEOWN M.P. and DARDEAU, E.A. JR. Utilization of filter fabric for streambank protection applications, Technical Report HL-80-12, U. S. Army Engineer Waterways Experiment Station, CE, Vicksburg, Miss. 1980.
2. FAIRLEY J.G. et al. Use of plastic filter cloth in revetment construction, Potamology Investigations Report 21-4, U. S. Army Engineer District, Memphis, CE, Memphis, Tenn. 1970.

Fig. 8. Stone riprap being placed over cushioning layer of fine sandy gravel; note filter fabric under gravel (photograph courtesy of Celanese Fibers Marketing Co.)

3. CALHOUN, C.C. JR. Summary of information from questionnaires on uses of filter cloths in the Corps of Engineers, Miscellaneous Paper S-69-46, U. S. Army Engineer Waterways Experiment Station, CE, Vicksburg, Miss. 1969.
4. ________. Interim report; Investigation of plastic filter cloths, Miscellaneous Paper S-70-2, U. S. Army Engineer Waterways Experiment Station, CE, Vicksburg, Miss. 1970.
5. ________. Development of design criteria and acceptance specifications for plastic filter cloths, Technical Report S-72-7, U. S. Army Engineer Waterways Experiment Station, CE, Vicksburg, Miss. 1972.
6. OFFICE, CHIEF OF ENGINEERS, DEPARTMENT OF THE ARMY. Guide specifications, Civil Works construction, plastic filter cloth, CE-1310 (superseded by CW 02215), Washington, D. C. 1973.
7. ________. Civil Works construction, guide specification, plastic filter fabric, CW 02215 (supersedes October 1975 specification), Washington, D. C. 1977.
8. ________. Final report to Congress, the Streambank Erosion Control Evaluation and Demonstration Act of 1974, Section 32, Public Law 93-251, Washington, D. C. 1981.
9. DARDEAU, E.A. JR. Use of filter fabric by Corps of Engineers Districts, memorandum for record, U. S. Army Engineer Waterways Experiment Station, CE, Vicksburg, Miss. 1978.
10. MILLER, S.P. Unsatisfactory performance of geotextiles at Corps of Engineers installations (in preparation), U. S. Army Engineer Waterways Experiment Station, CE, Vicksburg, Miss.

CONVERSION FACTORS

U. S. customary units of measurement used in this report can be converted to metric (SI) units as follows:

Multiply	By	To Obtain
Fahrenheit degrees	5/9	Celsius degrees or Kelvins*
feet	0.3048	metres
square feet	0.09290304	square metres
inches	25.4	millimetres

* To obtain Celsius (C) temperature readings from Fahrenheit (F) readings, use the following formula: C = (5/9) (F - 32). To obtain Kelvin (K) readings, use: K = (5/9) (F - 32) + 273.15.

13 Prototype tests of slope protection systems

K. W. PILARCZYK, MSc(Eng), Delta Department, The Netherlands

SYNOPSIS:In the past decade a rapid development in design and construction of artificial block and bituminous revetments has taken place. The fact that design rules are still limited in quantity has stimulated investigations in this area. The prototype tests are of great importance for verifying the results of desk and model studies. In this paper the scope and organisation of the prototype measurements as done actually in the Netherlands (1981-1983) are discussed. In addition a short review is presented of the protection systems involved in these prototype tests.

INTRODUCTION.

1. Numerous types of revetments have been developed in the past for shore and bank protection against erosion by waves and currents (i.e. rip-rap, blocks, asphalt, etc.). On the other hand, continued demand for relatively low-cost protection in estuaries and along the shores and navigation channels has stimulated investigations in the area of artificial block and bituminous revetments(ref. 1,8,9,10,11, 12) as well, in the area of geotextiles (ref. 3,4,5). The reason for this is the increase of the problem in respect to the defence of the shores and banks of navigation channels as well as the high cost and shortage of natural materials in some geographical regions. This demand has resulted, in inter alia, the rapid development of a great variety of artificial block-units and in the wider application of asphalt and geotextiles. At the same time, the quality of concrete blocks was gradually improving due to the improvements in the manufacturing process and the cost which diminished due to mechanical placing so that, as a result, concrete blocks of various sizes and shape are used satisfactorily in coastal protection and the protection of navigation channel banks under a variety of conditions.

2.For countries with their own sources of stones,dumped quarry stone is usually the cheapest material per ton,which can be used for revetments. However,dumped stone has a lower stability per unit weight against wave and current attack than the most concrete armour units. Because of these characteristics dumped quarrystone can serve as a reference by which the stability and cost of other revetments may be judged(ref.2).

3. The design of most of these revetments is still based more on rather vague experience than on the generally valid calculation methods. Therefore, however limited the existing knowledge on this subject may be,it is useful to systematize this knowledge and to make it available to designers (ref. 3,7).

4. In this paper, the actual research philosophy leading to the preparation of design guidelines on bank protection is briefly discussed (ref. 7, 8, 9). Next, a review of the organisation of the prototype measurements being a part of the total research programme, is also presented. Special attention is paid to the protective systems used in the prototype tests in the Netherlands (rip-rap, placed blocks, block-mats, gabions, sand-mattresses and open stone-asphalt). The advantages and disadvantages of the different types of revetments are compared and suggestions made regarding their practical application.

SCOPE OF THE RESEARCH PROGRAMME.

5. The developments in inland shipping in the Netherlands during the past decades are particularly related to the increasing number of high-power vessels and push-tow barges. Many problems arise from these developments, among them erosion of the banks. Depending on the local situation,ship-induced water motion or water motions generated by other mechanisms, e.g., wind, waves and currents may be decisive for the design of the bank protection. The design of bank protections is complicated and has no proper theoretical foundation yet. Consequently, at the Delft Hydraulics Laboratory, an extensive, fundamental research programme on ship-induced water motions and related design of bank protections and filter-layers (incl. geotextiles) is carried out by the commission of the Dutch Public Works Department (Rijkswaterstaat). This research is known as the M1115-research (ref. 8).

6. During the research programme hydraulic model studies have been carried out to a scale of 1:25. At this scale however, only very limited information on the stability of top-layers (especially regarding block-revetments) can be obtained. In addition, some scale effects may be involved in reproduction of the ship-induced watermotions. Therefore, it was decided to set up an extensive series of prototype-measurements (OEBES-project) to get better insight in the phenomena involved and to determine the scale effects by comparing the results of the prototype measurements with corresponding hydraulic model tests (ref. 6,7). The prototype measurements took place in the Hartel Canal (see figure 1) situated within an area of the Rotterdam-harbour (the first series in 1981 and the second one in 1983). The tests have been carried out

in close cooperation with the Delft Hydraulics Laboratory the Laboratory of Soil Mechanics,the Rotterdam Port Authority and the Ministry of Transport and Public Works.

7. The results of the prototype measurements, in combination with the model results and the calculation methods developed in the framework of the systematic research (M1115) on bank protection and systematic research (M1795) on dike protection extended with knowledge gained from practical experience,will lead to preparation of guidelines for reliable bank protection designs. The aim of the total research programme is to develop such design criteria that the amount of maintenance and construction costs of new revetments is minimized.

DESCRIPTION OF PROTOTYPE TESTS.

Location and test embankments.

8. The Hartel Canal proved to be a good location for prototype tests. The canal satisfies the set criteria: it has a straight fairway with a restricted width (bottom width ~75m, depth ~7m), continuous slopes of homogeneous subsoil, and little disturbance by shipping.
During the 1st series of measurmeents (1981) the Hartel Canal was closed by a lock navigation (i.e. constant water level was present). During the 2nd series (1983) the Hartel Canal was in an open connection-through tidal river Oude Maas - with North Sea (i.e. tidal flow fluctuation was present).
For the first series of measurements eight different 40m long testsections (slope 1:4) were constructed. Five others were added by different contractors in 1983.
The following test embankments, equipped with geotextile filter, were purposely constructed on slope 1 to 4 for the measurements campaigns (see figure 3):

Series 1981 and 1983.
1. rip-rap (5-40kg) on clay
2. blocks(0.15m height) on clay.
3. blocks on a layer of gravel, on sand.
4. blocks on sand.
5. basalton (0.15m) on sand.
6. rip-rap(5-40kg) on sand.
7. coarse gravel(80-200mm) on sand.
8. fine gravel(30-80mm) on sand.

Only 2nd series (1983).
9. basalton (0.12m) on silex/sand.
10. fixtone(0.15m) on sand asphalt on sand.
11. sand-mattresses(0.20m) on gravel/sand.
12. armorflex-mats(0.11m) on gravel/sand.
13. PVC-Reno mattresses (0.17m) on sand
14. ACZ Delta block-mats (0.16m) on gravel/sand.

Structure and organisation.

9. The test embankments (top-layer and subsoil), test-ships and wet cross-section of the test location were equipped with various instruments. The measurements were centrally directed from a shore-based "central" cabin(see figure 1). Just before entering the test-section of the canal by a test-ship, all instruments started operating simultaneously on a command given in the central cabin. Processing of the data followed

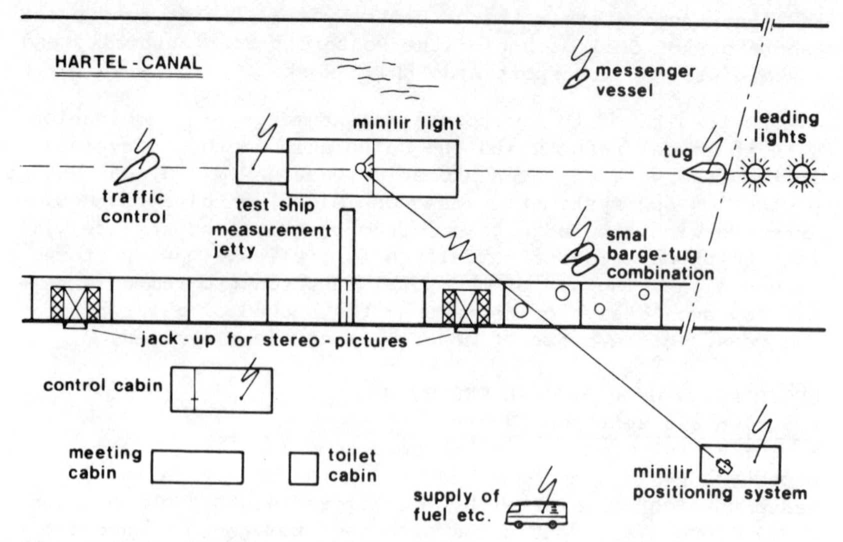

Fig. 1 Schematized set-up of prototype measurements.

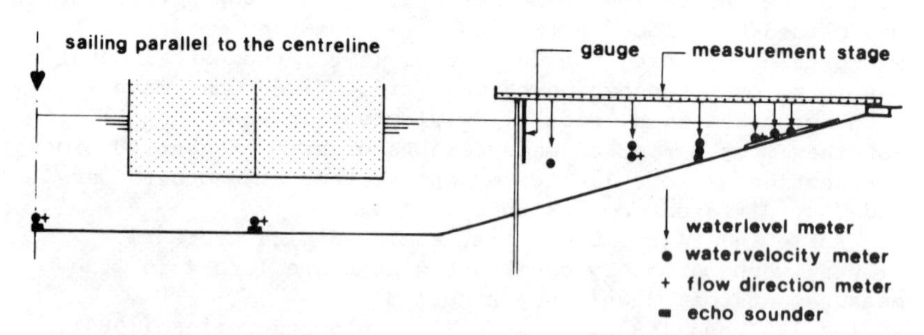

Fig. 2 Cross-profile at central measurement stage.

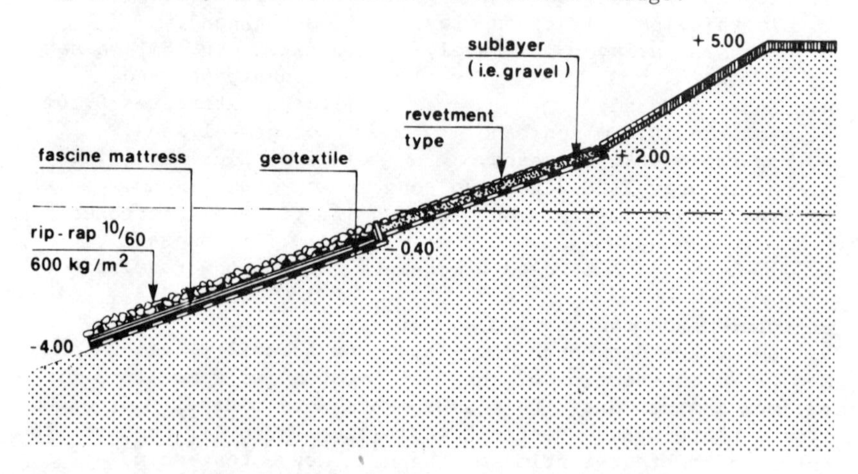

Fig. 3 Typical cross-section of prototype embankments.

immediately resulting in plots of the selected signals. By analyzing the plotted information in conjunction with erosion measurements,a well balanced selection of the next required test conditions could be made.

10. The extent of the measuring campaign required much of organisation. During the measurements the following groups were in charge: 1°) central cabin (data acquisition, signal handling and analysis group), 2°) test ship group, 3°) ships' position group, and 4°) erosion group, including a diving and two "Stereo-pictures" teams.
Next, a number of additionally required ships were in situ, viz.: 1°) a tug to assist the test-ship when manoeuvring, 2°) a barge-tug combination to assist the erosion-group, 3°) a liaison ship to collect the ship borne data after each run and to deliver the data to the central cabin, and 4°) a traffic control vessel.
During the execution of the measurements one universal frequency for communication was used.

Fig. 4. Ship-waves attack on test embankments.

Test-ships, test-runs and induced watermotion.

11. During the 1st series of measurements the following test-ships were used:
First series (1981): -small vessel for management
- pushing unit, 4500 hp, with barges (length each barge 76.5m, width 11.8m and draught 3.0m); four loaded barges in 2x2 formation, (see fig. 5), four empty barges and six loaded barges in 3x2 and 2x3 formation.
- tug, 700 hp.

During the 2nd series of measurements (1983) the following test-ships were used:

- pushing-unit, 5400 hp, four loaded barges in 2x2 formation (one test-run was also done with four empty barges).
- motor-vessel, 800 hp (80 x 9.5 x 2.5).
- tug, 1120 hp.

12. The number of test-runs with each test ship and the max. values of watermotion-components induced by test-ships are summarized below.

Test ship	Position	no. turns	V_R (m/s)	Z (m)	H (m)
1981					
2x2, loaded	centerline	19	1.27	0.75	0.40
2x2, loaded	toe of slope	21	2.02	0.85	0.40
2x2, empty	toe of slope	7	0.85	0.73	0.86
2x3, loaded	toe of slope	4	1.52	0.84	0.21
2x3, loaded	centerline	4	1.35	0.67	0.20
3x2, loaded	toe of slope	3	1.12	0.56	0.30
3x2, loaded	centerline	6	1.06	0.45	0.27
tug	toe of slope	2			
1983					
2x2, loaded	toe of slope	34	1.85	1.15	0.35
2x2, empty	toe of slope	1	0.75	0.55	0.75
motor-vessel	toe of slope	16	0.60	0.35	0.30
tug		38	-	-	0.80

V_R = velocity of return flow, Z=max. water-level depression, H= height of secundary waves.

Instrumentation.

13. Hydro instruments. To obtain a thorough insight into ship-induced water-motion, detailed measurements of wave heights, velocities and turbulence were carried out. Wave height meters, flow velocity meters and flow direction indicators were fixed to the measurement jetty(see figure 2). A wave run-up device was installed on the slope. Moreover, some other hydro measurements were carried out, viz.: measurements of pressures along the slope, measurements of vertical flow distribution (to determine development of the boundary layer), measurement of turbulence rate in the immediate vicinity of the rip-rap, and measurement of rate and direction of velocity, pressure and vertical distance to the ship (by means of echo-sounder) in the line of the ship-path.

14. Soil mechanical instruments. A number of water-pressure gauges were placed into the subsoil before the filter cloth and top-layers were put into place. These gauges were situated in such a manner that both horizontal and vertical gradients were measured. Beforehand, 6 concrete blocks and 3 basalton blocks had also been provided with pressure gauges on

Fig. 5. Loaded pushing unit at test location.

both sides of blocks which made it possible to determine the absolute pressure and the pressure difference exerted on the protection blocks when attacked by ship-induced waves. All hydro-, and soil mechanical instruments were directly connected to the central date-acquisition system.

15. A number of preparations and devices were applied to determine the beginning of motion and/or erosion of protective materials. These were: coloured sections of gravel and rip rap, flat cylindrical gravel-traps (to catch the gravel) and jack-up for stereo-pictures at rip-rap locations. When the ship had passed divers immediately inspected the embankments under the waterline, collected eroded rip-rap and surveyed the four gravel-traps. All trapped material was carefully analysed. From time to time the erosion holes were filled with the aid of a barge-tug combination.

16. On board of the test-ships the following variables were recorded: course and rudder-angle, trim of the barges, and torque and number of revolutions of the shafts. A gyrocompass was used to determine the course. All signals were recorded as a function of time. Transfer to a function of place is only possible when a very accurate position system is used. Data recorded on board during a run were directly collected by a liaison-boat and taken to the central cabin.

17. Positioning system. In the prolongation of the Hartel Canal leading lights were placed to show which course the test-ships have to maintain (centerline or toe of slope). The exact position of the test-ship was determined with a Minilir system in conjunction with an automatically operating distance-meter (Aga-120). The Minilir-system follows a lamp mounted on board of the ship (see figure 1.). In fact, the Minilir determines two angles: horizontally, which gives a bearing of the lamp and vertically, which gives the sinkage of the lamp. The accuracy of the Minilir is 0.5×10^{-6} rad. Through the continuous registration of the horizontal angle and the distance to the ship, the ship's position can be accurately determined. The position of the lamp in the fairway in combination with the location of the lamp on board and the momentaneous course of ship determines the position of the ship in the fairway. The sinkage of the lamp in combination with the trim indication of the barges determines the total sinkage and trim of the barges. These data were stored in the cassette and added to the central-acquisition system immediately after each test-run.

ANALYSIS.

18. With the results obtained, the insight in to hydraulic load- and damage mechanisms related to revetments when attacked by ship-induced water motion can be obtained (see figure 4). This will result in more reliable and universally applicable design rules for protection systems. The results of the first series of prototype measurements (1981),restricted mainly to the hydraulic load and stability of rip-rap,are partly published in ref.6 and 7. The results of the analysis of the 2nd series of prototype measurements (1983) will be available at the end of 1984.

BEHAVIOUR OF TEST EMBANKMENTS (see fig. 6).

Loose materials

19. Gravel on geotextile and sandy subsoil. Two gravel embankments, one with 30-80mm gravel and the other with 80-200mm gravel, were applied to verify the model relations describing the beginning of movement and transport of loose materials under ship-induced water motions. High gravel transports were only observed when push-tows and tugs sailed at high speed near the bank. In general, calculation methods based on model results give a proper approximation of the prototype values. It is interesting to note, that during the period between the two series of prototype measurements (October 1981-May 1983) an unexpectedly high transport of fine gravel took place as a result of a normal (rather low) shipping-intensity in the Hartel Canal. The probable explanation could be, that more vessels (particulary small vessels and tug-boats) are sailing nearer the bank than it was expected at the first sight. It also emphasizes the necessity of sufficient statistical data on behaviour of

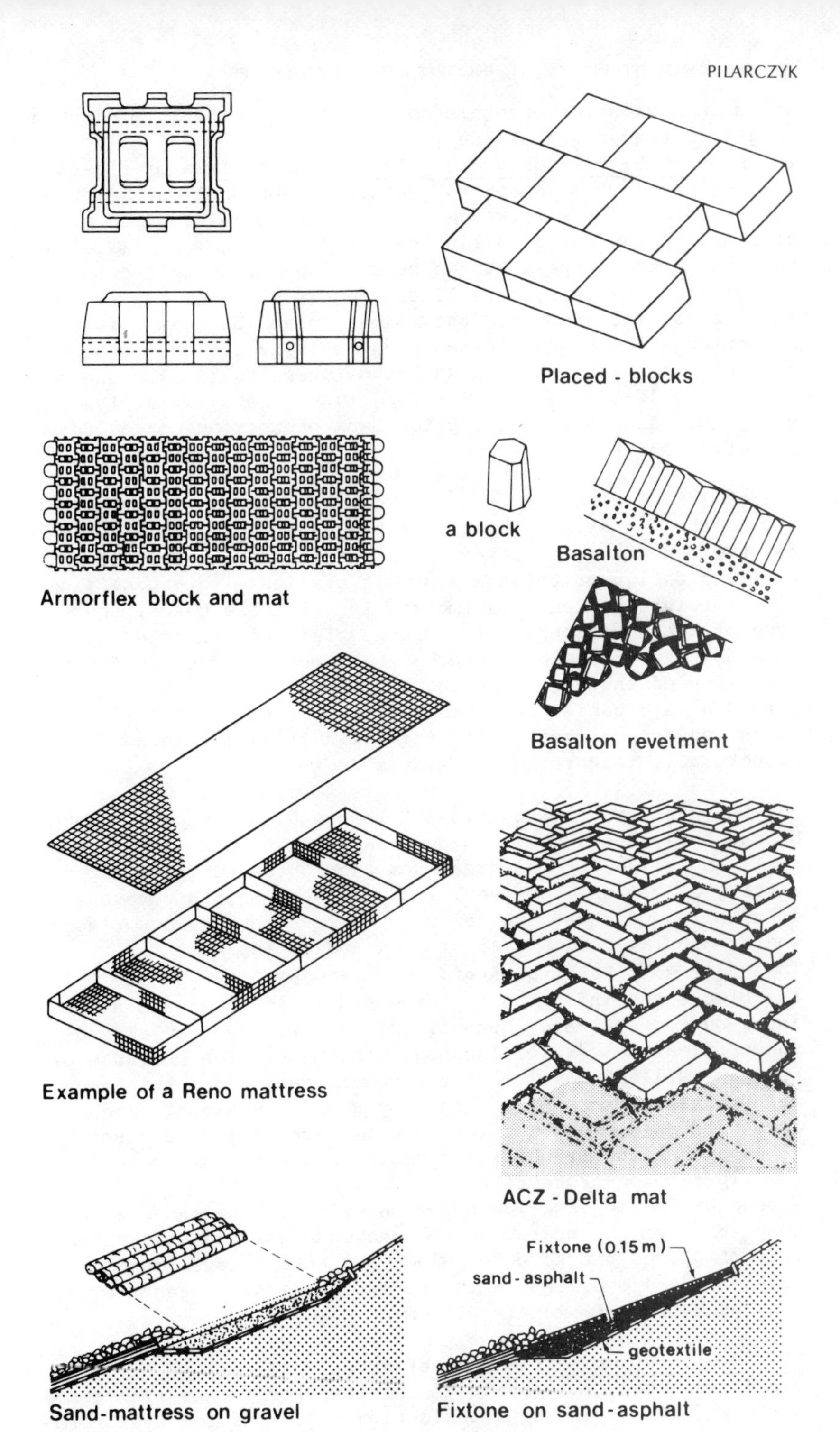

Fig. 6. Examples of constructions tested.

ships in navigation channels to be able to properly predict a long-term transport balance.

20. Rip-rap 5-40 kg on geotextile. For this class of rip-rap the beginning of movement was rather exceptional. It has been observed to occur mainly due to secondary waves induced by empty barges and tugs sailing very close to the bank (i.e. at the toe of the slope). It can be concluded that in the most practical cases this, or a little higher class of rip-rap would be satisfactory for the normal inland fairways. In the Netherlands the stone classes used normally for bank protection of channels with high shipping-intensity (incl.push-tow are of the 10-60kg and/or 60-300kg types. Detailed analyse of design procedures of rip-rap for bank protection can be found in ref. 7 and 8.

Placed (free) blocks.

21. In the Netherlands, concrete blocks are frequently used for the revetments of dikes, dams and banks. In general, no reliable design criteria are as yet available for these (and also for interlocked) revetments. In all these cases, the type of sublayer (permeable/impermeable) and the rate of permeability of the blocks are very important factors for the stability of these revetments. Sometimes, these blocks are threaded with cables or connected to geotextile by nails or nylon nooses (or even glued) forming a flexible and structurally integrated mat-system (ref. 3).

22. For prototype tests, 0.3 x 0.2m and 0.15m thick concrete blocks were used. Both sand and clay were used to form a subsoil. No failure of revetment has been observed (e.g. no uplifting of blocks). However, limited settlement and thus deformation of revetment has been observes at various places where underlayer consisted of sand and geotextile.
The analyse of the registrations of the pressure gauges below the blocks has indicated that the hydraulic gradients at the inter-surface of sandy subsoil and geotextile often exceeded the critical values for the beginning of erosion. Because of the geotextile the vertical transport was limited and the main transport took place probably along the slope. Some amount of sand could get lost due to lack of adequate sand-tightness at the transition from the block-revetment into the rip-rap toe-protection.
More detailed information hereabout will be available after completion of the hydraulic and geotechnical analysis. Stability of placed blocks under wind-wave attack is summarized in ref. 1 and 3 where the recent large-scale results from the Dutch Delta-Flume are also mentioned.

Flexible interlocked block revetments.

23. Three types of flexible interlocked revetments with three different principles of interlocking were used for the prototype tests, namely: grouted basalton blocks, cabled and grouted Armorflex-mats and noosed Delta-block mats (ACZ).

Fig. 7. Loaded pushing unit: 2x3 formation.

24. Grouted Basalton blocks (prisms). The excellent experience the Netherlands has had with natural basalt revetment (stone pitching) in conjunction with the high cost and shortage of natural basalt, have resulted in the development of artificial concrete prisms based on the shape of the natural basalt (patented as "Basalton"). This system is characterized by a polygon connection and consists of various shapes and different dimensions of prisms which allow even the construction of a random-shaped revetment. The blocks are produced in various sizes and with various densities. Lately they are also available as Basalton-mats. The area of the inter-block space equals about 20% of the total surface area (ungrouted). The blocks are slightly tapered vertically. Because of this, the prisms may sink lower if there is any settlement of the soil-body or erosion of the sublayer,which is immediately evident. Moreover, because of its tapered shape, the prisms have a firm position in the slope. The interstices between the blocks are filled with, for instance, graded broken stone silex or copper-slag (size 1-50mm). As a result the possibility of upward movement of the blocks is strongly reduced. In the Netherlands, the underlayer normally consists of graded broken stone or silex-stone (a waste product of the cement industry) of 0-60mm and about 30cm thick (i.e. for sea-dikes) It is also possible to place Basalton-blocks on sand-soil with a geotextile and broken stone in between. Recently full-scale tests on this type of revetment were performed in the large Delta-Flume of the Delft Hydraulics Laboratory (ref.12) As a result of grouting (= flexible interlocking) the

strength of the construction may increases up to $H_s/\Delta D \sim 10$ (slope 1:3) where H_s=significant wave height, Δ=relative density of block and D=block height. With such strong slope revetments, the stability of the filter-layer can become more critical.

25. The behaviour of the grouted Basalton revetment, used in the prototype tests in the Hartel Canal, was very satisfactory. During the first series of measurements (1981) only 15cm thick blocks were used. During the 2nd series (1983) 12cm high blocks were also applied. In both cases no stability problems occurred. The washing-out of the grouting material was, on average, restricted to a few centimetres in depth. As the inter-block space varies very little with the height of the block, the relative depth of washing-out (scour depth related to block height) increases in proportion to the decreasing of the block height. This means that, despite the high stability value, the height of the block must not be less than about 10cm to retain its stability performance.

26. Armorflex block mats. The purpose-shaped interlocking armorflex blocks are threaded with steel or nylon cables and bounded together, thus forming a flexible mat-system. A geotextile and/or graded filter are first spread over the slope to be protected, and than overlaid by the block-mat. Additionally, grouting material may be applied to the inter-block spaces to rigidify the mat once it is in place. Because of cabling, this system maintains its integrity in the event of subgrade deformation or severe dynamic loading upon a given exceedance of the design conditions for free blocks. The armorflex mat-system has been investigated extensively with respect to wave attack (ref. 11).

27. For prototype test embankments in the Hartel Canal the full and cellular type blocks of 0.11m height were used. Both systems were grouted with fine gravel. The performance of these systems was very satisfactory. Against expectation the washing-out of the grouting material was rather limited. When placing these mats special attention should be paid to the connection of the adjoining mats.

28. ACZ-Delta block mats. The system is characterized by a diagonal block-distribution and consists of rectangular purpose-shaped blocks (blocks 0.20x0.40m-underneath and 0.155x0.355m-outer surface). The blocks are poured onto the geotextile. The connection between blocks and geotextile is realized by four nylon nooses connected to the geotextile. The space between the blocks was about four centimetres wide. The block-mats (blocks 0.16m height) were placed on gravel and grouted with coarse gravel. The short term performance of this mat did not immediately lead to instability problems. However, due to the large interspaces, the grouting material was soon washed-out and the external hydraulic load was ac-

ting through the geotextile directly on the sand-sublayer. In the long-run it will lead to erosion of the sublayer and deformation of the revetment. Some settlement of the test revetment was already observed during the tests. Some improvement of this system is neededbefore it can be used in practice.
In general, the weak point of the block mats is how to repair them when the sublayer erodes and the underwater connection of the adjoining mats fails.

Sand-sausage mattresses (Profix).

29. Zinkon BV, a Dutch company specialized in bank and slope protection works, has developed a flexible erosion control system composed of filter-cloths and granular fill material, known as the Profix-system. Profix is a fast and relatively cheap method. Two tightly woven polypropylene mats are stitched together at regular intervals to give the design weight of at least 200kg/m^2 when filled with sand. Both filter-cloths must be sand tight. Moreover the outer cloth is stabilized against ultra violet radiation. It is provided with a felt layer to promote and develop vegetation providing extra protection against u.v. radiation. The cloths allow plant roots to penetrate into the subsoil thus providing extra stability to the construction. The required strength of the filter-cloths depends on the exerted loads, for instance, the design of slopes, the method of construction, the thickness and weight of the fill material. Sand and/or gravel are very suitabel as fill material, possibly mixed with cohesive additives. Mixing the fill with seeds can promote vegetation. The empty mattress is spreaded out at the top of the revetment and pulled out in stages, as they are filled, into the river or channel.Dry sand is blown through rubber hoses threaded in turn into flap covered openings every 5 to 10m along each tube of the mattresses.
These mattresses are actually used on a large scale for bank protection works in the Nigerian Delta area (since 1981). Special attention has to be paid to the risk of vandalism i.e. the mats may be cut away on purpose or damaged by pins through them. However, this has not been experienced. For areas which are not so densely populated (i.e. outside a serious risk of vandalism) this system may offer a good alternative bank protection.

30. The Profix mattress at the test embankment had an average thickness of 0.20m containing medium to coarse sand, between a flat laying filter cloth sticked together at intervals of 0.40m. The direction of the tubes was up and down the slope. The experience with this system obtained through the tests in the Hartel Canal showed the high importance of a sufficient degree of density (compaction) of the sand inside the mattress. Due to the breakdowns of the filling equipment this degree, unfortunately, was not obtained and migration of sand grains inside the tubes downward the slope could take

Fig. 8. Loaded pushing unit: 3x2 formation.

place. This as a result of the water movements caused by tidal action and passing ships. Consequently, the part of the mattress below the maximum water level was densified affecting the filling degree of the upper part. After refilling this part of the mattress the performance of the construction appeared to remain more satisfactory(during the 1st three months of observations).Future observations at this test location will supply additional informations regarding the behaviour of the Profix-mattresses under ship-induced load.

PVC Reno mattresses (Maccaferri Gabions).

31. Gabion is a large wire mesh basket coated with zinc or PVC (polyvinyl chloride) to ensure long life under adverse conditions. It is rectangular in shape, variable in size with diaphragms at certain intervals. These baskets, filled with relatively small rocks, are widely used in bank stabilisation and river training structures. The inherent flexibility of the gabions -the ability to bend without breaking- seems to be primary reason for their succes. Other important advantages are its permeability, stability, easy repair and relatively economy.

32. Mattresses used in the prototype tests were PVC-coated baskets 1.0 x 4.0m and 0.17m thick, filled with coarse gravel 70-130mm and placed on a geotextile on sand. The short-term performance was satisfactory. However, the placement was done with more than normal care. After the tests only a slight

swelling of the individual cells was observed. Also, only a limited number of stones has escaped throughout the wire-mesh of the mattress-baskets. Although the long-term performance has to give the final answer on the applicability of this system, it seems that gabions are a good protective alternative for these locations where vandalism is not a problem.

Fixtone (Bitumarin B.V.

33. Fixtone (open stone asphalt) is a new development in the construction of permeable asphalt revetment. Fixtone is prepared by mixing about 82% stones (16-56mm) with about 18% pre-mixed sand mastic (i.e. 64% sand, 16% filter and 20% bitumen 80/100), giving a material in which the stones are fixed firmly and form a stable, flexible and permeable (voids content 25% or more, pores up to 10mm) construction material. The lining generally consists of a layer of 15 to 20cm Fixtone on a filter layer. A more economical construction has recently been obtained with a layer of Fixtone on top of a fibre cloth which acts as a sand barrier. It is also produced in the form of Fixtone mats. Fixtone surface is resistant to currents up to 6m/s and waves up to 2m in height at least. Due to the complicated visco-elastic behaviour of asphalt mixes, which cannot be scaled down, the assessment of resistance to wave attack can only be carried out on actual scale. In order to provide a design tool for designers, the experien ce from several projects has been compiled into a "rule of thumb", reading:

$$D = C.H_s$$

in which: D = thickness of Fixtone layer, H_s = significant wave height and C = coefficient value being 1/6 in the case of Fixtone on filter cloth, and 1/10 on a sand bitumen filter This rule is also supported by the large-scale check-tests in the Delta-Flume of the Delft Hydraulics laboratory (ref. 12).

34. The prototype test embankment in the Hartel Canal consisted of a 0.15m thick toplayer of Fixtone (stone size 20-40mm) on an average 0.20m thick layer of sand-asphalt. The short-term peformance of Fixtone was rather satisfactory. Only a very small number of stones were loosened from the slope surface. These were small stones which had only single contact surfaces with the underlying Fixtone and had loosened because of rocking motions under current and wave attack. Due to high permeability of the "young" Fixtone there was no lifting of the construction as a result of upward pressure. However, further information on long-term behaviour of Fixtone, especially regarding the permeability, is still lacking. These conclusions are identical with the conclusions of the large-scale tests in the Delta-Flume where the wind waves up to H_{max} = 2,65m were generated (ref. 12). Actually (October '83) long term abrasion tests are being carried out in the large stream flume at Lith in the Netherlands. The

first results indicate that the surface erosion of Fixtone is rather limited even after a three weeks of continuous tests with water-current of about 4m/s.

END REMARKS.

35. After the completion of the short-term measurements in the Hartel Canal it was decided to keep all these prototype embankments for further studies on long-time behaviour in the coming few years. These data combined with the geotechnical analysis of the data obtained during the short-term prototype measurements will give more information on the practical applicability of the different protective systems.
The results of all these studies metioned above will bring designers closer to the solution of the typical problem of the choice of protective structure for bank protection works in respect to design load, the ability of materials, and desired function of construction.

REFERENCES.

1. DEN BOER K., C.J. KENTER and K.W. PILARCZYK. Large scale model tests on placed block revetment. Delft Hydraulics Laboratory, Publication no. 288, January 1983.
2. PILARCZYK K.W. and K. DEN BOER. Stability and profile development of coarse materials and their application in coastal engineering. Delft Hydraulics Laboratory. Publication no. 293, January 1983.
3. PILARCZYK K.W. Revetments (in "Closing Tidal Basins"), Delft University Press, 1983.
4. SPAN, H.J.TH. et al. A review of relevant hydraulic phenomena and of recent developments in research, design and construction of protective works. Dutch contribution to XXVth International Navigation Congress, Edinburg, 1981.
5. VELDHUIJZEN VAN ZANTEN, R. and R.A.H. THABET. Investigation on Long-Term Behaviour of Geotextiles in Bank Protection Works, 2nd Intern. Conference on Geotextiles,Las Vegas, 1982.
6. DELFT HYDRAULICS/SOIL MECHANICS LABORATORY. Prototype measurements in the Hartel Canal. Summary and conclusions 1st test-series (in Dutch), Report R1613, Sept. 1983.
7. New aspects of designing bank protection of navigation channels(in Dutch). KIVI-Symposium, Delft, 25 May 1983.
8. DELFT HYDRAULICS LABORATORY. Systematic research on bank protection of navigation channels, reports M1115, 1974-1985 (in Dutch).
9. DELFT HYDRAULICS LABORATORY. Systematic research on stability of block revetments under wave attack. Reports M1795, 1982-84 (in Dutch).
10. DELFT HYDRAULICS LABORATORY. Stability of basalton blocks (in Dutch). M1900, 1983.
11. DELFT HYDRAULICS LABORATORY. Stability of armorflex block slope protection mats under wave attack. M1910,1983.
12. DELFT HYDRAULICS LABORATORY. Fixtone: stability under wave attack. M1942, 1983.

14 Yugoslav experience in constructing revetments incorporating geotextiles

M. BOZINOVIC, MSc, M. MILORADOV, PhD and E. CIKIC, BSc, Jaroslav Cerni Institute for the Development of Water Resources, Belgrade

SYNOPSIS. The paper is a concise presentation of experience in Yugoslavia with respect to the design and construction of bank revetments with geotextile on the rivers and canals in Yugoslavia. It also gives some results of investigations performed on such bank revetments. It is concluded that the experience has so far been a positive one, but that further research is needed for clarifying certain aspects of the use of geotextiles, especially of their use as filters and separators.

INTRODUCTION

1. Extensive training works were begun on rivers in Yugoslavia towards the end of the 19th century and the beginning of the 20th century. Most of the works were made in order to improve navigation routes and protect against flooding on the rivers Danube, Tisza, Drava, Sava and others. However, a lot still remains to be done in the futureyears both on the construction of new training works and on the maintenance and reconstruction of the existing ones. The total length of rivers in Yugoslavia is about 110.000 km and of canals 900km, of which about 660 km are navigable.

2. Since bank revetments are the most expensive part of river training structures, cheap solutions are of a great economic interest. With the invention of geotextile, the possibilities for finding such solutions have been broadened and especially so since home-made geotextile became available in Yugoslavia.

3. There are at present two factories in Yugoslavia producing geotextile for construction purposes. Both factories produce polyestrous and polypropylene fibres by using the dry procedure, i.e. the needling method. Table 1 shows the properties of the two types of Yugoslav geotextiles that are most often used for bank revetments.

4. With the beginning of geotextile production in Yugoslavia, intensive investigation work was conducted in order to determine its possible application in hydraulic structures, especially for bank revetments. This investigation included the following:

- laboratory testing of geotextile fabrics

Table 1

Properties of "LIO-Filter Plastica" art.7010, type 300 production "LIO" - Osijek and of nonwoven textile "Politlak-300", production "TOZD-Filc"- Menges (Ref.1 and ref.2)

No.	Characteristics	Unit measure	"LIO" filter plastica	"POLITLAK-300"
1	2	3	4	5
1	Type of basic raw material-polymer	-	polyester	polypropylene
2	Mass of felt per unit of area	g/m^2	300$\pm$ 10%	250$\pm$ 10%
3	Thickness of felt	mm	2.6	3.0
4	Width of felt	m	2.40	2.20 and 4.40
5	Tensile strength a) according to length	kN/5cm	75	58
	b) acc.to width (DIN 53857)	kN/5cm	50	68
6	Tensile elong. a) acc. to length	%	70	140
	b) acc. to width	%	100	135
7	Water permeability felt (flow of water vertical to the surface of the felt; pressure on the felt 22 Pa	m/s	(1.5-2).10^{-3}	(3-4).10^{-3}
8	Absorption of water at 21°C and 65% relative air humidity	%	0.3 - 0.4	0.01 - 0.1

- field observation of bank revetments where geotextile has been used
- examination of samples of geotextile taken from bank revetments
- preparation and publication of standards for the use of geotextile in bank revetments

5. This paper reports on the results of these investigations and on the experiences in the application of geotextile for different types of bank revetments of rivers and canals in Yugoslavia.

Proposal of standards for the use of geotextile in bank revetments

6. The first proposal has been published under the title: "Temporary Standards for the Application of Geotextile in the Construction of Bank Revetments" (ref.3). In this document the principles and techniques developed at BAW-Bundesanstalt für Wasserbau, Karlsruhe, W.Germany, were adopted to a great extent (ref.4). The methods and procedures used for the attestation of geotextile as recommended by the mentioned Institute provide the designers with the necessary information, especially when using geotextile as a protective filter in the reversible flow of water.

7. However, a great many problems in this field still remain to be solved as geotextile is a relatively new material while the possibilities of its application are very numerous and different.

Applied techniques for bank revetments with geotextile

8. The following guidelines are recommended for the design of bank revetments with geotextile: (ref.5,6and 7)

- in a bank revetment geotextile should serve primarily as a separator and filter, and if need be, as a drain and structural reinforcement;
- a bank revetment should be simple and adapted to the local conditions;work should be mechanized as much as possible
- the structure of a bank revetment should resist the loading and other influences both in the course of its construction and its use. Special attention should also be paid to the analysis of wave effects and to the washing out of fine soil particles from the banks,
- when choosing the material for the construction of bank revetments, the changes that can occur in the material from the time of its production till the end of the life span of the structure should also be considered. For example, in permanent structures, geotextile must be protected against the direct effect of sunlight as well as against other effects which can influence its durability.
- the cost of construction and maintenance of bank revetments with geotextile should be lower than they would be if classical materials were used for the same purpose.

9. Taking all this into consideration, several types of bank revetments have been designed and built on Yugoslav rivers and canals. A short description of some of them will now be given.

10. Fig.1 shows a bank revetment on the river Sava exposed to relatively high waves where geotextile serves both as a filter and separator. The geotextile that was used was made in Yugoslavia. It is the so-called "LIO" filter plastica which is made of non-woven fabric made of polyestrous fibres. The thickness of the geotextile (felt) is 2,6 mm while the unit mass is 300 gr/m^2.

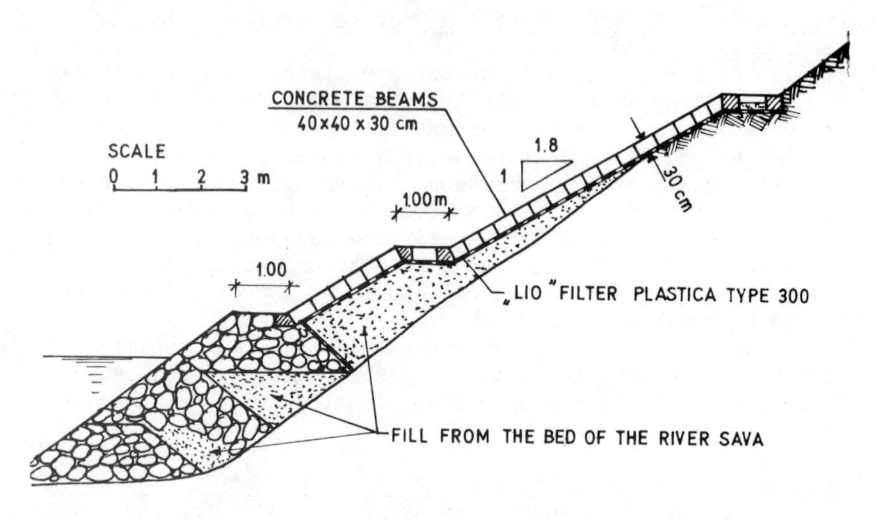

Fig.1 Bank revetment on the left bank of the river Sava at Baric

11. The revetment shown in Fig.2 was used on the river Drava for preventing intensive bank erosion (ref.8) and the geotextile here serves both as a filter and reinforcement of the rip rap. This bank revetment is actually a flexible mattress made of geotextile and fascines places crosswise and loaded by rip rap. The mattress is anchored to the bank by placing the upper end of the fabric in a small trench and loading it with rip rap and by fixing the fascines with wooden poles. The mattress is prepared on a floating platform while rip rap is placed by using a crane. This type of mattress is also used as a foundation for groynes and other training structures on river beds consisting of material prone to settling and erosion.

12. The type of bank revetment shown in Fig.3 was used on the river Tisza (ref.9). The porous geotextile serves for separating the gravel from the very fine material in the river bank. The geotextile thus protects the gravel filter and drain from the intrusion of the very fine particles and being porous, it also acts as an auxilliary filter.

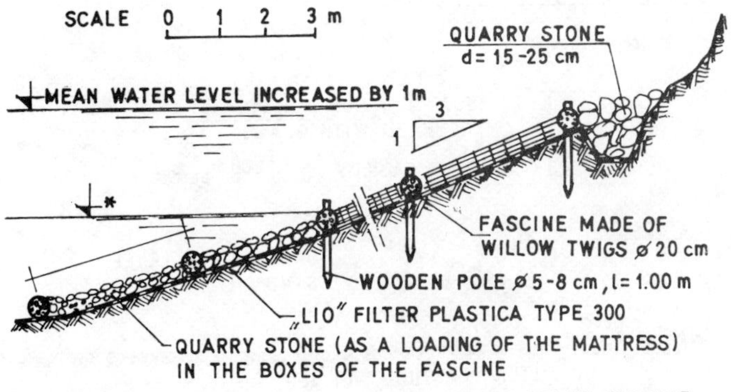

Fig. 2 Bank revetment on the right bank of the river Drava at Bistrinac

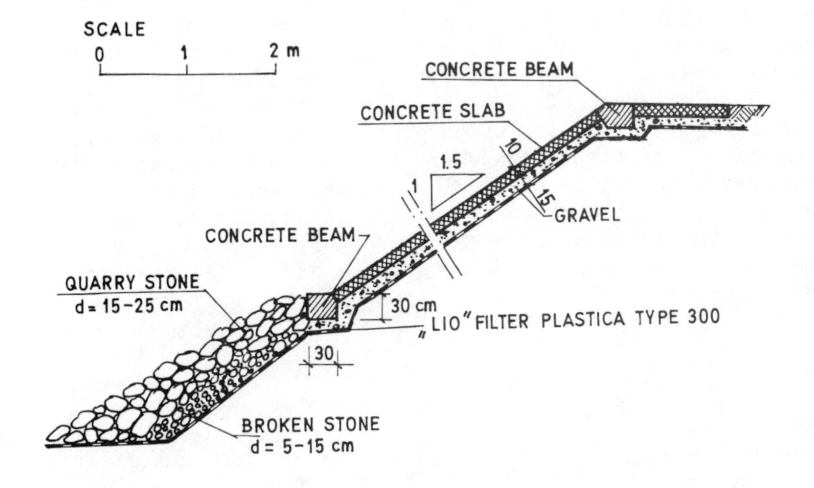

Fig. 3 Revetment on the right bank of the river Tisza at Kanjiza

13. Fig.3 shows the protection of a slope of a flood protection dike on the river Kolubara. Since the body of the dike is made of very fine granular material, geotextile here too serves as a protective filter as it prevents the removal of very fine particles of material from the body of the levee through the joints on the lining of the slope of the levee. Cincc the floods here occur only rarely and are of a short duration, only the bottom part of the slope of the levee was protected by using concrete slabs while the top part was covered with humus and grass.

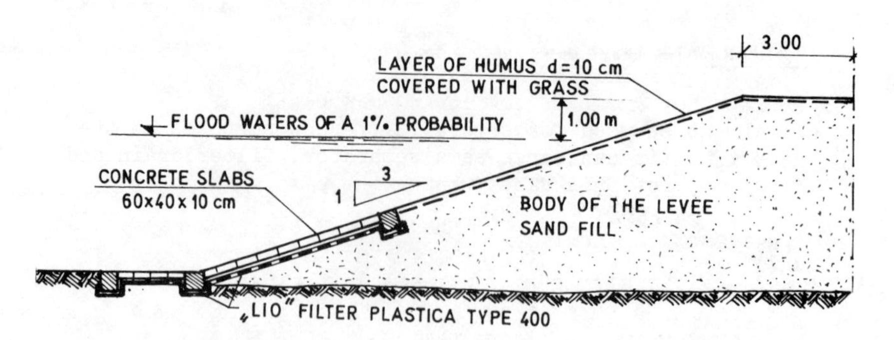

Fig. 4 Protection of the slope of the levee on the right bank of the river Kolubara at the mouth of the river

14. Geotextile is also used for repair works on damaged banks of canals and fish ponds. Fig.5 shows repair works on a canal bank damaged by waves. In this case, geotextile was used in combination with the "Netlon" net and anchored by wooden poles. This type of repair works has been used successfully on a length of 15 km at canals of the Danube-Tisza-Danube system. Similar types of bank protection were also used for some fish ponds where geotextile serves as a filter and separator while the Netlon net and anchored wooden poles ensure the static stability of the structure. Although this protection is of a temporary character, it can be made to last longer if there is vegetation along the contour of the bank revetment.

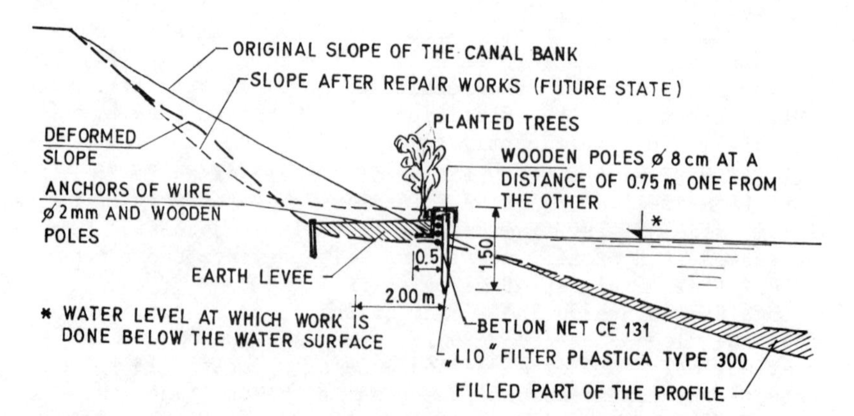

Fig.5 Repair works on the deformed slope of the canal Danube-Tisza-Danube by using LIO filter plastica and Netlon nets

Experience and results of investigation concerning the use of geotextile in the construction of bank revetments

15. The fact that geotextile has after a relatively short period of time been accepted for use in the construction of bank revetments in Yugoslavia is due to the following reasons:

- besides being relatively cheap, geotextile can in a bank revetment be used for several different purposes at the same time (it can serve as a separator, filter,drain and reinforcement)this rarely being the case with other materials;
- the industrial production of geotextile guarantees a standard quality of the product which makes it possible to use its mechanical and other characteristics to a greater extent and this in turn ensures more rational technical solutions.

16. The relatively small weight of geotextile reduces transport, storage and manipulation costs, makes the use of mechanization easy saving thus man-power in construction.

17. Geotextile can be successfully applied in bad weather and is easier and safer for building in under water than is the case with other materials. For example, it is very difficult to construct a multi-layer filter of gravel under water, which is not the case with the placing of an underwater filter made of geotextile.

18. In many cases, geotextile can be used as a substitute for more expensive materials. It also makes it possible to use material available on the spot for the construction of bank revetments, even if the material is of a poor quality. And finally, the proper use of geotextile can improve the bearing capacity and other characteristics of poor quality soil.

19. Based on the results of investigation of unused new samples of geotextile as well as of samples of geotextile already used on ban revetments, the following can be concluded:

- there is no significant change in the coefficient of filtration on the samples of geotextile taken after 3 to 5 years from the bank revetments on the river Drava (Fig.6) (ref.10) . However, the results of laboratory tests(ref.11) have shown that the coefficient of filtration can very quickly and significantly be reduced if the filter is clogged up by suspended particles. For example, exposed to the impact of waves of a height of 0,3 m and a concentration of suspended sediment in water of 3,5 gr/l, geotextile made of polyestrous fibres (250gr/m^2) lost within 12 hours around 50% of the initial value of the coefficient of filtration.
- the grain size distribution of soil samples taken directly from under the geotextile (in the bank revetment on the river Drava) changed insignificantly over a period of 5 years after the bank revetment was constructed. (Fig.7) (ref.10)

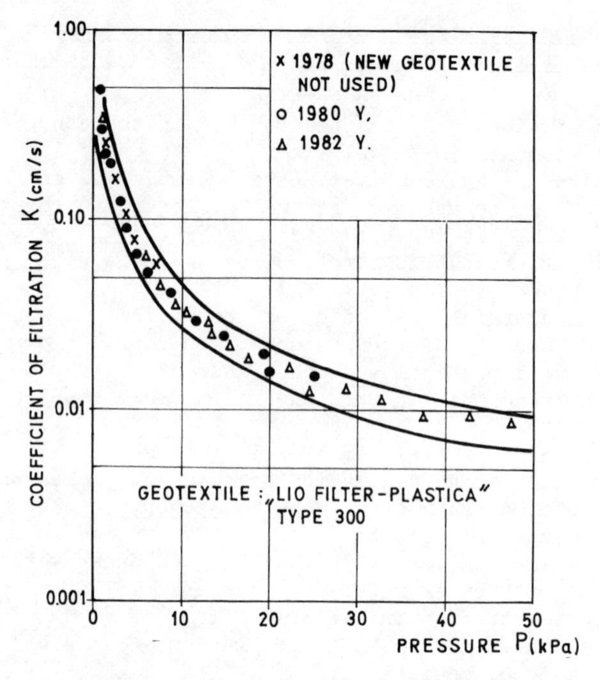

Fig. 6 Relation K=f (P) for samples of geotextile taken from the bank revetment on the river Drava at Bistrinac

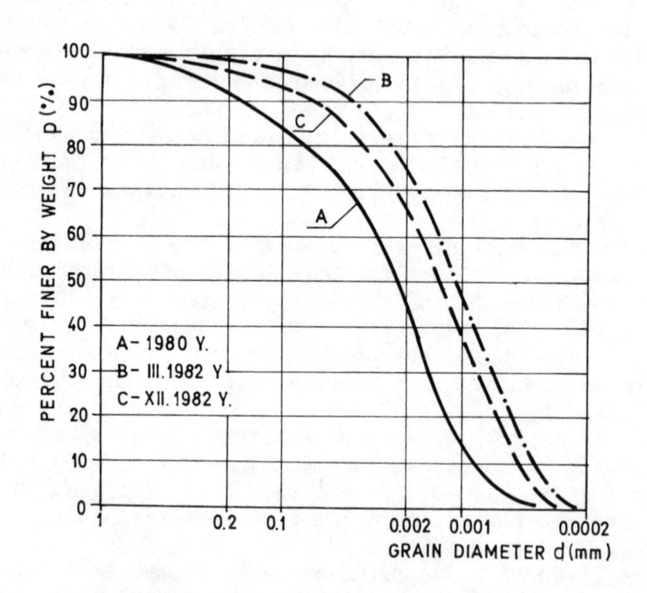

Fig.7 Change in the grain size distribution of samples of the material immediately under the geotextile on the bank revetment on the river Drava at Bistrinac

- the coefficient of friction between geotextile and the soil on which it is placed is relatively high and varies between 0,4 and 0,8. (ref.10). However, it has been noticed that the friction is greatly reduced if geotextile is saturated with a slimy and slippery colloidial substance. This must not be overlooked when designing bank revetments on sewer canals..
- a reduction of 15% of the original tensile strength of geotextile was found after five years of service.

CONCLUSIONS

20. Althpugh geotextile has been used for bank revetments in Yugoslavia only since a few years ago, the experience has proved to be positive both from the point of view of reducing the costs and the duration of construction. However, the lack of detailed specifications, standards and guidelines negatively influences the wider use of geotextile in everyday practice.

21. Advances in using geotextiles for bank revetments and other structures can be expected when more information on its behaviour in different types of structures and the results of laboratory investigations become available.The most important aspects to be investigated are the performance of geotextiles as filters and separators in the bank revetments.

REFERENCES

1. NOVOTA A. The Application of Fabric Filter Plastics in Some Fields of the Construction Industry, Symposium on the Application of Filter Plastica in Hydraulic and Road Construction, Belgrade, 1979.
2. ZMAVC J., VILHAR M., The Use of Fabric Foundations - Needless Felt in the Construction Industry, Textile-Ljubljana, TOZD-Filc-Menges, Institute for Textile Technology and Yugoacryl-Ljubljana.
3. BOZINOVIC M. Temporary Guidelines and Standards on the Use of Fabric Filters for Bank Revetments, Institute for the Development of Water Resources, Belgrade and"LIO"Textile Industry, Osijek, 1981.
4. LIST H.J. Woven and Nonwoven Fabric Filters in Waterways Engineering, Tests and Dimensioning , C.R.Coll, int.Sols Textiles, Paris 1977.
5. BOZINOVIC M., M.MILORADOV M. The Use of Filter Plastica for Training Works, Symposium on the Use of Filter Plastica in Hydraulic and Roadway Construction, Belgrade, 1979.
6. MILORADOV M. BOZINOVIC M . River Training Works, Civil Engineering Calendar, 1981, Union of Civil Engineers and Technicians of Yugoslavia, Belgrade, 1981.
7. BOZINOVIC M. On the Use of Nonwoven Fabrics in Hydro Construction, Journal "Waters of Vojvodina 1983", Novi Sad, Yugoslavia, 1983.

8. KRISTOFOROVIC J. Use of Plastics in River Engineering, Symposium on River Engineering and Its Interaction with Hydrological and Hydraulic Research, belgrade, 1980.
9. MILORADOVIC T. Protective Levee on the River Tisza at Kanjiza, Journal "Waters of Vojvodina'1982", Novi Sad, Yugoslavia, 1982.
10. Reports on the Investigation of "LIO" Filter Plastica in 1981 and 1982" "Jaroslav Cerni" Institute for the Development of Water Resources, Belgrade, 1981 and 1982.

11. MUSKATIROVIC D. BATINIC B. JOVANOVIC M. Some Aspects of Hydraulic Analysis of Synthetic Filters, International Symposium on River Engineering and Its Interaction with Hydrological and Hydraulic Research, Belgrade, 1980.

15 Theoretical basis and practical experience-geotextiles in hydraulic engineering

Ing. M. WEWERKA, Chemie Linz AG, Austria

SYNOPSIS. To choose the right filter material in hydraulic engineering avoids unpleasant surprises and high repairing costs. To get optimum results, all different stages of the construction and function have to be analysed in order to be informed about the maximum possible stress, which must be used for the design of the geotextile, therefore it can be different for each project. The following report tries to analyse these stages exactly, dealing with necessary properties of the geotextiles and their testing methods in close regard to practice.

I. INTRODUCTION

For some decades geotextiles have been used in hydraulic engineering. Mainly 3 different types are used:

Mechanical bonded nonwovens
Thermical bonded nonwovens
Wovens

For the majority of the projects their primary task is to act as erosion protection, alternative for, or part of a mineral filter. Whereas during the last years their effectivity has been discussed, today there is a tendency to create criterias, which enables the responsible engineer to choose the material best suitable for certain conditions. Basis for such criterias is the knowledge, what kind of forces are acting on the geotextile in a project.

Mostly there are three stages of stress:

1. During and after laying the geotextile
2. During placement of riprap
3. After completion of the revetment

The geotextile must resist each stage, otherwise its function cannot be guaranteed.

II. LAYING THE GEOTEXTILE

1. Laying under dry conditions

Laying the geotextile under dry conditions is relatively simple. The material is mainly delivered in rolls on site. It can be placed by rolling off over the area that has to be protected. Jointing may be effected by overlapping 0,50 m -

1,00 m, depending on subsoil conditions and riprap. It must be guaranteed that overlapped geotextiles are prevented from becoming separated during placement of any stones. This problem can be avoided by welding (Fig. 1), sewing or by means of clamps.

Fig. 1. Welding of geotextiles on site

2. Laying under water

Laying under water can cause troubles because of current and waves. The geotextile can easily be driven out of the designed position if it is not fixed properly. The hydrophobic behaviour of the fibres of geotextiles (polypropylene (PP) and polyester (PES) are the mainly used raw materials) makes it difficult for water to penetrate into the pores. Therefore loading is necessary to achieve sinking, or at least the end of the layers must be fixed to the ground. The number of overlaps, which must be made under water, shall be low, as they are difficult to control. This is the reason why layers as wide as possible (or welded or sewn layers) shall be used.

3. Occuring loads

During this stage the geotextile has to take up low tensile forces (but this is usually no problem) and by UV-rays of the sun. These UV-rays destroy all geotextiles (5) therefore a protection is necessary under all circumstances, which is provided by the riprap and water anyway. The critical period is between laying the geotextile and placement of stones. Especially stabilized PP and PES are sufficient stable raw materials, that can resist usual periods without any problems. Practice has shown, that the reduction of the strength is depending on several factors (geographical position, humidity and dust content of air, growing of plants ...) therefore exact instructions can hardly be given (Fig. 2).

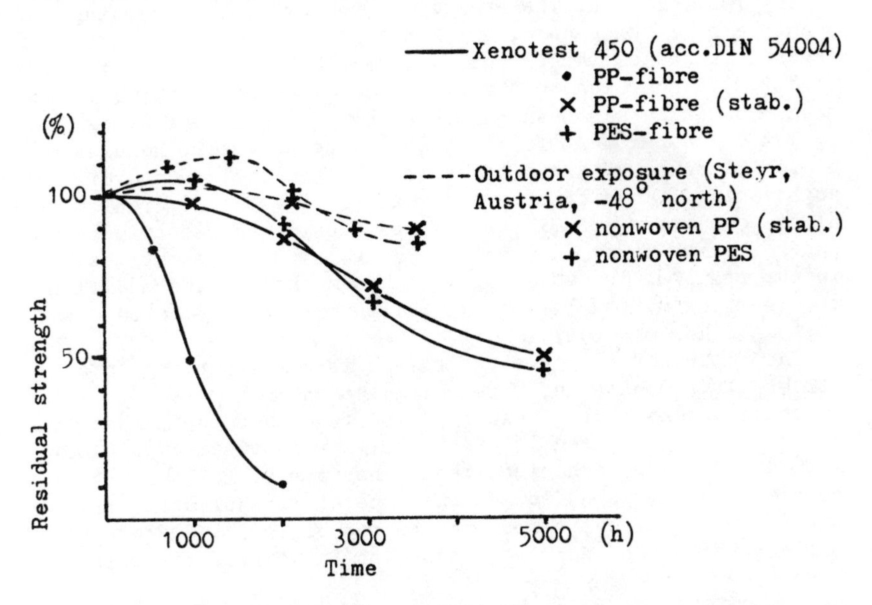

Fig. 2. Tensile strength retained versus UV-exposure time.

III. PLACEMENT OF RIPRAP

1. General indication and occuring loads

Placement of riprap means first of all high mechanical stress to the geotextile. The level of stress is depending on the type of placing (throwing, hand placing) and shape and size of stones. Peak values of stress are certainly obtained by throwing stones. The energy of the falling stone is concentrated on a rather small area, this can easily lead to damage. To avoid this the geotextile must have mechanical properties which allow transmission of the energy into subsoil without damage, or any protection layer has to be placed (e.g. gravel layer), what of course increases project costs.

In principle it is possible to place stones weighing some tons directly onto the geotextile, if the puncture resistance is high and the subsoil consists of sand, silt or clay. In the reach of wave action, problems may arise if the distances between the stones are too big. The geotextile is not fixed to the ground firmly. The constant to and fro movement caused by waves looses the texture of fibres in the course of time, which leads to deteriorated retention behaviour. Geotextiles consisting of endless fibres have advantages compared to those consisting short (staple) fibres, as they cannot be removed completely out of the texture. In the reach of wave action, stones bigger than 0,30 m - 0,50 m should therefore be placed individual as close together as possible, dumping cannot be recommended. If this is not possible a compensation layer consisting of smaller stones or gabions has to be placed beforehand.

2. Tests and properties

Information about the quality of geotextiles concerning their puncture resistance under dynamic loads can be obtained by the testing method of the Technical Research Centre (TRC) of Finland (10). According to that test the diameter of the hole is measured, which has been punctured by a 1 kg cone falling down 0,50 m on the geotextile which is laying on water. Results of up to 10 mm are obtained by geotextiles suitable for hydraulic engineering. A modification of this testing method, replacement of the cone by a pyramid, should be considered, as this shape can rather be found in practice and which gives sometimes different results. The advantage of this method is the addition of a supporting material (water), while this is not the case with most of other methods.

The necessary puncture resistance is corresponding to a stress-strain behaviour from which the necessary tensile strength automatically results. Experience has shown that the lower limit is between 14 - 15 kN/m (strip tensile test acc. ASTM D 1682) for mechanical bonded nonwovens. But this is of course strongly dependent on the type of riprap. Using e.g. gabions the mechanical stress can be rather low, therefore the geotextile design is primary based on its filter efficiency.

IV. FUNCTION

After completion of the project the filter efficiency of the geotextile is of primary importance. The mechanical stress results from water action and current, which can be taken up by the geotextile without any problems if the above mentioned recommendations are considered.

To provide filter criterias for geotextiles a lot of investigations have been done, but mainly for drainage systems. It has to be distinguished between criterias for laminar flow and alternating turbulent flow. For waterway engineering permeability under laminar flow and retention capability under turbulent alternating flow is of special interest.

1. Permeability

1.1. General observations In isolation, geotextiles have almost always higher permeability than the suboil and would therefore be suitable in this regard as filter material. However, in practice, soil particles are deposited at and in the geotextile. The decrease of permeability arising because of that is depending on the amount and size of the soil particles. The amount again is depending on the available space and type of water flow. Geotextiles pressed tightly onto the subsoil make rearrangements of the soil hardly possible, the composition of subsoil adjacent to the geotextile is changed only slightly (supposing adequate retention capability of the geotextile).

Testing the filter efficiency the Federal Institute of Waterway Engineering (BAW - Karlsruhe) uses an equipment that simulates wave action (4). The conditions of this test are close to practice, although the interpretation and therefrom deduced recommendations are partly not justified. A pot, of which the bottom is formed by a geotextile, is filled with a certain test soil. In intervals of 30 seconds it is dipped into water and pulled out again. Although there is a considerable portion (up to 30 %) of grain sizes smaller than the effective pore size D_W (acc. to Franzius Institut Hannover) of a mechanical bonded nonwoven only appr. 1,7 % of the test soil passed through it within 34 hours. Additional the amount passing through is decreasing in the course of time. Water permeability of the soil-geotextile system scarcly changed after a short increase at the beginning of the test (Fig. 3). Under this condition clogging can hardly be expected, as all accumulations of fine particles are disturbed again and again by the alternating flow (exception see 22.3.) Therefore the critical area concerning the permeability is in the area where water does not change its direction too often (e.g. flow of ground water).

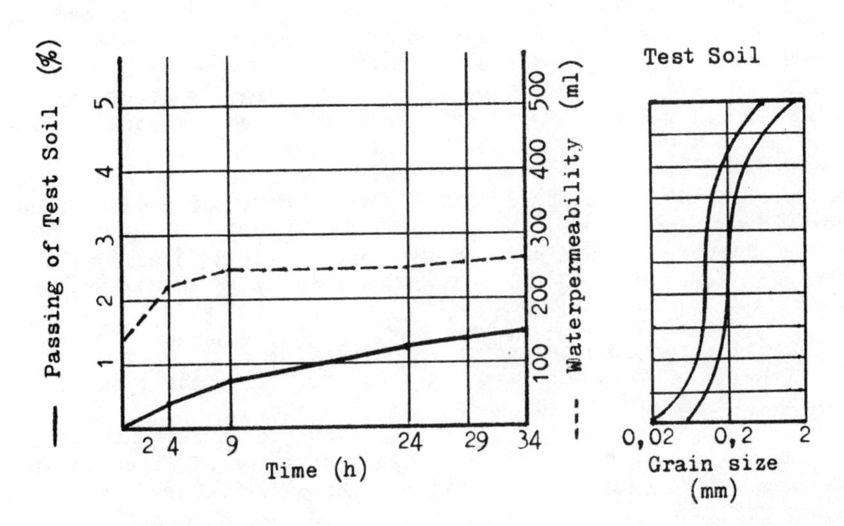

Fig. 3. Waterpermeability and passing of soil through the geotextile (D_W = 0,1 mm) under turbulent, alternating flow versus time.

1.2. Existing criterias In different investigations (1,2,3,4) there is a demand for the coefficient of permeability k_g of the geotextile to be 10 - 1000 times higher than the coefficient k_s of the soil.

$$k_g/k_s = 10 \div 1000 \qquad (1)$$

Especially high values are demanded for woven products to avoid the blocking effect (blocking of the open area of wovens by soil particles).

2. Retention capability

2.1. Existing criterias. While the hydraulic filter efficiency has to be observed critically under laminar one-directional flow, the retention capability has to be examined primary under turbulent and alternating flow. Different recommandations (1) give a correlation B_{50} between the pore size D of the geotextile and the grain size d_{50} of the soil wherein

$$B_{50} = D/d_{50} = 0{,}4 \div 2{,}2 \qquad (2)$$

2.2. Theoretical basis for criterias. The test of filter efficiency under turbulent flow as described above (IV.1.1.) showed almost no passing of soil particles, if there is a direct contact between soil and geotextile and a certain amount of soil particles is bigger than the effective pore size D_w.

Due to turbulent and alternating flow soil particles which are smaller than the opening size are passing through the geotextile at the very beginning. Consequently there is an accumulation of soil particles bigger than the pore size. On one hand this accumulation reduces more and more kinetic energy of waves, on the other hand penetration of finest particles is impeded and further migration is stopped. A natural filter is formed founded on the geotextile. The amount of particles passing through is depending on several factors:

2.2.1. Geotextile. Reduction of kinetic energy is depending on k_g and thickness (3). The pore size is decisive for soil particles, which are retained from the very beginning. With increasing thickness there is an increasing retention capability because of the prolonged filtration length (6) and the higher amount of deposited soil particles.

2.2.2. Subsoil. Grain size is in close relation to pore size of the geotextile. The coefficient of uniformity

$$C_u = d_{60} / d_{10} \qquad (3)$$

influences the formation of the natural filter; if C_u is high the formation takes more time. At cohesive soils the risk of erosion is reduced by the amount of cohesion.

2.2.3. Contact geotextile - subsoil. Cavities underneath the geotextile can lead to increased transportation of fines towards the geotextile (instable condition under water). Very small pore sizes and low kinetic energy of water can then form a filter cake of very low permeability (clogging).

2.2.4. Riprap. The les kinetic energy can pass through, the lower is the risk of erosion.

2.3. Retention criteria. As there are so many factors influencing the filter efficiency in practice, it is difficult to create criterias based on laboratory experiments alone, which of course cannot take all these factors into consideration. Based on practical experience, different test results and the above made considerations the relation

$$D_w / d_{80} \leq 1 \qquad (4)$$

seems to be justified as criteria for the retention capability to cohesion less soils of mechanical bonded geotextiles used in waterway engineering. With uniform soils (C_u 5) a quick formation of a natural filter is possible, passing of fine particles through the geotextile will be very limited. For non-uniform soil ($C_u > 5$) with a grain size $d_{80} < 0,06$ mm cohesive forces can start to change their behaviour. The effective pore size D_w of the geotextile can be larger than determined by the above mentioned relation. In practice good results have been obtained by

$$D_w \leq 0,1 \text{ mm} \qquad (5)$$

with mechanical bonded nonwovens with a thickness greater than 2 mm (under a pressure of 0,2 N/cm2).

These criterias are not valid for special types of soil (e.g. dispersive clay, suffosion) and systems where absolute no passing of soil particles is allowed.

V. SUMMARY OF CRITERIAS

The properties of the geotextile, that should be used in waterway engineering have to be evaluated in steps in accordance with progress of work.

1. Laying

Resistance against UV-rays until to placement of riprap (for prolonged periods stabilized PP or PES are recommanded).

2. Placement of riprap

Puncture resistance has to be choosen according to the type of riprap and its placement.

3. After completion

Tensile strength must be according to the kinetic energy of water. Permeability (k_g) of the geotextile has to be at least 10 times higher than that of the soil. Effective pore size D_w has to be equal or smaller than d_{80} of the cohesionless soil. If there is any cohesion D_w shall be equal or smaller than 0,1 mm.

VI. LONGTERM BEHAVIOUR

Geotextiles are relatively new building materials. It is difficult to evaluate their behaviour in advance for some decades. But different investigations showed good results (2,8, 9) in this regard, provided that the placement has been done correctly. High stress occurs during placement and the resulting alteration of the mechanical properties must not be too high in order to have sufficient reserves to act successfully as filter layer. In any case geotextiles have to be protected against UV-rays.

VII. PRACTICAL EXAMPLES

The possibilities of placement of geotextile and revetment is depending on the project itself and the working conditions (availability of building materials, manpower, technical equipment). The following examples are only representative for many other possibilities.

The geotextiles used in these examples have been mechanical bonded nonwovens. Other geotextiles could behave different and cause some problems as reported by F.B. Couch Jr. (7). (When using woven materials problems concerning the puncture resistance and filter efficiency arose.)

1. Due to recommendations of BAW a geotextile with especial good retention capability had to be used for the stabilisation of the slopes of the Elbe-Lübeck canal in Germany. This requirement as well as other necessary properties called for a geotextile weighing more than 700 g/m2. Revetment was continuously assembled on a pontoon (Fig. 4). When moving the pontoon foreward the revetment (gabions on the geotextile) slided directly on the slope of the canal.

Fig. 4. Slope protection assembled on a pontoon.

At the following projects a nonwoven made out of endless PP-fibres, weighing 240 g/m2 and 280 g/m2 resp., with the trade name Polyfelt TS has been used.

2. To avoid erosion due to wave action geotextiles have been used in the course of the development of the Danube in Austria. Laying of the geotextile was done under dry conditions, but was relatively difficult because of the non-cohesive subsoil and steep slope. Larger widths were prepared before on flat ground by welding. Then they were rolled down the slope. Rather heavy stones were placed as close as possible directly on the geotextile without any precautions (Fig. 5).

3. For slope protection of the Saone (France) the vertically placed geotextile was fixed by means of steel nails on the upper and lower end. Jointing of the geotextile was done by fixing the overlap with steel nails. The riprap consists of non-uniform stones ensuring good bond and close bedding (Fig. 6).

Fig. 5. Slope protection in the reach of wave action.

Fig. 6. Stabilisation of the banks of the Saone.

4. The stabilisation of the Nuberia canal in Egypt was done under dry conditions after digging a new slope parallel behind the original bank (Fig. 7). The geotextile was unrolled parallel to the canal and the riprap was hand-placed directly on it. After completion of the revetment the earth dam was removed between canal and new slope.

Fig. 7. Hand-placing of riprap direct onto the geotextile.

REFERENCES

1. TEINDL H., Filterkriterien von Geotextilien Bundesministerium für Bauten und Technik, Straßenforschung Heft 153, Wien 1980

2. HEERTEN G., Geotextilien im Wasserbau, Anwendung, Prüfung, Bewährung, Mitteilungen des Franzius-Institutes für Wasserbau und Küsteningenieurwesen der Universität Hannover Heft 52 (1981)

3. GRABE W., Mechanische und hydraulische Eigenschaften von Geotextilien. Sonderdruck aus Heft 56 (1983) der Mitteilungen des Franzius-Institutes für Wasserbau und Küsteningen nieurwegen der Universität Hannover.

4. LIST H.J., Woven and Nonwoven Fabric Filters in Waterway Engineering - Test and Dimensioning. Intern.Conf.on the Use of Fabrics in Geotechnics, Paris 1977

5. HÜTTNER G. and SCHNEIDER H., Müssen Geotextilien lichtbeständig sein? Allgemeiner Vliesstoff-Report 6-1982

6. WITTMANN L., Soil Filtration Phenomena of Geotextiles. 2nd Intern.Conf.on Geotextiles, Las Vegas 1982.

7

7. COUCH F.B. Jr., Geotextile Applications to Slope Protection for the Tennessee-Tombigbee Waterway Divide Cut. 2nd Intern. Conf.on Geotextiles, Las Vegas 1982.

8. SCHNEIDER H., Die Langzeitbeanspruchung der Geotextilien im Bauwesen. 8. Nationale Technisch-Wissenschaftliche Leichtindustrietagung, Bukarest 1981.

9. SCHNEIDER H., WEWERKA M., Long-Term Behaviour of Geotextiles Geotex 3/82

10. RATHMAYER H., Experiences with "VTT-GEO" Classified Non-Woven Geotextiles for Finnish Road Constructions, 2nd Intern.Conf.on Geotextiles, Las Vegas 1982.

16 Bidding procedure and placing operation of geotextile filter layers

Dipl.Ing. H.-U. ABROMEIT, Bundesanstalt für Wasserbau, Karlsruhe, West Germany

SYNOPSIS. Geotextiles are used as a special filter layer for bank revetments of waterways in W.-Germany for more than 15 years. Because of the empirical values obtained in this period of time Bundesanstalt für Wasserbau has set up terms of delivery for geotextiles fixing certain general and specific demands geotextiles have to satisfy if they are used for standard constructions of bank revetments. Besides it terms of delivery determine the procedure of permission to start the placing operation of geotextiles and the number of control tests of deliveries. Details concerning placing operation and settlement of account of the delivered geotextiles are stated in supplementary technical specificatons.

Standard constructions for bank revetments

1. The following standard constructions are applied to bank revetments of waterways of the European category no. IV (1350-t-ship) or smaller and to a slope inclination m ≤ 1 : 3. Regarding the design of geotextile layers we only have to distinguish fundamentally three various construction methods:

2. Rip-rap. The geotextile filter layer lies on the in-place soil and is covered with a rip-rap layer german stone-classification II (Ø 15 - 25 cm) or if traffic load is heavy class III (Ø 15 - 45 cm), minimum thickness of rip-rap 0,60 m (Fig. 1). Nearly 35 % of stabilized banks of channels and 90 % of natural waterways are protected in this way. The qualities of a geotextile filter layer have to be designed to grain size and to permeability to water of the in-place soil, to the high strain caused by casting the stony material and to abrasion caused by movement of stones.

3. Rip-rap sealing compound. The geotextile filter layer is placed on the in-place soil and covered first with rip-rap class II. Afterwards

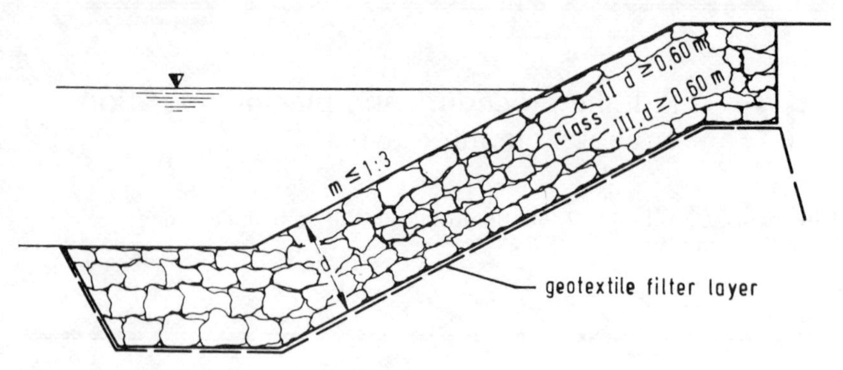

Fig. 1. Rip-rap

grouting with sealing compound has to be done. The thickness of covering layer must be minimum 0,40 m if rip-rap are partially and 0,35 m if it is completely grouted (Fig. 2).
Concrete or mastic asphalt are used for sealing compound.
Nearly 40 % of stabilized banks of channels and 5 % of natural waterways are protected in this way.
The qualities of geotextile layer have to be designed to grain size, to permeability to water of the in-place soil and to the high strain caused by casting the stony material and also to high temperatures to 170° C in case of hot laying of bituminous material.

4. Permeable, holohedral layers (concrete blocks, slabs or mattings, etc.). Permeable, holohedral layers used for slope surfacing on banks of waterways are concrete blocks, mattings or concrete slabs etc. Concrete blocks are only suitable if they are vertically and horizontally indented. Sufficient permeability to water has to be proved. Because of high costs only a small part of stabilized banks of waterways is protected in this way.

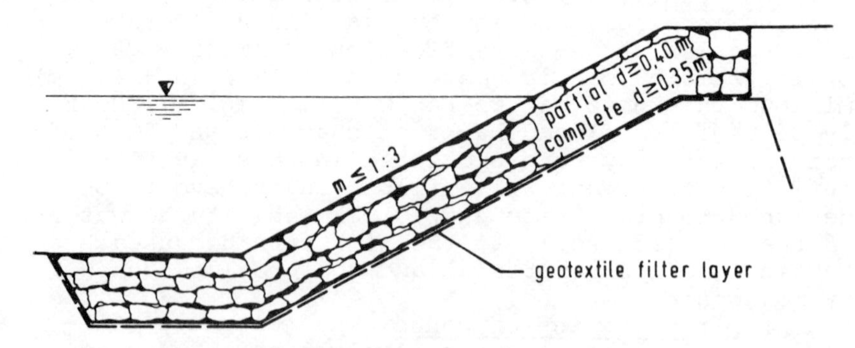

Fig. 2. Rip-rap sealing compound

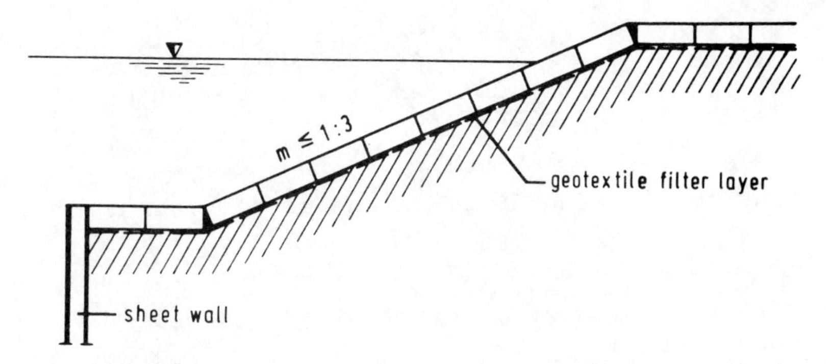

Fig. 3. Concrete blocks, slabs or mattings etc.

The geotextile layer is placed between ground and surface layer (Fig. 3). Forces caused by placing operation of covering layer are not worth mentioning.

Biding procedure of standard revetments

1. The bid documents of standard revetments comprise the list of bid items and quantities, and from now on the terms of delivery and supplementary technical specifications. Standardized texts can be used for the list of bid items and quantities. In the pay item "geotextiles" the type of soil must be designated. The minimum values demanded for the type of soil and the provided type of standard revetment are shown in a table appended to the terms of delivery.

2. The succesful bidder has to prove the qualification of the provided geotextile by a valid testreport.

Terms of delivery

1. General demands. Geotextiles used as a filter layer have to be oil-, seawater-, frostresistant and innocuous for ground-water. They must have enough long-term behaviour.
The actual experience has shown that the following raw materials: polyacryl (PAC), polyamid (PA), polyester (PES), polyethylen (PE), polypropylen (PP) satisfy these general demands. If new raw materials are used these demands have to be proved by a qualification testreport.

2. Specific minimum values. The specific minimum values for geotextile filter layers depend on type of bank revetment and in-place soil.
They are fixed for the aforesaid standard constructions in the following manner:
- thickness of filter layer:
  a) $d \geq 4{,}5$ mm on sand and no abrasion strains

b) $d \geq 6,0$ mm on cohesive soil or abrasion strains
- thickness of a supplementary roughness layer (only demanded on soil endangered to motion of soil to toe of slope):

$$d \geq 10 \text{ mm}$$

- tensile strength in longitudinal and transvers direction and joints:

$$F \geq 1200 \text{ N/10 cm}$$

- efficient size of openings of supplementary roughness layers:
  a) Dw = 0,32 - 1,5 mm on cohesive soil
  b) Dw = 0,5 - 2,0 mm on sand
- hydraulic properties (measured on the soil penetrated geotextile)
  a) $k > 10 \times k_{soil}$ on sand
  (k = coefficient of Darcy)
  b) $k > 10^2 \times k_{soil}$ on cohesive soil
- soil-particle retaining ability to a defined standard type of soil (Fig. 4).

Permissible total volume of soil penetration in the course of test

a) type 1 - 3 : <25 g/34h/225 cm^2

b) type 4, :<300 g/150 min/225 cm^2

The test methods are different for a) und b). Most of cases the in-place soil can be adjoined to one of the 4 types of soil. If it is not possible the test can be done with the in-place soil.

- resistance to rupturing strength
  W > 600 Nm (stones class II)
  W > 1200 Nm (stones class III)
- abrasion resistance:
  The minimum values of tensile strength and the permeability to water have to be preserved after the abrasion test of geotextile.
- resistance to high temperatures to 170° C:
  After heating the geotextile to 170° C and the following cooling down the minimum values of tensile strength and permeability to water have to be preserved.

3. Control tests. Control tests must be done as well with the first delivery as with the following deliveries.

In this connection must be checked up the minimum values of
- layer thickness
- tensile strength
- size of openings

and the following characteristic values of the geotextile

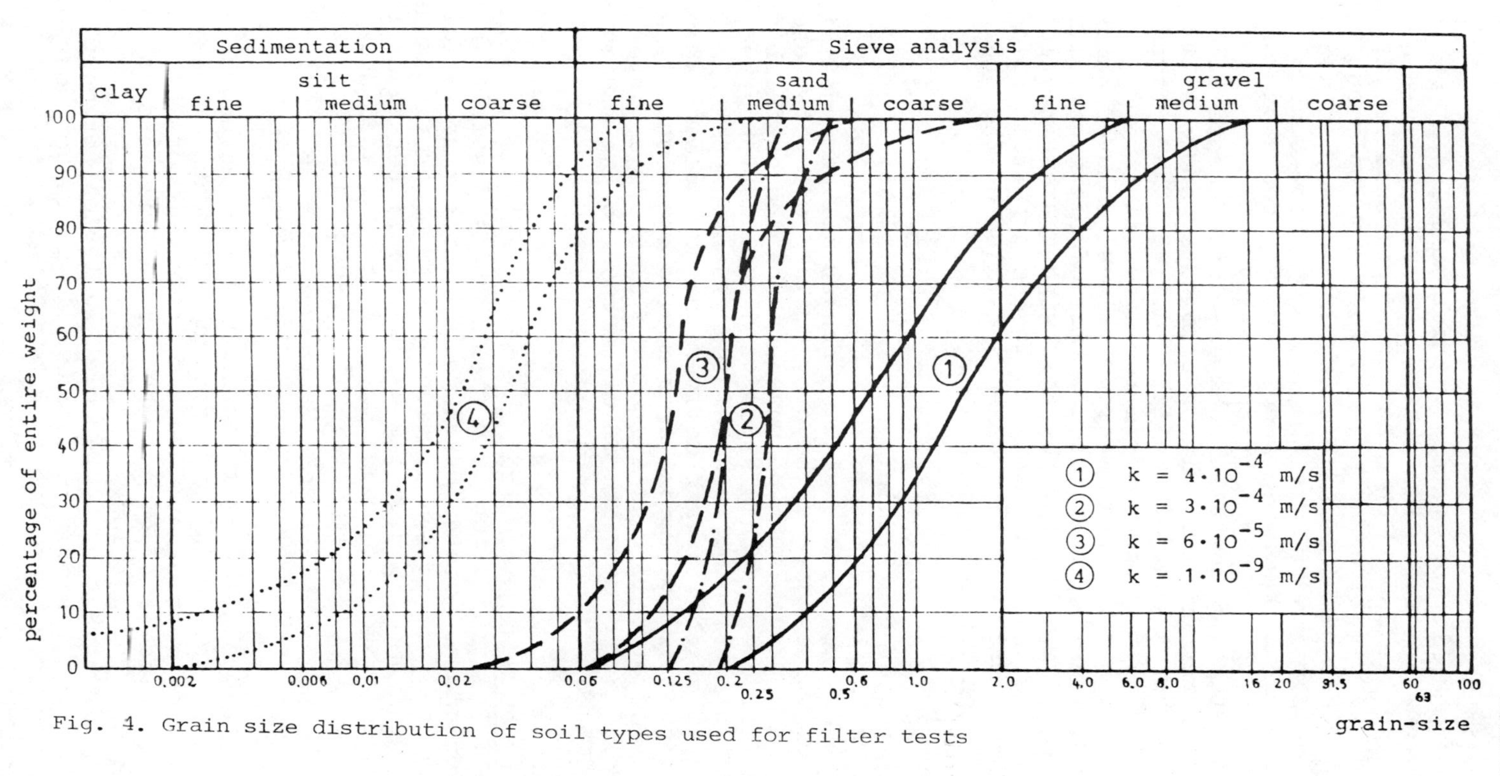

Fig. 4. Grain size distribution of soil types used for filter tests

- weight
- permeability to water (without soil).

The number of control tests done with the further deliveries depend on quantity of geotextiles. If there are deliveries of max. 20.000 m^2 only one control test per 5000 m^2 has to be executed, in cause of deliveries of more than 20.000 m^2 one control test per 10.000 m^2 is necessary. The costs of control tests are payed by the employer.

4. Starting the placing operation of geotextile filter layers will only be allowed by employer if the control test of the first delivery has proved the identity with the offered material.

It is necessary that minimum values of layer thickness, tensile strengths and size of openings must be observed. Besides the following characteristic values of material must lie inside a certain tolerance compared with the average value of the qualification testreport:

weight:

single layer geotextile:	± 10 %
multilayer geotextile:	± 15 %
permeability to water:	± 50 %

If these tolerances will be passed all demanded minimum values have to be proved. The costs of these extra controls must be payed by the successful bidder. If one of the minimum values is not observed, the deliveries will be refused. If it happens in course of the further deliveries the employer keeps reserve for the demand of removing the geotextiles already placed or to extend the time of guarantee according to the size of difference to the minimum value. Geotextiles which are not yet placed have to be refused.

Supplementary technical specifications

1. The connection of geotextiles is possible in overlaps or joints. Overlaps or joints have to follow the inclination of slope. Width of overlaps must be minimum 0,50 m if the geotextile filter layer is placed above water and 1,0 m if it is placed under water. If the filter layer is placed on slopes partly under water geotextiles have to overlap minimum 1,0 m if the quantity placed under water is greater than in the air.

2. Geotextiles must lie smoothly on the ground. Driving over with construction equipments or heavy vehicles must be avoided. If it is not possible a protective layer of non angular material has to cover the geotextiles.

3. Treatment by ultraviolet radiation for more than a week should be avoided. If it is not possible a protective layer must be brought up.

4. Stones may not be thrown down on geotextiles from a height of more than 2 m.

5. Material properties of geotextiles change for the worse if temperatures are falling below the freezing-point. Therefore placing operation is only allowed if it is possible without damage of geotextiles.

6. Accounting. Deliveries and placing operation of geotextiles will be accounted on the base of the covered area.

T3 Experiences in the use of geotextiles in the water construction field in Finland

J. JUVONEN, MSc(Eng), Roads and Waterways Administration, Finland

SYNOPSIS. The effects of ultra-violet light, salt, water, freezing etc. on geotextiles was studied in a disposal area of dredging works in Finland. Some tensile strength tests were carried out with these geotextiles at the Technical University of Helsinki. From these and other experiences recommendations were given for the use of geotextiles in the slopes of disposal area dams and canals in Finland.

INTRODUCTION

1. Geotextiles have been used in Finland in the water construction field since 1970. In 1981 8 different types of geotextiles, of which one was domestic, were marketed in Finland. About 5 million m2 of geotextiles were used in the construction field in 1980.

2. The majority of the geotextiles used in the construction field are non-woven. The fabrics used in soil and road construction were mainly bonded by heat sealing and those used in water construction are mechanically bonded by needling.

3. The Roads and Waterways Administration of Finland has used geotextiles in the water construction field mainly in the slopes of canals and in the inner slopes of dams surrounding disposal areas of suction dredging works.

THE VAASA PROJECT

1. In Finland the first time a geotextile was used in the inner slope of a dam surrounding a disposal area of suction dredging works was in 1979. The project was the deepening of the incoming channel leading to the port of Vaasa in the Gulf of Bothnia.

2. The function of the geotextile was to let the water out of the disposal area and to keep the dredged material in. The disposal area, about 2.7 hectares, was surrounded by a dam constructed of blasted rock. The disposal area has formed a new island in the Vaasa archipelago.

3. During the dredging works the water flowed in and out of the disposal area depending on the water level of the sea outside the dam. This helped to keep the geotextile from clogging.

Picture 1. The geotextile was placed on the inner slope of the dam surrounding the disposal area

4. The geotextile was a heat sealed, UV-stabilized fabric weighing about 300 g/m2. During the first summer the strength of the geotextile seemed to be quite adequate but in the fall holes begun to appear in the geotextile. The holes were most frequent around water level and above water level in the area which had been exposed to ultra-violet rays.

5. In september 1979 and March 1980 a series of tests (plane strain tensile tests) was carried out at the Technical University of Helsinki. The objective of the tests was mainly to check the influence of ultra-violet rays, sea water and freezing on the strength of the geotextile. The results obtained from the tests are shown in Table 1.

6. When the test results were analyzed it became apparent that the sea water had very little effect on the strength of the fabric (70% polypropene, 30% polyetene) or the heart sealing. The sunlight, inspite of the UV-stabilisation weakened the breaking point of the fabric with about 10%. The UV-rays weakened the fabric also so that after the breaking point of the fabric the geotextile broke completely, whilst with the new fabric the geotextile broke only partly and lost resistance only gradually. The breaking point of the geotextile that had been in the dam through the winter and had been frozen for about 3 months had weakened about 30 % compared to the new geotextile. Also this geotextile broke completely after the breaking point.

7. Next spring after the ice had melted the geotextile in the dam of the disposal area was practically useless. A new geotextile was installed over the old one so that the dredging

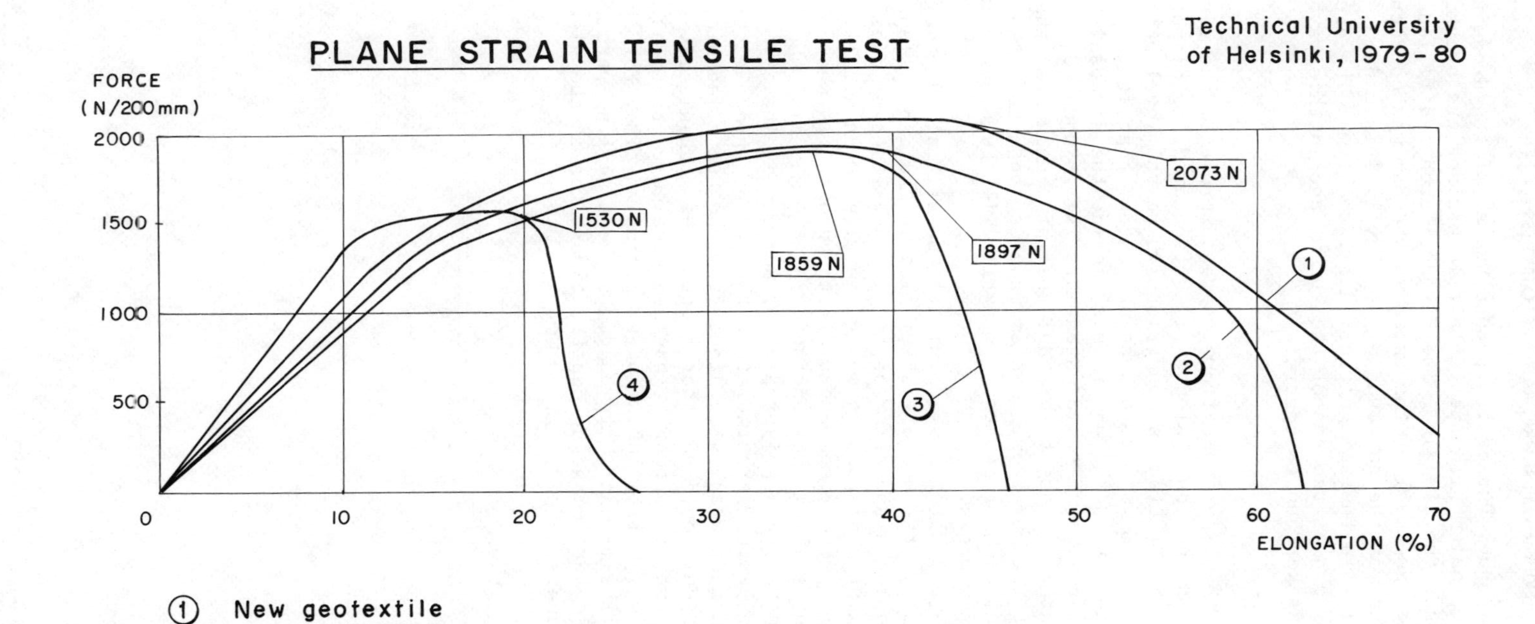

(1) New geotextile
(2) Geotextile that had been in seawater for 100 days
(3) Geotextile that had been influenced by sunlight for 100 days
(4) Geotextile that had been in the dam over winter (frozen for about 3 months)

Table 1. Results of tests carried out with geotextiles in the Vaasa project

works could continue. During the second summer of the dredging works the main problem with the geotextile was not the breaking of the fabric but the clogging. The disposal area had filled up so much that the level of the water in the dam was consecutively above that of the surrounding sea level. This caused the water to flow in the same direction all the time. The dredging was slowed down at the end of the summer and in this way the clogging was kept under control.

RECOMMENDATIONS

1. Based on the experience from the Vaasa project, the use of geotextiles in the inner slopes of suction dredging disposal areas should be restricted to conditions where the level of the water outside the dam varies to such an extent that it is occasinally higher than the water level inside the disposal area, if the grain size of the dredged material is under 0.06 mm. Also the area of the geotextile should be large enough so that the velocity of the out flowing water is close to zero. If these conditions are not met the geotextile will be clogged in a matter of a few weeks.

2. The Roads and Waterways Administration of Finland made recommendations for the use of geotextiles in the inner slopes of dams surrounding disposal areas and in the slopes of navigable canals. In the disposal areas geotextiles mechanically bonded by needling were recommended. If the geotextile was produced by heat sealing the manufacturing raw material should be polyester rather than polypropene or polyetene. The UV-stability of polyester is better that that of polypropene or polyetene. From test results obtained from the State Technical Research Center one could see that geotextiles mechanically bonded by needling are stronger than heat sealed geotextiles and keep their durability for a longer period of time. For this reason geotextiles mechanically bonded by heat sealing were also recommended for demanding construction sites in the slopes of navigable canals.

Discussion on Session 4: Materials

Dipl.-Ing. H. U. Oebius, Versuchsanstalt für Wasserbau und Schiffbau

Many authors have referred to the return flow. As I understand it, they define it as flow parallel to the channel axis caused by the water system of the ship. This definition is only correct if a completely symmetrical ship is cruising exactly in the centreline of a fully symmetrical channel. Only under these conditions will the return flow be symmetrical. The intensity of the return-flow is a function of the blockage effect and the ship speed.

If the ship is not symmetrical, or not cruising in the centreline of the channel, the resistance field, which expresses itself in the front wave (not that at the bow), will be asymmetrical, with gradients not parallel to the centreline of the channel (see Fig. 1). The velocity field then will have a current direction from the place of higher resistance to that of lower resistance. The consequence will be an inclined velocity field down the slope of the embankment. You will therefore have not only components parallel to the centerline of the channel, but also components acting normally to it. These currents are responsible for the erosion of the toe of revetments, and have been investigated by the Versuchsanstalt fur Binnenschiffbau (Inland Ship Research Station) in Duisburg at Germany.

I wonder if this component has been observed by other authors and whether it has any influence on the stability of revetments.

I would like to add that this hydraulic procedure is very common in channels with two-way traffic, as you can see from Fig. 3. of Paper 3.

Dr J. Fowler, Us Army Engineer Waterways Experiment Station

There are problems associated with armoured revetment failure caused by the improper use of geotextiles with the wrong EOS and open area. After the geotextiles with an EOS of 70 to 100 and an open area of 10 was replaced with an EOS of 30 to 40

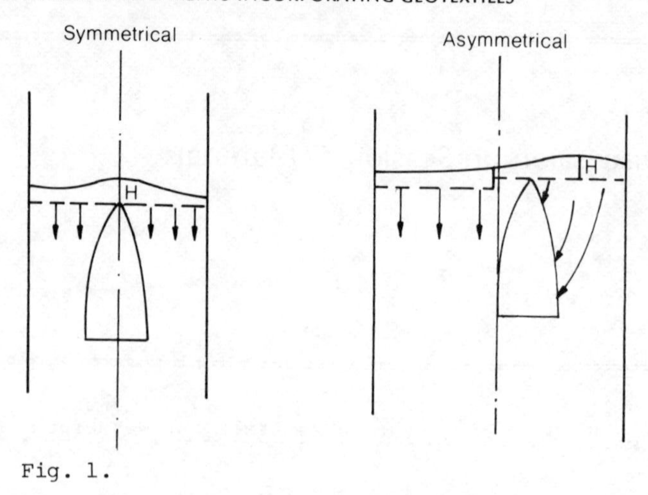

Fig. 1.

and an open area of 25 to 30 there have been no failures under similar stormwater conditions.

Dr G. Heerten, Naue-Fasertechnik, Germany

I would like to give some results of field investigation testing the permeability of dug up fabrics. The results given in Paper 11 show the very sufficient long-term permeability of needlepunched nonwoven fabrics, in contrast to the poor permeability of woven fabrics.

Some reasons for the very good behaviour of needlepunched nonwoven fabrics are:

(a) the high porosity (normally, needlepunched nonwoven fabrics have about 90% pores and 10% fibre mass, in contrast to soil structures with only 2 to 50% pores and 50 to 75% grain material)

(b) the unexpected stable thickness in contact with sand and finer soils (all dug up needlepunched nonwoven fabrics show the original thickness of the virgin fabric, and have not been compressed according to actual revetment load; a spontaneous grain-fibre-grain interaction prevents any compression during installation).

The results given are valid for revetment structures of sea dikes and waterways on silty sands and sands. The grain size distribution of the soils are given in Fig. 2.

In the discussion of revetment design and construction I think that it is very important to give information about:

(a) the subsoil (e.g. grain size distribution, cohension)

(b) the kind of fabric (e.g. nonwoven or woven or multi-layer fabric)

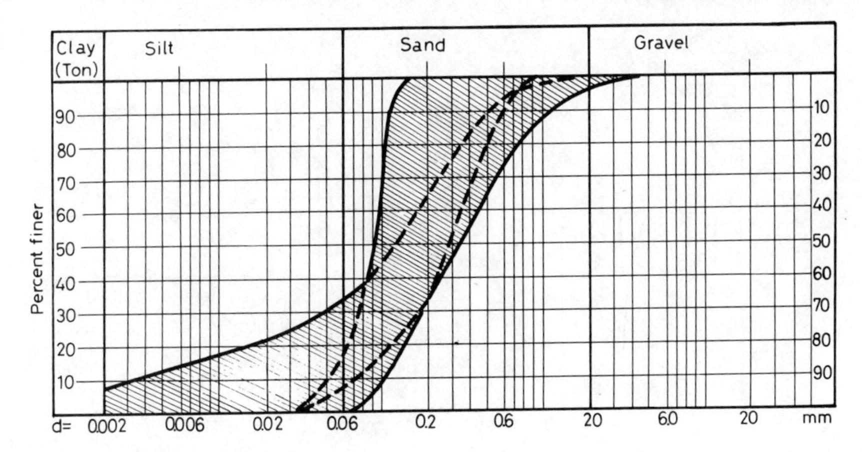

Fig. 2. Grain size of the subsoils at the sampling locations

(c) the load conditions (e.g. current and wave spectrum)

(d) the revetment armour layer (e.g. weight, kind of interlocking)

to make sure that we are not comparing apples and eggs.

Mr J. G. Berry, Bertlin & Partners, Surrey

The opportunity to check whether geotextiles have been damaged seldom arises. I was able to do this about 10 years ago with continuous heat banded filiment fabric.

About 1.5 m of rubble, maximum size 300 mm, was placed with care on the fabric to form a wedge of free draining material behind a sheet pile wall in order to reduce the earth load (the wall was leaning). The soil was silt. Fig. 3 shows a cross-section.

Because the depth of excavation was found later to be inadequate the rubble and fabric was removed. The fabric was found to be punctured in several places, although I had personal knowledge of how difficult it was to cut with a knife during the installation process.

I believe the punctures were due to the fabric being under tension (which is a fact referred to in Paper 12), as the fabric was held up at front and rear and the base of the excavation was soft silt. This should not be a problem with a simple revetment slope.

I would be interested to know if other delegates have had a similar opportunity to examine a fabric after it has been covered with rubble.

The need for test panels to be constructed and taken apart as part of a construction contract is strongly advised (see Paper 12).

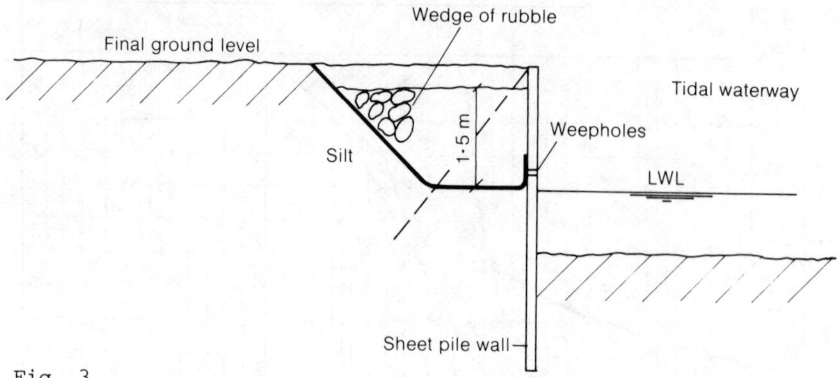

Fig. 3.

I see a good future for composites, where a net is used to give strength and the filter fabric, bound to it, can be of ordinary strength below it.

In order to avoid tension, it is presumed that the best way of laying rip rap is to start at the base and work up the slope, the fabric being allowed to slide down the slope if needed. I have not seen this matter referred to.

The use of a cushion of fine material over a fabric before placing rubble or rip rap is a good way of reducing the incidence of damage to fabrics. This is referred to in Paper 12. Presumably, with time the finer material will be leached out and upwards. This may not necessarily affect the performance of the revetment.

Dr Fowler showed slides of a highway protected by a revetment using perforated concrete facing blocks. On the foreshore there was rubble from remains of previous revetments. I should like to know if rubble thrown up by wave action caused excessive abrasion of the blocks.

Mr P. Rankilor, Manstock Geotechnical, Manchester

Manstock is presently conducting a ten-year weathering programme of geotextiles in different climates and different geotechnical locations: in soil; on the coast; buried; exposed to ultra-violet light; north of the Arctic Circle; in the Saudi Arabian desert and coasts, and in Europe and the Java coastline. Weathering is one of the most important items needing current research. Weathering also stimulates changed failure modes which are important to bear in mind at the design stage. This is not just a simple loss of strength.

Dipl.-Ing. H. U. Abromeit

If gabions are used for bank revetments on slopes lying

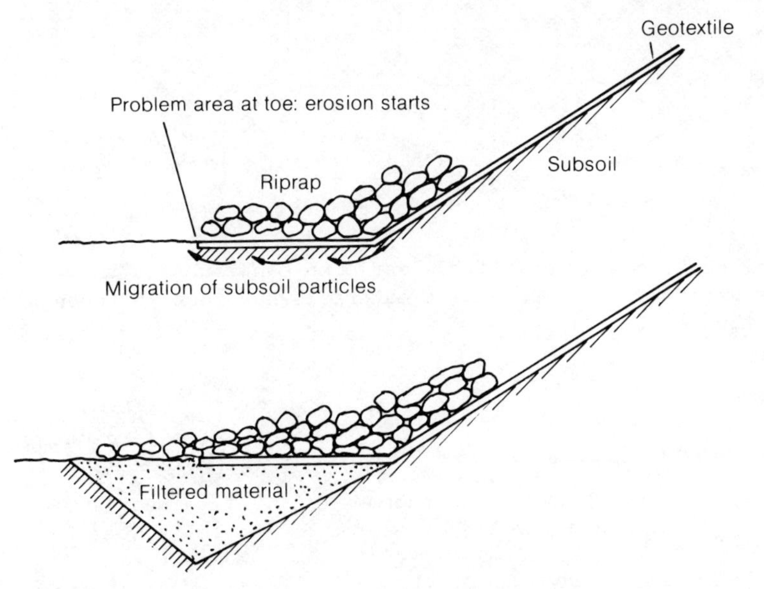

Fig. 4.

partially under water, what possibilities exist, in cases of damage, to repair the destroyed gabions under water-level (not locally, but on greater lengths)?

Mr M. Wewerka

To repair destroyed gabions on greater lengths under water it is necessary to know the cause of destruction. In case of corrosion of the wire we propose to construct new gabions on top of the old ones (without using a geotextile). Another possibility is to cover that area with concrete slabs.

Mr M. E. S. Bjerkan, Sivilingenior M. Bjerkan A/S, Norway

The erosion of subsoil underlying geotextiles is an effect due to the toe design (Fig. 4). I assume most problems with erosion of subsoil are due to migration of subsoil particles towards the toe, causing subsidence of the embankment. This can be presented by using filter material at the toe. More attention should be given to toe design.

Mr M. P. Keown

Current design guidance specifies that the revetment and fabric extend below mean low water to prevent erosion of a bank under attack due to streamflow action. Key trenches can help to prevent loss of the revetment toe if wave attack is

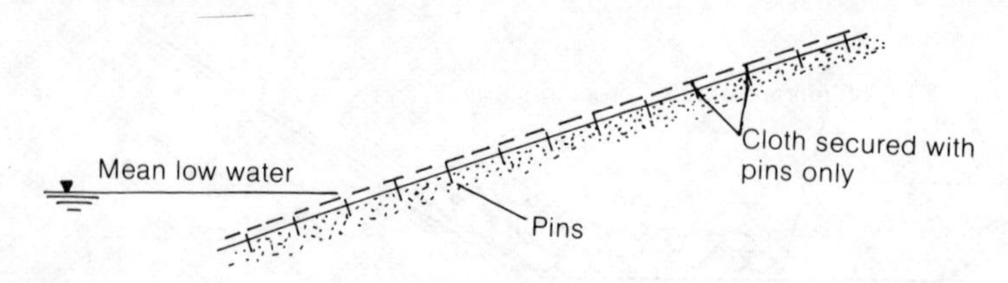

Fig. 5. Placement of filter fabric on bank subject to streamflow action. Revetment materials have not yet been placed on the fabric

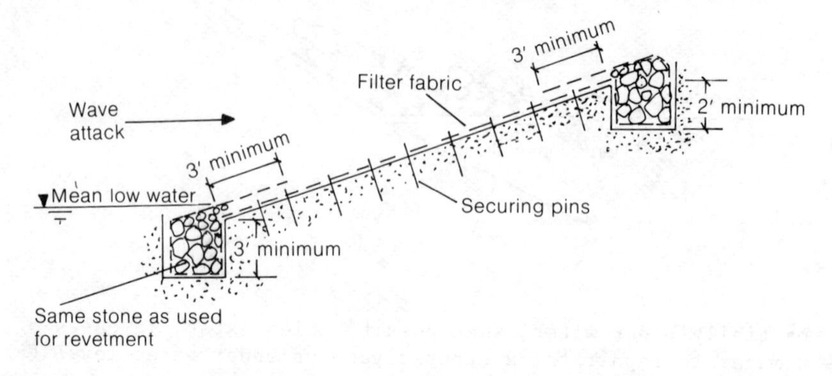

Fig. 6. Filter fabric on bank subject to wave attack showing placement of vertical wall key trench at toe and top bank. Rvetment materials have not yet been placed on fabric

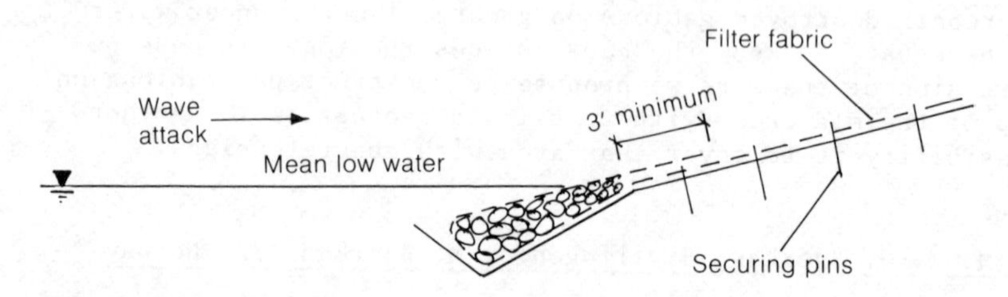

Fig. 7. Key trench design used when soil conditions do not permit construction of vertical walls

the principal erosion mechanism. Good engineering practice may also dictate that a key trench is needed to maintain revetment stability, if during streamflow attack the currents parallel to the bank are anticipated to be excessively strong or if eddy currents are set up near the revetment. These points are amplified in the following quote from "Utilization of filter fabric for streambank protection", US Army Engineer Waterways Experiment Station Technical Report HL-80-12, M.P. Keown and E.A. Dardean, 1980.

"To prevent soil leaching, the filter fabric strips should be properly joined together by seams and/or overlaps. The strips should be placed with the longer dimension parallel to the current when used along streams where currents acting parallel to the bank are the principal means of attack. The upper strip of fabric should overlap the lower strip (as roofing shingles are commonly placed) and the upstream strip should overlap the downstream strip. To avoid having long sections of continuous overlap, the overlaps at the ends of the strips should be staggered at least 5 ft. The revetment and fabric should extend below mean low water to minimize erosion at the toe (Fig. 5).

When the revetment materials and fabric are subject to wave attack, the customary construction practice is to place the fabric strips vertically down the slope of the bank. The upper vertical strip should overlap the lower strip. The fabric usually needs to be keyed at the toe to prevent uplift or undermining (Fig. 6). It is important that the key trench be below mean low water to prevent erosion of material adjacent to the trench and the subsequent loss of the trench. When it is not possible to maintain vertical trench walls, the fabric may have to be keyed as shown in Fig. 7. Here the trench is excavated and the walls are allowed to assume a stable slope. The key at the top (shown in Fig. 6) is usually not necessary unless wave action is expected to reach that elevation or overbank drainage is anticipated. A key trench at top bank should also be used for streambank protection works subject to current attack where there is an overbank drainage problem."

Mr V. E. Mosca, Pauling PLC, London

We have been shown pictures of a large rock being dropped onto a geofabric with no puncturing. Paper 16 states that rocks should not be dropped more than 2 m. Another photo has shown rip rap being placed with the aid of a batter rail stacked through the geotextile. Can anyone explain from what height rocks can safely be dropped? Also, if the material is so resistant to puncture, how does the labourer hammer in the stakes for the many batter rails invariably required (Fig. 8)?

Mr M. Wewerka

The height from which rocks can safely be dropped depends on the following factors:

(a) weight of the rock
(b) shape of the rock (angular, round)
(c) resistance against deformation of subsoil
(d) tensile strength of geotextile
(e) elongation at break of the geotextile
(f) distance between already placed rocks or other fixings.

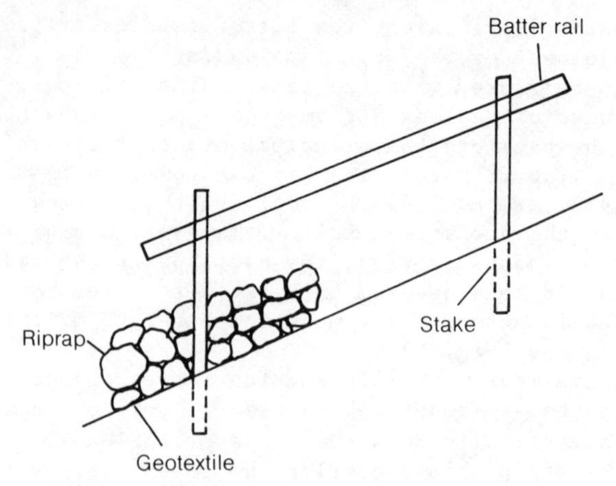

Fig. 8.

Due to the energy of the falling rock, the geotextile and subsoil will be deformed. As long as the elongation of the geotextile is high, subsoil will take up most of that energy without destroying the geotextile (as this does not reach its elongation at break). Wherever the elongation at break is low, the strength of the geotextile has to be high enough to take up the bulk of the energy, as the resistance of the subsoil cannot be taken into consideration.

The factor of safety against puncturing is much higher for materials with high elongation at break (mechanically bonded nonwovens) than for those with high tensile strength but low elongations. To hammer in stakes can be done by cutting the geotextile beforehand by means of a knife.

Mr P. Rankilor

I have observed unexpected failures of geotextiles under site conditions. These include the unravelling of a needle-punched geotextile during marine construction works by wave oscillation action. Another was the brittle impact failure of a strong polypropylene woven geotextile under site conditions.

Mr E. Loewy, Sir William Halcrow & Partners, London

The session has included six papers on materials, which is in many ways the most important aspect of the conference. We are seeking new products and materials, relatively untried in the construction and civil engineering industries but required to have properties and characteristics such as strength, durability, ease of hauling, 'greenability' in the environmental impact, all at lower first and maintenance costs than the traditional materials and techniques.

All current methods of laboratory accelerated weathering

have great limitations and are of value mainly in comparing one product with another. They are not able to simulate actual working conditions over an extended period. Hence the vital role of full-scale prototype trials, carefully designed and monitored as described in the papers from such organizations on the US Corps of Engineers, Rijkswaterstaat, and others. The active support of manufacturers of geotextiles, of research bodies and consulting and contracting engineers is equally essential. The identification of environmental acceptability parameters needs to be related, in such prototype trials, to cost and to technological efficiency. The various papers show quite differing attitudes to these matters.

Civil engineers working in these fields have to familiarize themselves with the essential physical and chemical properties of the various new geotextiles, much as they long have done with traditional materials. They must establish new parameters and test methods to enable the engineering properties to be identified and quantified. Such work had to be done with soils before soil mechanics could be properly established.

Dipl.-Ing. P. Giorgi, Dipenta SpA

In the case I shall describe here, a reinforcement realized with a polyester mesh and an elastomeric additive to the bituminous mixture have satisfied the strict specification requirements. The prefabrication of the revetment executed in slabs on the canal banks by a standard road finisher is achieved by means of a reinforcing mesh, which in addition to its role in the operational stage allows for easy handling and placing of the elements.

Dipenta SpA executed in 1978, as technical and financial partner, the lining of the main canal of Nekouabad, which is part of the hydraulic works managed by the Iranian Ministry of Water and Power in the Esfahan region. The function of the canal is to divert the water of the Zayandeh river for the irrigation of approximately 200,000 acres of land. The Engineer for the project was Sogreah Grenoble. Total length of the project was 10 km, and the total surface to be lined was 150,000 m^2. Works included the fabrication and placing of bituminous concrete slabs with a polyester mesh as reinforcement and an addition of an elastomer of the butadyene styrene type to the bituminous mixture.

Technical specifications:

Slab thickness	25 mm
Specific weight	2270 kg/m^3
Weight percentage of bitumen + additive	10.5%
Percentage ratio of voids to the total volume	2.5%
Compressive strength resistance (test LCPC B 14), minimum	100 bar

Fig. 9.

Fig. 1O.

Fig. 11.

As above but after 7 days under water, minimum	90%
Penetration test Din 1996	3.5 mm
On slope stability (H3/V2) at 80°C after seven days maximum	0.1 mm
Ageing index of elastomerized bitumen minimum	85%

General condition of contract: 10 years guarantee

Description of the material employed:

(a) Composition of the mixture

sand 0-5 mm	970 kg/m^3
gravel 5-12 mm	660 kg/m^3
filler (cement)	575 kg/m^3
bitumen (60-70 penetration)	182 kg/m^3
elastomeric additive (Europrene Sol T-ANIC)	9.6 kg/m^3

(b) Geotextile polyester mesh, Structofors type

size	18 x 18 mm
tensile strength	20 da N/cm
percentage of elongation at maximum strength	10%

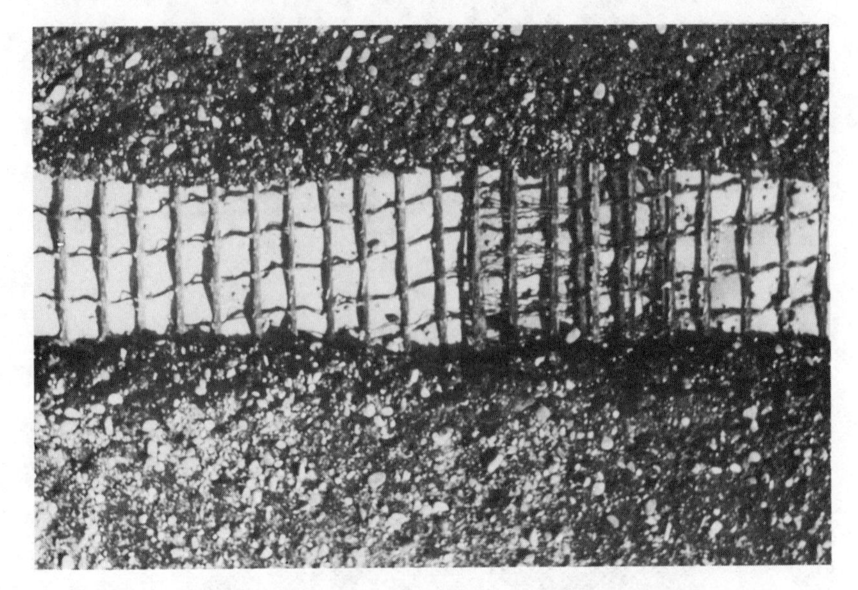

Fig. 12.

Fig. 13.

The mixture, prepared by an asphalt mixing plant, was transported by a heating and mixing truck to the canal banks (fabrication site), and the temperature in the mixture was in this way maintained at 230-245^{0}C. The spreading was achieved with a standard road finishing machine equipped in the front with a roll carrying chassis (Fig. 9). The unrolling of the geotextile roll of reinforcement was obtained by the advancing of the road finisher itself. Spreading and levelling of the hot mix on the unrolled geotextile completed the operation (Fig. 10).

Figure 11 shows a detail of the side sliding form. Fig. 12 shows the discontinuity in the spread mix to allow for the separation cut of different slabs. The cut lengths of slabs were put into position in the canal by means of an hydraulic crane equipped with a special outrigger, itself equipped with hydraulic jacks (Fig. 13). On-site production outputs with two teams spreading and two teams placing reached 1400 m^2 per day.

The most interesting feature of the revetment in question is the stability to the severe climatic conditions (excursions from -15°C to +70°C are foreseen). In addition, the rheological behaviour of the slabs obtained with the use of the elastomeric additive allows for all deformations which occur during placing.

For the short duration load application (placing stress), the combination of the bituminous concrete with the geotextile reinforcement in polyester mesh, the deformation modulus of both being within the same order of magnitude, allows for a good co-operation within the slab. On the other hand, for the long duration stress (slab stability during operation time) the polyester reinforcement guarantees a good dimensional stability.

Mr P. Rankilor, Manstock Geotechnical Consultants, Manchester

This technical contribution briefly describes some of the principles of design for an impermeable lining system for rivers over areas potentially subject to subsidence.

In the case of a lining scheme specifically required for a mining situation, the loss of water per se would not be important, but the consequences of the disappearance of that water into the active mine workings below would be catastrophic.

In view of the economic nature of mining projects, and the psychological factors associated with the world-wide investment in the mining industry, it is considered that a lining proposal carries not only the requirement that the river should be effectively lined to prevent the possibility of water ingress during mining activities, but also that any scheme should be so well engineered and so substantial that it can clearly be seen that the possibility of catastrophic water entry into the mine does not exist. It is therefore to be expected that, if a lining system be commissioned, it would be of the highest technical quality and best material quality available. We also would expect it to be thoroughly researched and properly designed.

It is often the case that general/regional ground-water levels have fallen as a result of general mining activities in any mining area. This establishes a prima facie case that mining could induce a situation such as that shown in Fig. 14. In this diagram the opening-up of joint systems above the mine workings is postulated.

Figure 15 shows the situation where the introduction of an impermeable lining will act as an obstruction to the collecting function of the river. Therefore, (a) hydrostatic imbalances must be catered for at the river edges, (b) hydrostatic uplift forces must be catered for on the river bed, and (c) water-flow nets must be checked and a suitable sized collector designed with vertical inflow capability and horizontal transporter capability to allow 'over-topping' in a limited vertical space.

A different situation involving an impermeable lining is described in Fig. 16. This shows the hydrological situation where the rock mass above the mine void has become dry as a result of the opening of fissures through subsidence. It is to be expected in this situation that the impermeable lining will effectively contain the river, but that the local ground-water table will fall as ground-water passes down into the mine workings. Consequently, there remains the possibility of soil piping as a result of ground-waters transporting soil particles down into the mine workings, but the limited source of ground-water volume, once the river has been isolated, is very small and the risk is therefore no greater than in any other normal mining situation. The layered functions of a suitable lining are described in Fig. 17.

The chosen bank geometry which I consider to be suitable for most general purposes is shown in Fig. 18. There are three fundamental aspects to the slope geometry: the slope

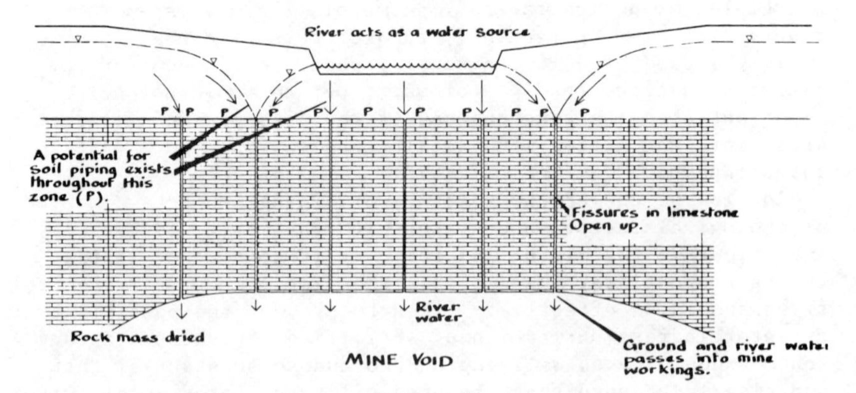

Fig. 14. Hydrological situation without impermeable lining if leakage occurs into mine

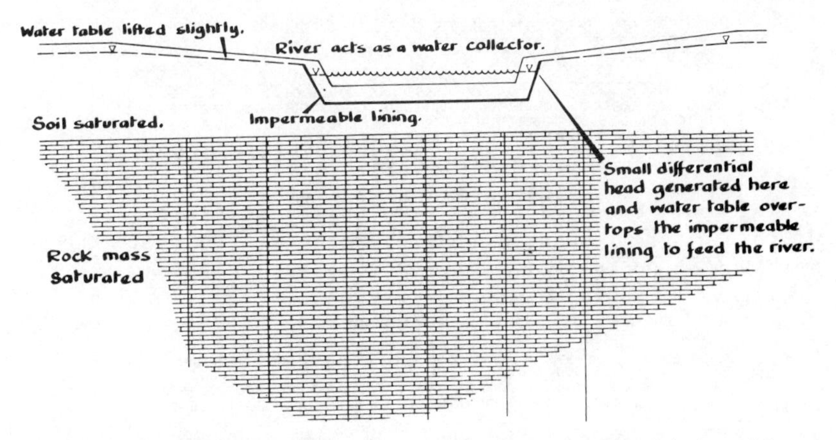

Fig. 15. Hydrological situation with impermeable lining if no leakage occurs into mine

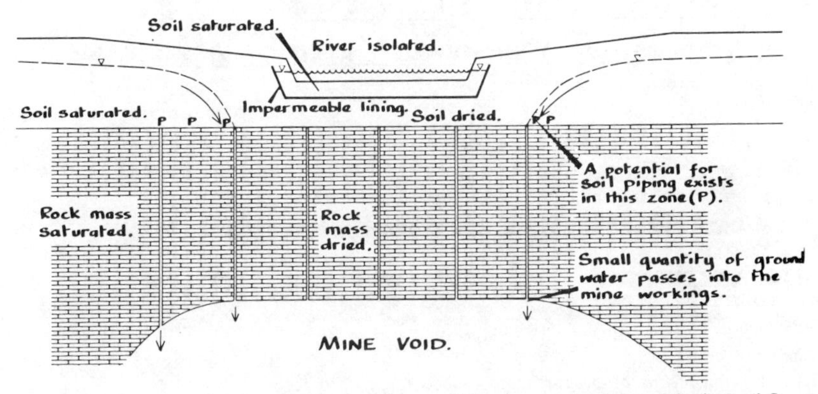

Fig. 16. Hydrological situation with impermeable lining if leakage occurs into mine

angle; the bank height relative to the river bed; and the size of rocks necessary to resist the rolling effect of the river flow.

The criteria affecting the slope choice are firstly, practicality of laying with regard to the membranes, and secondly the friction coefficients between membrane and soil and between each membrane. If the slope is too steep, then it would be difficult to lay the fabrics and they will tend to slide down the slope.

The height of the bank is chosen in relation to the possible peak flow, and will be modified in terms of a final design, both in view of the ultimate peak flow expected and the centrifugal force of the water causing an upsurge on the outer sides of bends. The height of the slope is also vertically extended by the appropriate amount, in order to cater for future calculated mining subsidence at any point along the channel. The rock size will also be finally chosen

PROPOSED SPECIFICATION RANGES

(Final Specification Yet To Be Defined)

VERY STRONG PERMEABLE MEMBRANE OR WEBBING MAT. PARAWEB OR EQUIVALENT. YOUNG'S MODULUS AND ULTIMATE TENSILE STRENGTH TO BE CONSIDERABLY HIGHER THAN FOR THE UF AND LF BUT LOWER THAN THOSE OF LPM PROBABLY ABOUT 25 TONNES/M. JOINTING SYSTEM MUST DEMONSTRATE ADEQUATE STRENGTH TRANSFERENCE HIGH U.V. RESISTANCE. HIGH ABRASION RESISTANCE. DETERIORATION RESISTANCE. DEGRADABILITY TO BE TAKEN INTO ACCOUNT IN STRENGTH CALCULATIONS.

A THICK FELT WITH YOUNG'S MODULUS AND ULTIMATE TENSILE STRENGTH TO BE CONSIDERABLY HIGHER THAN FOR THE CM. PROBABLY A POLYESTER OR POLYPROPYLENE NEEDLE-PUNCHED MATERIAL APPROX 5-10 mm THICK. TERRAFIX OR EQUIVALENT.

AN IMPERMEABLE MEMBRANE

PROBABLY <1 mm THICK WITH EXTENSIBILITY >100 % AND HIGH ULTIMATE TENSILE STRENTH. MUST HAVE EXCELLENT AND PROVED JOINTING SYSTEMS WITH HIGHER BOND STRENGTH THAN THE SINGLE MEMBRANE. PROBABLY BUTYL RUBBER OR POLYETHYLENE.

THICK FELT: PROPERTIES AS FOR UF. TERRAFIX OR EQUIVALENT.

LINEA COMPOSITES LTD. HEAVY DUTY PARAWEB OR EQUIVALENT. PERMEABLE MEMBRANE: PROPERTIES AS FOR UPM, BUT UV RESISTANCE NOT CRITICAL YOUNG'S MODULUS AND ULTIMATE TENSILE STRENGTH TO BE CONSIDERABLY HIGHER THAN UPM IN ORDER TO ABSORB. MAIN STRUCTURAL STRESSES. PROBABLY OF THE ORDER OF 50 100 TONNES/LIN METRE IN BOTH DIRECTIONS WITH LOW CREEP CHARACTERISTICS.

THICK FELT: TO ACT AS A SACRIFICIAL ABRASION PROTECTION BELOW THE LPM WHERE LPM IS PLACED DIRECTLY ON ROCKHEAD AND DRAG WOULD BE EXPECTED UNDER DIFFERENTIAL MOVEMENT CONDITIONS. TERRAFIX OR EQUIVALENT.

PROPOSED CROSS-SECTION OF COMPOSITE IMPERMEABLE LINING MEMBRANE.

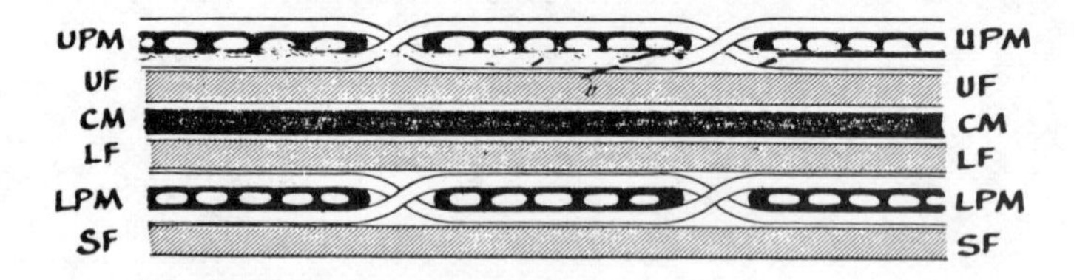

PROPOSED FUNCTIONS

UPPER PROTECTIVE MAT (STRONG)

TO TAKE THE DIRECT PLACEMENT OF BOULDERS AND ROLLING OF BOULDERS; TO RESIST ANY ACCIDENTAL SURFACE DAMAGE (MAN MADE); TO ABSORB THE EXPECTED SUBSIDENCE-INDUCED TENSION AWAY FROM THE FELT AND IMPERMEABLE MEMBRANE, PERMEABLE

UPPER FELT (THICK)

TO RE-DISTRIBUTE POINT LOADS FROM ABOVE; TO ACT AS A SMOOTH 'LUBRICATOR' BETWEEN THE MAT ABOVE AND THE CENTRAL MEMBRANE; PERMEABLE TO ALLOW WATER THROUGH TO ASSIST IN LUBRICATION AND REDUCE SUBSIDENCE-GENERATED PRESSURE DIFFERENTIALS; TO ACT AS A SECOND SAFETY BARRIER IN THE EVENT OF FAILURE OF THE IMPERMEABLE MEMBRANE, SINCE IT'S PERMEABILITY IS LOW, AND IT WILL CLOG FURTHER AND BECOME SELF-SEALING IN THE EVENT OF PIN-HOLE LEAKS; TO ACT AS A CUSHION LAYER BETWEEN THE JOINTS IN THE UPPER MAT AND THE CENTRAL MEMBRANE.

CENTRAL MEMBRANE (IMPERMEABLE)

TO HAVE EXCELLENT JOINTING CHARACTERISTICS; TO HAVE HIGH STRENGTH AND EXTENSIBILITY, TO LAST THE REQUIRED LENGTH OF TIME WITHOUT DETERIORATION.

LOWER FELT (THICK)

TO RE-DISTRIBUTE POINT LOADS FROM BELOW; OTHER FUNCTIONS AS FOR UPPER FELT.

LOWER PROTECTIVE MAT (STRONG)

TO ABSORB POINT LOADS FROM BOULDERS BELOW OR FROM DIRECT ROCKHEAD PLACEMENT, TO ABSORB SUBSIDENCE INDUCED TENSION AWAY FROM THE UPM, THE FELTS AND CENTRAL MEMBRANE; TO RESIST DAMAGE FROM BEING DRAGGED OVER UNDERLYING MATERIAL; PERMEABLE. TO ACT AS PORE WATER PRESSURE DISSIPATOR.

SLIDING FELT (THICK)

REQUIRED BENEATH LPM TO ALLOW THE LATERALMOVEMENT OF THE LPM WITH MINIMUM ABRASION i.e. TO ACT AS A LUBRICATOR BETWEEN THE MAIN STRUCTURAL ELEMENT AND THE ROCK BELOW, AND TO ACT AS A FINE FILTER ELEMENT.

Fig. 17.

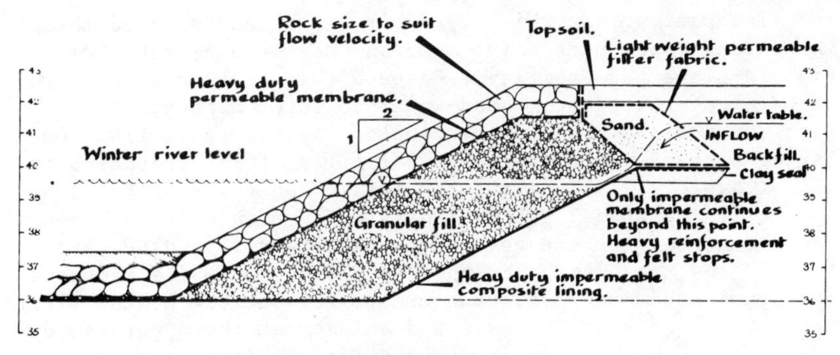

Fig. 18. Winter flow before subsidence

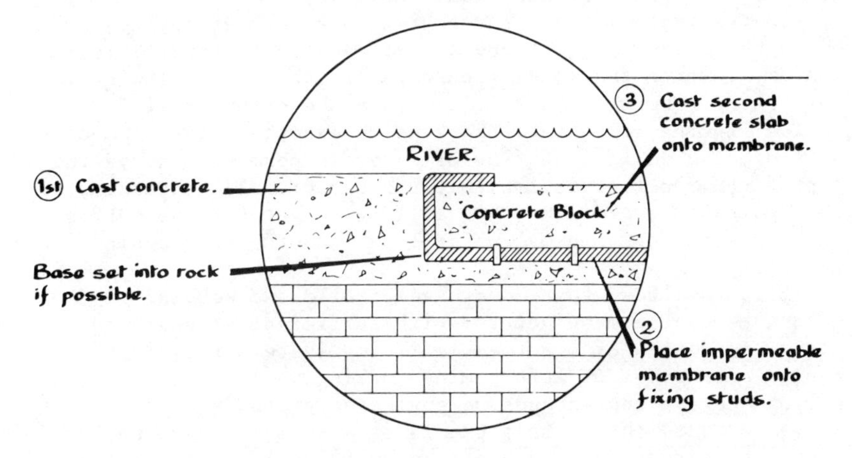

Fig. 19. Detail (not to scale)

in relation to the ultimate expected edge velocities.

The object of making the structure out of granular material is to ensure its flexibility under extreme subsidence conditions. In this regard, the overall structure should be substantially repairable after subsidence has taken place. Simple stone infill and re-levelling should suffice to repair this damage and no work should be necessary beneath the structure unless it were felt that a particular void had been generated which could be suitably backfilled by grout. It is apparent that once the mine workings are abandoned and flooded, then there would be no problem generated even if the impermeability of the membrane became impaired subsequently.

As can be seen in the figures the upper edge of the impermeable membrane is bedded into a clay seal. This is in order to prevent direct hydrostatic connection developing between the ground surface and the back of the impermeable membrane. Even a small connection could generate high hydrostatic heads which should be avoided.

It is expected that hydrostatic pressures could build up beneath the impermeable membrane if allowance is not made for the dissipation. In this regard it can be appreciated that the strong lower protective membrane has been selected with a high in-plane permeability. Beneath this is a thick filter felt which will allow ground-water to enter through it at a rate in excess of the percolation of the natural soils. Once having passed through the external filter felt, ground-water will travel laterally along the interwoven voids of the lower protective membrane, and will dissipate outward at each end of the structure. It can be seen on the lower left-hand part of Fig. 19 that the design brings the entire composite membrane up through the concrete structure to the surface of the river bed. Pore-water pressures can dissipate at these points, and the exact permeabilities and hydraulic gradients are calculated as part of the final design.

The drawings illustrate the principle of overtopping, by which the river can continue to act as a collector. This is done by lifting the local ground-water table and allowing it to overtop through a sand filter into the granular fill.

The figures illustrate that, as an example, after 1 m of subsidence a nominal outflow of water is permitted above the top of the impermeable lining. The use of a 2 m thickness of sand guarantees a minimum hydraulic gradient for the outflow, and a concomitant minimal water-flow into the surrounding ground.

It is concluded that modern geotextile and webbing membranes can be manufactured with sufficient strength to allow the design of impermeable lining systems for rivers or canals over areas of future mine workings. In order to meet all of the varying demands of envisaged ground conditions, it is recommended that a multiple-layer system is used forming a composite structure. Each of the individual layers is chosen to contribute towards the overall performance of the composite lining, giving the engineer considerable scope for design specification.

17 Case histories using filter fabric underneath revetments in lower Louisiana

L. E. DEMENT, MS, US Army Engineer District, New Orleans, and J. FOWLER, PhD, PE, US Army Engineer Waterways Experiment Station

SYNOPSIS. Flexible articulated mattresses used in conjunction with geotextiles or filter fabrics are a viable method of erosion protection. However, several case histories in Louisiana have shown that selection or proper specifications of filter fabric along with adequate weight of armor unit are essential factors in achieving an acceptable design.

INTRODUCTION

1. The objective of this paper is to present the "Louisiana Experience" on several major construction projects along navigable waterways. The authors will illustrate key factors that should be considered when designing revetments incorporating geotextiles.

REASONS FOR USING GEOTEXTILES OR FILTER FABRICS

2. There are many case histories where filter cloth has been shown to be a valuable component. Private interests, state agencies, and the New Orleans District of the Corps of Engineers have used filter fabric on many projects and on occasion they have built isolated test sections. Cost estimates show that filter fabric is far cheaper than conventional rock filter layers. Control of filter installations, both underwater and in the dry, is more efficient and cost effective with filter fabric and it assures positive coverage. Filters or filter layers should be considered as integral parts of a typical dike, breakwater, jetty, or revetment where the dynamic forces of water such as wind-waves, currents, and ship forces interact on all components of the structure. The percentage of the cost of the total structure attributed to filter fabric is small, but the returns or benefits can be large. There are many filter fabrics on the market today and the engineer must be careful in writing his specifications. There are filter fabrics which cost less than 10 cents a square foot, and others that cost over one dollar per square foot (0.0929 m^2). Products with tensile strengths of 200 pounds per inch would not be appropriate to locations with shear forces of 1000 to 4000-pound per inch

(179 to 714 kgs/cm) tensile fabric would be suited. The engineer should continue to investigate the advantages of using certain types of filter fabrics and he should challenge the industry to provide better fabrics that will improve the performance of structures.

TYPES OF STRUCTURES USED FOR ARMOR PROTECTION IN FLEXIBLE REVETMENTS AND STRUCTURES

3. The industry is becoming very competitive and new products have reached the market to help resolve the erosion problems. A brief list of these structures is given below:

a. Cellular concrete blocks.
b. Fascine mattresses.
c. Light weight concrete mattresses.
d. Articulated cable connected concrete mattresses.
e. Interlocking concrete block mattresses.
f. Rock used in various ways with filter fabrics to form flexible revetments and coastal structures such as breakwaters and jetties.
g. Grout and sand filled flexible structures.

LIST OF CASE HISTORIES

4. The following paragraphs give a brief description of projects where geotextiles have been used. This is not a comprehensive list of all projects, but it illustrates several different applications (**Fig.1**).

5. Mississippi River Outlets (ref.1). In 1979 several test sections using filter fabric were incorporated in offshore navigation jetties constructed with shell core and rock armor. These projects are located 90 river miles (145 kilometers) south of New Orleans and are described in detail later in this report.

6. Southwest Pass, Mississippi River. In 1981, three 1000-foot-304.8 m-long test sections were built using lightweight concrete armor mattresses in conjunction with filter fabric. These structures are revetments or dikes built to protect the river banks and are located approximately 110 river miles (177 kilometers) south of New Orleans and are also described later in this report.

7. Mississippi River Gulf Outlet (MRGO). In 1982 six revetments were built along the south bank of the MRGO to investigate various concepts in flexible revetments including concrete armor mattresses and conventional rock revetments. The project is located 20 miles (32 kilometers) east of New Orleans along a deep water navigation channel. See detailed description later in this paper.

8. Grand Isle. Four projects using filter fabric have been built on this barrier island located approximately 80 road miles (129 kilometers) south of New Orleans.

9. In the early sixties the first phase of the east terminal jetty (or groin) was built adjacent to Barataria Pass.

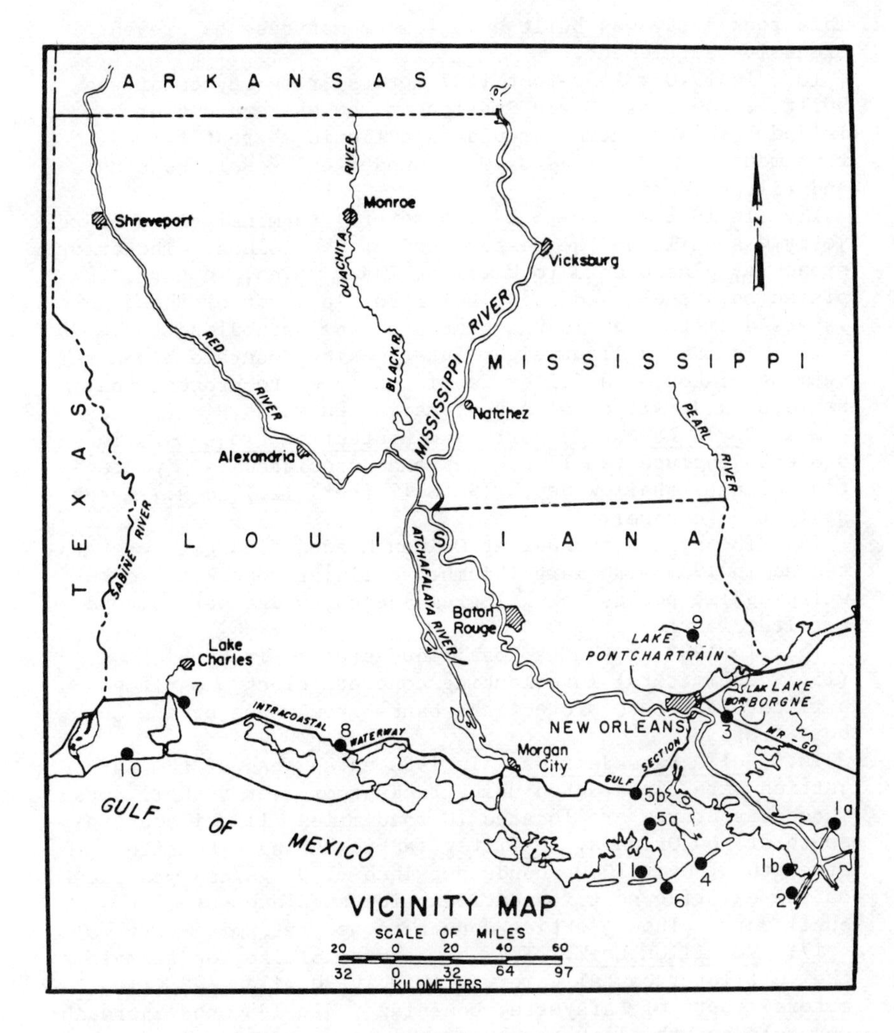

LEGEND

1. Mississippi River Outlets
   a. Baptiste Collette Jetties
   b. Tiger Pass Jetties
2. Southwest Pass, Mississippi R.
3. Mississippi River Gulf Outlet (MRGO)
4. Grand Isle
   a. West Jetty
   b. East Jetty
   c. Coast Guard Revetment
   d. Pirates Cove Marina
5. Bayou Lafourche Hurricane Protection Project
   a. Golden Meadow Floodgate
   b. La Rose Floodgate
6. Belle Pass Jetties
7. Calcasieu Lock
8. Vermilion Lock
9. Fontainebleau State Park
10. Holly Beach, La.
11. East Timbalier Island

Fig. 1. Vicinity map

This rock jetty was built on a lumber mattress to prevent excessive settlement.

10. In 1970 a 1400-foot (427 meters) rock revetment was built at the Coast Guard Station on the eastern end of the island. This revetment replaced a flexible grout filled revetment. The rock jetty was founded on a shell bedding and filter fabric.

11. In 1972 a 3000-foot (914 meters) terminal groin or jetty was built on the western end of the island. The armor stone was placed on a rock filter layer which, in turn, was placed on a shell bedding. Prior to placement of these layers a filter fabric was placed on the natural bottom.

12. In 1983 a lightweight interlocking concrete block revetment placed on a filter fabric was used to protect the banks or side slopes of a boat marina known as Pirates Cove.

13. Bayou LaFourche Hurricane Protection Projects. These projects include two floodgates and associated bank protection along a shallow depth (9 to 12 feet) (2.7 to 3.7 meters) navigation channel.

14. The bypass channel at Golden Meadow floodgate was protected in 1983 with large "jumbo" cellular concrete blocks weighing 115 pounds (52 kilograms) each, which were placed on a filter fabric.

15. In 1983 at the La Rose floodgate a 6-inch-thick (15.2 centimeters) interlocking concrete block was placed on a filter fabric to protect the banks from waves created by boat traffic.

16. Belle Pass Jetties. In 1980 extensions of the parallel jetties were constructed using rock armor over a shell core. These structures are located 70 road miles (113 kilometers) south of New Orleans. A filter fabric having a tensile strength of over 1000 pounds per inch (179 kgs/cm) was placed on the existing soft foundation prior to placement of the shell core. These jetties appear to be performing very well.

17. Vermilion Lock. This structure is also located along the Gulf Intercoastal Waterway (GIWW), 30 miles (48 kilometers) south of Lafayette, Louisiana. In 1984 the approach channels and the lock chambers will be protected with rock and filter fabric similar to the Calcasieu Lock protection.

18. Fontainebleau State Park. This project was completed in 1979 and was part of the National (Section 54) Low Cost Shoreline Erosion Control Demonstration Project. It is located on the north shore of Lake Ponchartrain, 30 miles (48 kilometers) north of New Orleans. Filter fabric was used extensively on this project which included several different type structures including concrete block revetments and timber pile used tire breakwaters.

19. Holly Beach (ref.2). This project is located on the Gulf of Mexico adjacent to Louisiana Highway 82, approximately 40 miles (64 kilometers) south of Lake Charles, Louisiana. Small concrete blocks were used in conjunction with filter fabric and performed well for approximately 10 years since 1970. Several recent storms caused severe damage to this

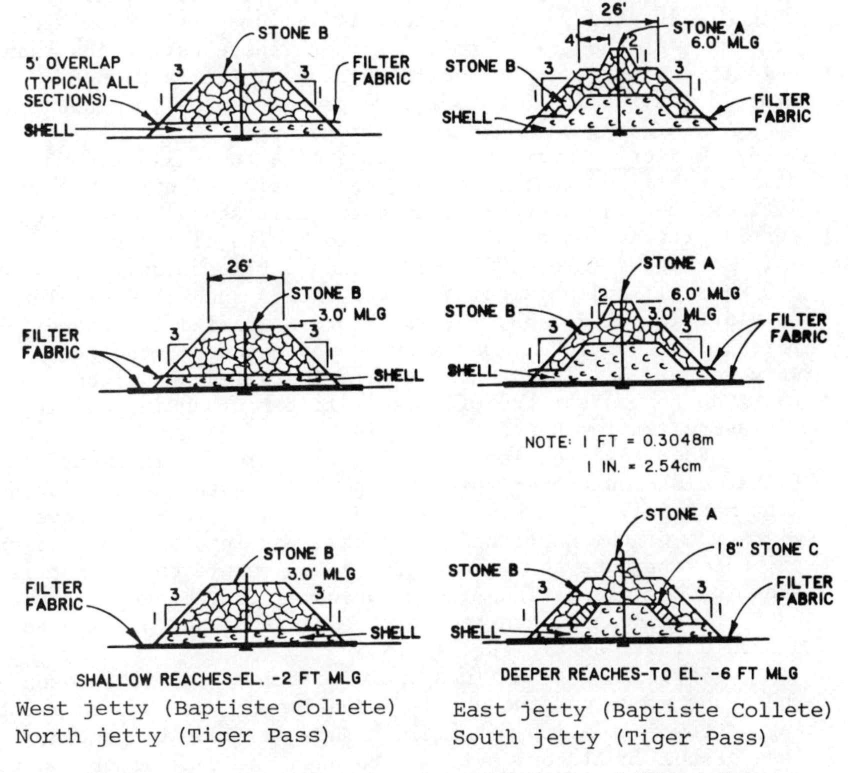

Fig. 2. Miss. River outlets, Baptiste Collete and Tiger Pass

revetment and a major 3 mile (4.8 kilometers) reconstruction of this project was begun in 1983. Several new concepts in concrete block revetments will be tested at this site. All of these blocks will be much heavier than the original blocks and will use filter fabric.

20. East Timbalier and Timbalier Islands. These projects are located approximately 60 air miles (96 kilometers) southeast of New Orleans and incorporate the use of filter fabric with rock armor revetments to protect the shorelines of offshore barrier islands.

SELECTED CASE HISTORIES

21. Three of the above mentioned projects using filter fabrics will be discussed in detail to illustrate lessons learned. These projects are the Mississippi River Outlets (Baptiste Collete and Tiger Pass Jetties), Southwest Pass of the Mississippi River, and the Mississippi River Gulf Outlet (MRGO).

Mississippi River Outlets

22. Location and description. This project includes two minor outlets or distributaries of the Mississippi River which are known as Baptiste Collette and Tiger Pass. They

intersect the main stem of the river at Venice, Louisiana, which is located 90 river miles (145 kilometers) below New Orleans. At the gulf terminus of both these navigation channels it was necessary to build twin shell fill and rock armored jetties in 1979 to assure proper depths to minus 16 feet (4.9 meters) mean low gulf (MLG).

23. Objective. In order to determine the validity of utilizing filter fabric in the construction of offshore jetties, several test sections were built at each jetty to test different concepts. Filter cloth with different tensile strengths (200, 400, 1000 pounds per inch (36, 71, 179 kgs/cm)) and physical characteristics such as equivalent opening size (EOS) (35, 70, 100) were specified for the different test sections. One section was nonwoven, the rest were woven. Settlement plates were installed along the center line to monitor the effects of filter cloth in reducing excessive settlement of the jetties.

24. Four test sections (see Fig.2) included placement of filter cloth on the existing bottom underneath the shell bedding material. This has a twofold purpose, i.e., to prevent migration of the shell and the rock armor into the foundation and to distribute the load of the jetty more evenly. Settlement in poor foundations cannot be completely prevented; however, filter fabric can reduce the magnitude and provide for a more uniform settlement.

25. Several test sections included filter fabric between the stone armor and the shell foundation or core and on existing bottom foundation materials. This was done to prevent loss of the shell core material through the rock armor and to distribute the load of the rock uniformly over the shell.

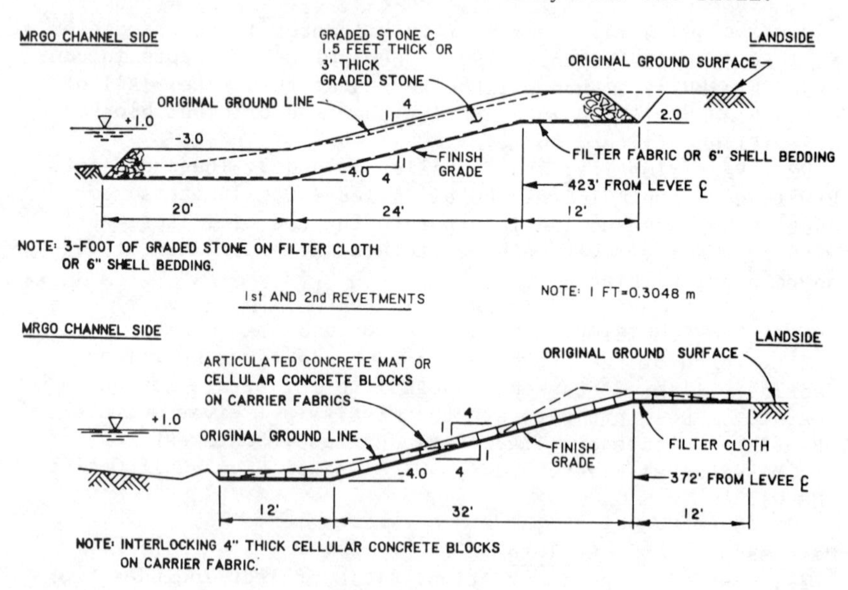

Fig. 3. Southwest Pass, Miss. River

26. Design criteria. The design wave used for rock armor was 4.5 feet (1.4 meters) with an associated wave period of 4.5 seconds. This wave criteria was based on a storm surge elevation or stillwater line of 4.0 feet (1.4 meters) above mean low gulf (MLG). Higher storm surges would tend to ride over the structure and be less critical to the jetty.

27. Soil conditions. Borings taken along the Baptiste Collette Bayou and Tiger Pass Jetty alignments indicate that marsh deposits, consisting of very soft clays, were encountered from ground surface to approximately elevations - 10 feet (3 meters) and - 20 feet (6.1 meters) NGVD, respectively. These marsh deposits were underlain by interdistributary deposits of very soft to medium clays with a few silt layers to the maximum depth of borings.

28. Settlement analysis. Based on the results of consolidation tests performed on undisturbed samples, it was estimated that the total foundation consolidation settlement would be approximately 4 feet (1.2 meters).

29. Performance of project and lessons learned. Jetty repairs were scheduled for 1983 to rebuild the jetties to design elevations which were 6 feet (1.8 meters) NGVD on the most exposed jetties and - 3 feet (0.9 meters) NGVD on the leeward jetties. Although settlement plates were included along both jetties, it was difficult to assess the total settlements due to conflicts in survey information. It appears that jetty settlement since 1979 varied between 1 to 3 feet (0.3 to 0.9 meters). How much is attributable to regional subsidence and how much is due to construction is questionable. All of lower Louisiana is subject to significant subsidence and it is difficult to ascertain the correct elevations of the benchmarks. Methods of placing the filter fabric on the existing bottom can be improved. Fascine mattressess (built with filter fabrics and willow chambers) such as used in Holland can be floated into position and sunk on site (by using high tensile strength filter fabric and providing rigidity). This approach should considerably improve the performance of jetty structures--constructed on soft foundations. Filter fabric can also be used to make the jetties more impervious on the land side to prevent loss of dredged and natural materials. Decreasing settlement by using filter fabric needs to be further evaluated by the construction of more detailed test sections.

Southwest Pass, Mississippi River

30. Location and description. Three separate test sections using lightweight concrete armor protection in place of typical rock dikes revetments were built in 1981 along the middle reaches of Southwest Pass. Southwest Pass is the major distributary of the lower delta, which connects the main stem of the Mississippi River with the Gulf of Mexico. Southwest Pass is located approximately 100 river miles (161 kilometers) south of New Orleans. The lower delta is experiencing major subsidence and losses of approximately

3000 acres (12,141 square kilometers) per year. The purpose of the dikes or revetments parallel to the river is to prevent erosion of the banks. The three test sections consist of a shell core overlain with filter fabric and armored with three types of concrete block mats.

31. The first section consists of 4-foot-wide (1.2 meters) mats, 20 feet (6.1 meters) long, attached or glued to a carrier strip of filter fabric. The individual concrete blocks are 8 inches square and 4 inches (10.2 cm) thick and weigh approximately 13 pounds (5.9 kilograms) each. This revetment weighs 30 pounds (13.6 kilograms) per square foot and the blocks have an open area of approximately 20 percent.

32. The second test section consists of a similar cellular concrete block which is attached to a carrier strip of filter fabric to form 4-foot-wide (1.2 meters) mats, 20 feet (6.1 meters) long. These blocks are considerably larger, having dimensions of 24 inches (61 cm) long by 16 inches (41 cm) wide and 6 inches (15 cm) thick. Each block weighs approximately 115 pounds and weighs approximately 45 pounds per square foot (2.2 kilopascals).

33. The third test section consists of 4-foot-wide (1.2 meters) mats that are fabricated in 25-foot (7.6 meters) lengths. These mats consist of solid concrete blocks that are 48 inches (122 cm) long by 14 inches (36 cm) wide by 3-5/8 inches (9.2 cm) thick and weigh approximately 140 pounds (64 kilograms) each. These blocks weigh approximately 30 pounds per square foot (1.4 kilopascals) and have been traditionally used by the U. S. Army Corps of Engineers to form large articulated concrete mattresses for deep underwater placement. They are usually placed by special barges that build continuous mats of interconnected blocks which are tied together with copper wire. See Fig.3 for typical cross sections illustrating the three test sections used in Southwest Pass.

34. Objective. Traditionally, rock dikes have been used in Southwest Pass; however, recapping or rebuilding the dikes to their original design elevations is often required. This maintenance can be significant in areas of soft in situ foundations. The purpose of these test sections was to test the viability of using a light armor revetment in lieu of the heavy 3-foot-thick (0.9 m) rock armors to reduce the settlement and consequent maintenance costs of the dikes.

35. Design criteria. The critical factor used to design these revetments was ship waves. The design wave was 4.0 feet (1.2 m) high which was based on observations and curves for predicting wave action in navigable waterways.

36. Specifications for filter fabric used in these test sections were EOS between 30-70 and a minimum tensile strength of 200 pounds per inch (36 kgs/cm).

37. Soil conditions along Southwest Pass vary; but, they can generally be described as marsh type soils consisting of very soft clays. No specific borings were taken at the test sites. The river stage varies from 1.0 to 4.0 feet (0.3 to

1.2 m) above NGVD.

38. Performance of project. Shortly after construction of the 13-pound (5.9 kilograms) cellular concrete block revetments, the mats were displaced by ship forces. All of the test sections were displaced by ship waves; however, the smaller blocks experienced the most severe damage and the most displacement. The toes of all three test sections were constructed to - 3.0 feet (0.9 m) below NGVD. Most of these test sections experienced uplifting of the toes and whole mat sections were often overturned. Mats of these types had often been used successfully along shorelines with waves approaching 4 feet (1.2 m), therefore other facts must be considered as having caused the significant damage. The concrete articulated Corps mats performed best which was attributed to the positive interconnection of the blocks with copper wire to form continuous mats above the water line. The larger cellular block (115 pounds) (52 kgs) performed considerably better than the small cellular blocks (13 pounds) (5.9 kgs), and this can be attributed to the larger unit weight. If the mats had been heavier at the toe or had been extended further below the water line (a minimum of two wave heights), damage could have been reduced.

Mississippi River Gulf Outlet (MRGO)

39. Location and description. Six separate revetments were built in December 1982 (along the south bank of the MRGO) in St. Bernard Parish approximately 20 miles east of New Orleans to test different types of armor. Fig.4 for typical cross sections.

40. The first revetment consisted of a 400-foot (122 m) reach of 36-inch-thick (91 cm) rock armor placed on a minimum 6-inch (15.2 cm) shell bedding layer.

41. The second revetment consisted of a 400-foot (122 m) reach of 36-inch-thick (91 cm) rock armor placed on large pre-sewn filter fabric panels 50 feet (15.2 m) wide and 56 feet (17.1 m) long. The filter fabric had an equivalent open size (EOS) of 50 and the percent of open area was less than 10 percent. Tensile strengths exceeded 300 pounds per inch (53.6 kgs/cm).

42. The third revetment consisted of a 500-foot (152 m) reach of 4-inch-thick (10.2 cm) interlocking blocks placed on the pre-sewn filter fabrics panels described above.

43. The fourth revetment consisted of a 400-foot (122 m) reach of 18-inch-thick (45.7 cm) rock armor using Class C stone or quarry stone gradation varying in weight from 5 to 400 pounds (2.3 to 182 kgs), with 50 percent lighter by weight varying between 18 and 100 pounds (8.2 to 45.4 kgs). This rock armor was placed on a shell bedding layer with a minimum thickness of 6 inches (15.2 cm).

44. The fifth revetment consisted of a 400-foot (122 m) reach of 18-inch-thick (45.7 cm) rock (Class C) placed on a large pre-sewn filter fabric panels. All of the test sections using filter fabric had the same physical characteristics.

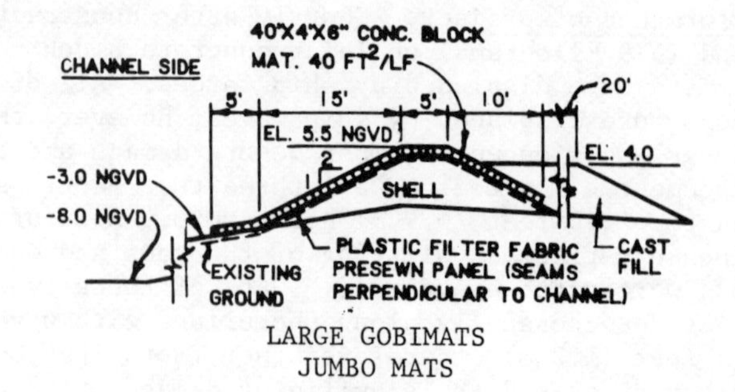

FORESHORE TEST SECTION TYPE "A"

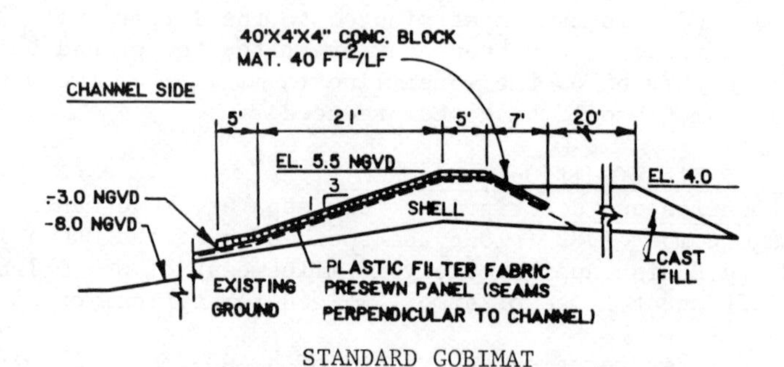

FORESHORE TEST SECTION TYPE "B"

NOTE: 1 FT=0.3048 m
1 IN.=2.54 cm

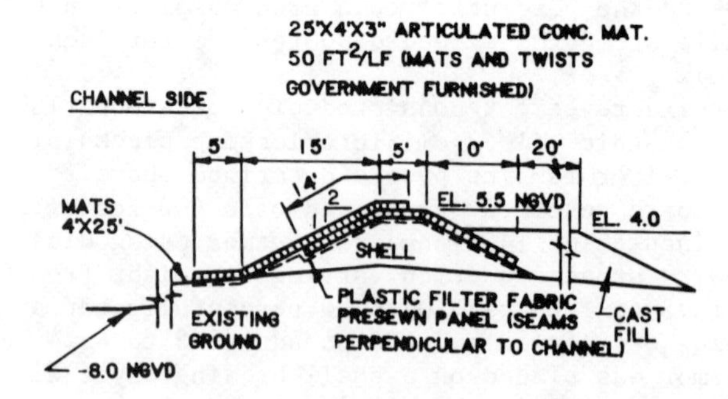

FORESHORE TEST SECTION TYPE "C"

Fig. 4. Miss. River Gulf Outlet

45. The sixth revetment consisted of a 500-foot (152 m) reach of 4-inch-thick (10.2 cm) articulated concrete mattress (ACM) furnished by the Corps of Engineers. This mat is similar to that placed along the Mississippi River as bank paving except that at this location it was not placed in a continuous width from the mat laying plant. The mat consisted of individual 25- by 4-foot (7.6 by 1.2 m) mats connected together above the water line. This mat was placed on filter fabric and was similar to that used on the test sections of Southwest Pass.

46. Objective. The purpose of the test sections was to test alternative designs to improve the overall costs of future wave wash protection for the MRGO. The length of the future protection project is approximately 20 miles (32 km), and is located adjacent to the present Chalmette hurricane protection levee. Due to soft in situ soils, significant maintenance costs have been experienced in nearby projects along the MRGO. The principal reason for high maintenance costs was attributable to settlement, therefore alternatives use of lightweight armor layers required investigation.

47. Design criteria. The mean tide level along the MRGO is about 1.0 foot NGVD (0.3 m) with a mean tide range of approximately 1.2 feet (0.4 m). Northerly winds can depress the Gulf of Mexico and cause stages as low as - 2.0 feet NGVD (0.6 m). The highest observed stage along the MRGO was in 1965 during Hurricane Betsy when 10 feet NGVD (3 m) was experienced. A recent winter storm on 20 January 1983 resulted in stages of about 8 feet NGVD (2.4 m) at the test sections. A design wave height of 4.0 feet (1.2 m) was selected for designing the rock armor and concrete block revetments.

48. Performance of project. Shortly after completion Dec 82 of the test sections, significant damage was experienced on all of the test sections with the exception of the first two which were built with 36-inch-thick (91 cm) rock armor. Damage was caused by rapid drawdowns and return flows and ship waves from large ships navigating the narrow channel. The width from bank to bank is approximately 1000 feet (305 m) and the channel is 500 feet (152 m) wide and 40 feet (12 m) deep.

49. The 36-inch (91 cm) rock test sections have performed well as far as displacement due to ship waves and return flows. An area of concern, however, is the extent of settlement which may occur over time. Settlement initially approached 1 foot (0.3 m) and occurred within two weeks after construction, indicating settlement into the soft foundations or loss of shell fill. The rock revetment appears to be protecting the immediate shoreline against wave attack. The bank behind the first two sections is relatively high which may have contributed to its better performance. Wave reflection and return flow could eventually cause significant scour of the underwater toe. Future monitoring of comparative cross sections will determine whether this is significant.

50. The interlocking concrete blocks suffered extensive

damage and many mats were overturned. The drawdown of ships caused an upward pressure to "float" the mats and the return flow allowed a "wall" of water to get underneath the edges of the mats and filter fabric and to further uplift the mats in the direction of the fast moving water. The concrete block revetment had approximately a 20 percent open area; this, however, could not relieve the uplift pressures due to lack of permeability of the filter fabric which had an open area of only 4 to 6 percent. Another major factor was probably the light unit weight of the blocks. It should be noted that damage was initiated at the toe of the mats which suggests the need for extending the toe deeper than - 4.0 feet NGVD (1.2 m) (recommend 2H below). Mats should be suitably anchored along all edges using heavier armor at and below the water line and a more pervious filter fabric should be used.

51. The 18-inch-thick (45.7 cm) Type C stone revetment suffered significant damage and this was due to inadequate thickness of armor layers and insufficient weight of the rock armor. Rock of this type should not be used where ship drawdown and return flows are significant. Drawdown and return flow were estimated to vary between 3 to 5 feet (0.9 to 1.5 m) and possibly higher. Many rocks in the revetment were displaced landward from 50 to 100 feet (15.2 to 30.5 m).

52. The 4-inch-thick (10.2 cm) articulated Corps mattress suffered extensive damage and many mats were overturned at the water's edge, indicating that uplifting forces initiated damage at the toe of the structure or at - 4.0 feet NGVD (1.2 m). This suggests that the toe should be extended deeper or that a heavier mat should be used. Mats weighing 50 to 70 pounds per square foot (2.4 to 3.4 kilopascals) are available and may provide sufficient stability for similar design conditions. Construction details such as burying the edges of the mat and anchoring the toe well below the water line would improve the performance of these structures.

SUMMARY

53. This paper briefly discussed case histories in Louisiana in order to examine factors that are important in using geotextiles in conjunction with flexible revetments. Space does not permit an exhaustive discussion of all the problems one encounters in the field. The following is a brief list of lessons learned on a few projects.

Mississippi River Outlets

54. Measurement of settlement in order to determine the performance of structures is often difficult in areas where subsidence and soft soils are common. Detailed monitoring should take place before, during, and after construction.

55. Fascine mattresses with high tensile strength filter fabrics should be investigated for supporting jetty structures. Simply placing filter fabric on soft foundations will not guarantee success.

56. Filter fabric can be used in conjunction with structures

(jetties and groins) to interrupt the movement of littoral material through pervious structures such as rock jetties. This can be instrumental in reducing maintenance of navigation channels.

Southwest Pass, Mississippi River

57. Lightweight blocks should not be used where ship drawdown and return flows are significant (exceed 3 feet) (0.9 m). Model tests can be used to determine required weights. It is estimated that flexible revetments should weigh a minimum of 50 pounds per square foot (2.4 kilopascals).

58. Filter fabric should perform as "filters" and not impervious membranes in revetment designs. Water permeability should be high. Other ways to achieve this is to specify percent open areas between 15 and 30, and EOS between 20 and 50.

59. Proper instrumentation should be designed to measure the drawdown and return flow along with waves created by large ships.

60. Quality control is important to insure that the mats are properly anchored and can articulate properly with the design forces encountered. Cables and interlocking methods should be adequate.

61. Toes of flexible revetments should extend to two wave heights below the water line or to depths that are sufficient to prevent uplifting of the mats by ship waves and forces or be anchored properly or use of heavier blocks or rocks along the toe.

Mississippi River Gulf Outlet

62. Small rock (quarry stone gradation) or thin rock armor layers do not perform well where ship forces in narrow channels are significant.

63. Rock designs without filters or filter fabric do not articulate as mattresses. In soft foundations this can often cause rock to settle nonuniformly and can destroy the integrity of the revetment, causing large maintenance costs.

64. Experience in 1969 at Holly Beach and other similar projects demonstrated that filter fabric should have a high retention for the local soil and that both filter fabric and armor cover should have a high ratio of open area in relation to the bank soil material under it; i.e., the percent open area should exceed the porosity of the bank soil material to relieve hydrostatic pressures.

BIBLIOGRAPHY

1. U. S. Army Corps of Engineers. 1978. "Mississippi River Outlets Vicinity of Venice, La.," General Design Memorandum, Supplement No. 1, "Jetties Design," U. S. Army Engineer District, New Orleans, La. (March).

2. Dement, L. E. 1976. "Two New Methods of Erosion Protection for Louisiana," (June).

18 The use of geotextiles impregnated with bitumen in situ as bank revetment completion of a section of the Milano-Cremona-Po inland waterway

G. DELLA LUNA, Eng., Consorzio Canale Milano-Cremona-Po, Cremona,
D. A. CAZZUFFI, Eng., Research Center for Hydraulics and Structures–ENEL, Milano, and M. CEPORINA, Eng., Geotextiles Group O.R.V., Padova

SYNOPSIS. The paper deals with the works made for the bank revetment completion of a section, about 1,000 m long, of the Milano-Cremona-Po inland waterway: in this area a particular geotechnical situation involved some specific problems. The waterproofing and the continuity of the banks revetment are ensured by a system of geotextiles impregnated with bitumen "in situ", by means of a simple equipment purposely arranged. The construction of the flexible armoured revetment including geotextile has given some very interesting technical-economic results.

INTRODUCTION

The inland waterway from the Po river to Milano

1. The city of Milano has been engaged with the design of a waterway connection with the Po river during all the centuries of its long history. In the ancient times, a communication was obtained by means of the tributaries of the Po (Adda, Ticino and Lambro rivers) and of artificial inland waterways the historical "Navigli". built between the city and the above rivers.

2. In the present time, this design has been undertaken again and foresees an inland waterway, 63 km long, which originates from the Po river just upstream of Cremona and ends at the South side of Milano. The difference of level between the Po and the port of Milano is of about 60 m and will be overcome by means of 8 navigation locks. The inland waterway has to cross the Adda 18 km from its starting point by a canal-bridge, which represents the most engaging structure of the whole design. Construction of ports is planned along the inland waterway at Cremona, Pizzighettone, Casalpusterlengo, Lodi and Milano.

3. So far, the first 15 km of the inland waterway have been built or are under completion between the Po and Adda rivers, with the first two locks of Cremona and Acquanegra. The port of Cremona was built too and the port of Pizzighettone is in progress.

4. The inland waterway waterproofing. One of the most important characteristics required by the Milano-Cremona-Po inland

waterway is a complete waterproofing: the design of bank and bottom protections has always a primary role in the inland navigation fairways design, in which cost-benefit analysis plays an important function too (ref. 1).

5. In the considered waterway the complete waterproofing has become necessary for the two following reasons:
a) to reduce water consumption;
b) to prevent any interference with the surface flow, on which the production of the richest agricultural land in Italy depends.
In port basins, where the area covered by water is greater, waterproofing is obtained by means of peripheral concrete seepage cutoffs, on which the toe of the bank revetments is connected. The whole cross-section along the inland waterway is lined instead with waterproofing revetments.

6. Typical sizes of the waterway cross-section in the two so far built sections are as follows:

- water level width	m	38.50 ÷ 41.50
- bottom width	m	28.00
- depth	m	3.80
- bank slope		2/3 ÷ 1/2
- wet section	m^2	120 ÷ 128

7. The waterproofing revetment in the first inland waterway section (about 8 km long) was made of bituminous concrete (ref.2), spread in three layers consisting of:
- a first layer of pervious mixed bituminous gravel (thickness 0.08 m);
- a second layer of impervious bituminous concrete (thickness 0.05 m);
- a third layer of impervious bituminous concrete (thickness 0.04 m).

The waterway cross-section in the first section and a detail of the revetment are shown in Fig.1.
The impervious layers content was 8 parts of bitumen on 100 parts of aggregate by weight. The percentage of voids by volume after compaction was lower than 3%.
Many experimental attempts were performed in order to realize a sealing coat on these revetments. The following processes were tested specifically:
- treatments with bituminous emulsions;
- treatments with different types of tar mastic and pitch;
- treatments with epoxy resins.

Little encouraging or quite unsatisfiable results were obtained from all these treatments. Bituminous concrete revetments have given fairly good results. However, they showed stability limits when placed on banks with great slope (2/3) as well as a very lowered strength to effects due to uplifts.

8. The second section of the inland waterway, about 7 km long, was built from 1979 to 1983, with geometrical characteristics similar to those of the first section. A different type of revetment has been used for this second section, because of the material cost variation occurred after the first section construction and of experiences made during the said works (ref.3). Banks were lined with concrete slabs, 0.15 m thick,

reinforced with electric welded steel meshes,or with prestressed concrete slabs, 0.05 m thick, placed on the concrete base (see Fig.2). For the bottom waterproofing, a bituminous concrete layer, 0.06 m thick, was spread on a compacted gravel base by means of road finishers.

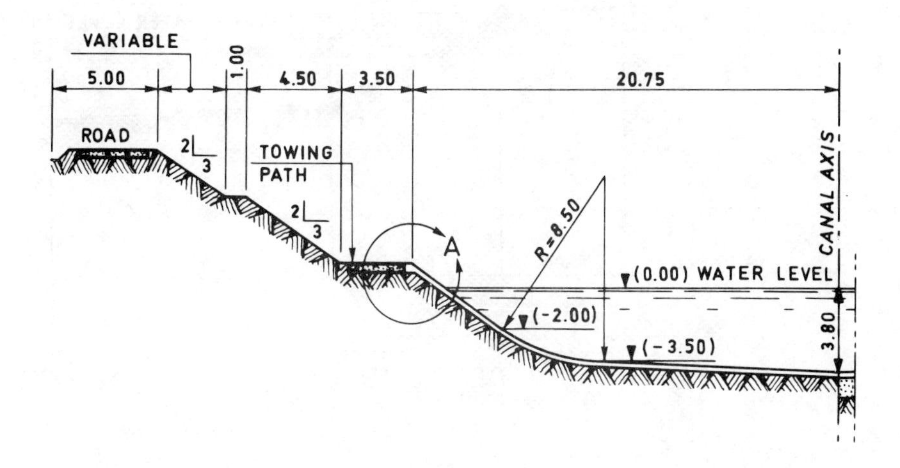

DETAIL "A"

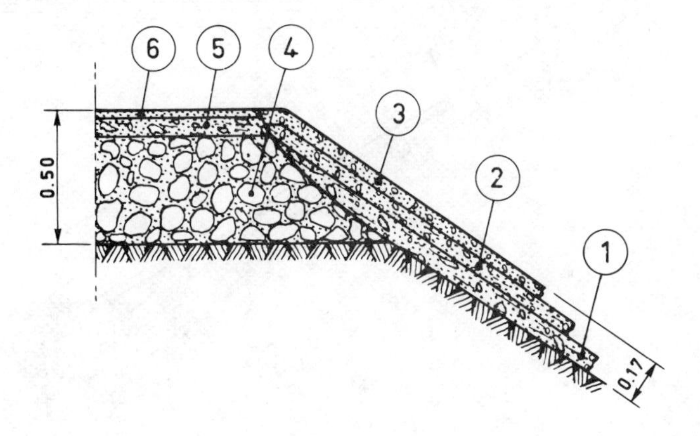

Fig.1. The inland waterway cross-section in the first section (above) and a detail of the revetment (below).

1. Pervious mixed bituminous gravel (thickness: 0.08m)
2. Impervious bituminous concrete (thickness: 0.05m)
3. Impervious bituminous concrete (thickness: 0.04m)
4. Compacted gravel base (thickness: 0.04m)
5. Pervious mixed bituminous gravel (thickness: 0.08m)
6. Impervious bituminous concrete (thickness: 0.02 m)

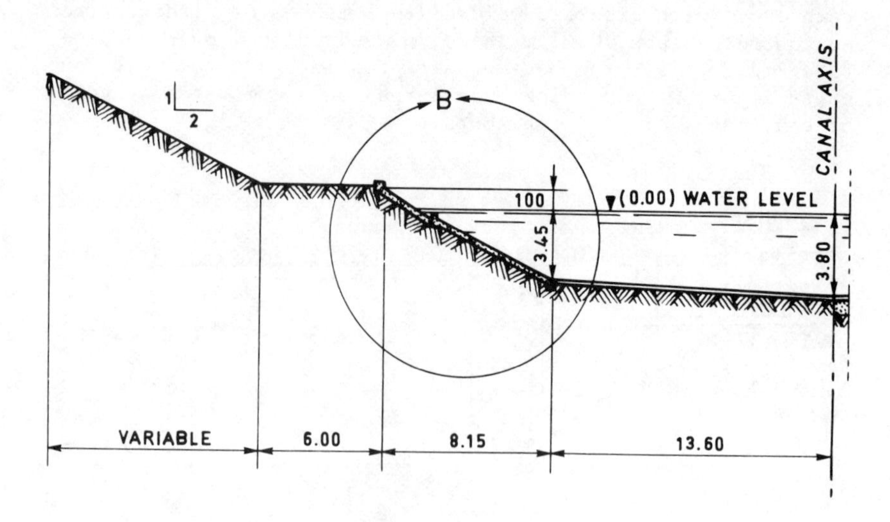

DETAIL "B„

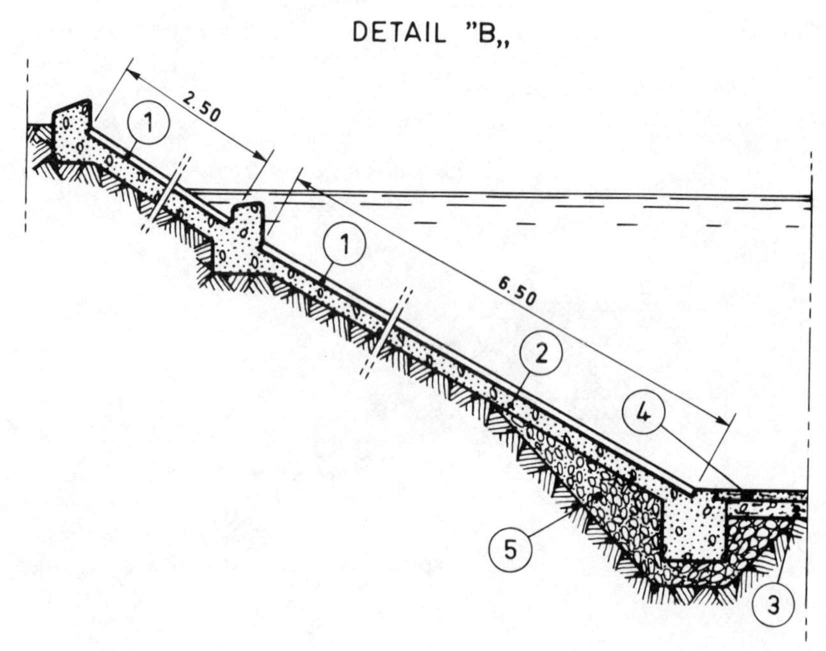

Fig.2. The inland waterway cross-section in the second section (above) and a detail of the revetment (below).

1. Prestressed concrete slabs (thickness: 0.05m)
2. Concrete base (thickness: 0.15m)
3. Compacted gravel base (thickness: 0.15)
4. Impervious bituminous concrete (thickness: 0.06m)
5. Granular material

THE CONSIDERED INLAND WATERWAY SECTION

1. The inland waterway section, where the application of flexible armoured revetments including geotextiles was made, is placed just upstream the double lock of Acquanegra, includes the outer port of the same lock and therefore presents cross-sections with variable width.

2. This section was wholly realized by an embankment from 1968 to 1973. Excavation volumes of the waterway section placed downstream the double lock of Acquanegra, built during the same years, were used. Embankment height in respect of country level ranged from 6.00 to 6.50 m at bank level and from 1.00 to 1.50 m at bottom level.

3. During the same period, a part of the bank and bottom revetment, specifically the first two layers consisting of bituminous concrete, 0.08 and 0.05 m thick, was built too according to the first waterway section (see Fig.1).

4. The problems causing a delay in completion of the bank revetment in this section were quite only of a geotechnical nature.

THE GEOTECHNICAL PROBLEMS

1. The considered waterway section presents a rather uniform stratigraphy along a development of 1,000 m. It consists of the following layers:

a) from the country level to a depth of about 0.50 m: silty clay, meanly hard, with the presence of organic substances;
b) from 0.50 to 5.50 m: silty peat, very soft;
c) beyond 5.50 m: sand.

Groundwater level is at a depth of about 0.90 to 1.00 m.

2. In the presence of such a stratigraphy, geotechnical problems obviously arose because of the settlements due to peat bed consolidation during the embankment construction.

3. The results obtained from consolidation tests made at the Geotechnical Laboratory of the Technical University in Milano, gave a vertical consolidation coefficient c_v for peat equal to:

$$c_v = 2.10^{-7} \ m^2/s \qquad (1)$$

whereas unconfined compression test showed a very low initial cohesion value equal to:

$$c_u = 5 \ kPa \qquad (2)$$

4. Therefore, it had been necessary to build the embankment by degrees, in successive layers, so that it was possible to take advantage from shear strength development due to peat consolidation during the previous load increase.

5. The consolidation process of soil bed was speeded up by means of drainage trenches, 5.00 m distant. In this way, a drainage of the peat layer was performed towards the outside too, by trenches connected one another on the upper side by a

sand bed forming the embankment base.

6. The embankment completion was reached by 5 successive load increases, at an average interval of about 70 days one from another (ref.4). This interval corresponds to the theoretical time necessary for reaching 50% of the primary consolidation obtained by applying the well known formula which gives the time of consolidation t depending on the time factor T_v, the half layer height H to be consolidated and the coefficient of consolidation c_v:

$$t = T_v \cdot \frac{H^2}{c_v} \tag{3}$$

In the case under examination, since drainage of the peat bed was possible upwards and downwards (H = 2.50 m), it was found:

$$t_{50} = 0.2 \cdot \frac{2.50^2}{2.10^{-7}} \text{ sec} \simeq 72 \text{ days} \tag{4}$$

7. The first embankment layer was 1.10 m high on the whole width of the inland waterway. This value corresponds to the maximum height of embankment which can be built without causing cracks on a peat bed of 5.0 m with a cohesion value of 5 kPa, calculated according to the method proposed by Jakobson (ref.5). The next layers were built according to the following progressive heights: 1.80 m, 3.10 m, 4.60 m and 6.00 ÷ 6.50 m.

8. The application of such methods enabled embankments to be built in their whole height without causing any failure of the peat bed. During the embankment construction, settlements were measured. For this purpose, 39 levels (three for each section), the plates of which was embedded in the embankment base sand, were placed. A constant survey of settlements made it possible to find some important differences from one point to another and to follow the slow course of this phenomenon.

9. After about 300 days from the embankment completion, the primary settlement of the peat was practically ended. However, the secondary settlements, exhaustion of which was foreseen after many years only, were not negligible because of the peat bed presence.

10. At the end of embankment construction, it was decided to carry out a soil preload in the middle side of the inland waterway, so that the most of settlements due to a waterway filling, planned of 4 m, occurred in advance. In fact, the preload application, kept for two years (1969-71), made it possible to obtain the whole primary settlement corresponding to the further water filling; the secondary settlement effect, particularly important under the banks, was to be taken into account.

THE BANKS FLEXIBLE ARMOURED REVETMENT

1. After the embankment completion and the further preload application, the construction of bottom and bank revetment was

started; at the beginning it consisted of three bituminous concrete layers, according to the same design of the first waterway section (see Fig.1).

2. It was prudently decided to postpone the construction of the last impervious revetment layer to the beginning of waterway filling because of the importance of the secondary peat bed settlements. In fact, it was expected that the development of secondary settlements and the presence of differential settlements at the concrete structure level would have caused some cracks.

3. The experimental revetment. At the end of the '70, when settlement progress had already shown some crack systems, more important at the points of conjunction to engineering structures, it was decided to perform an experimental lining.

4. The aims to be reached were defined as follows:
- waterproofing and continuity of the revetment;
- absorption capacity of small tensile stresses to prevent destructive strains (as for bituminous concrete);
- saving in construction and maintenance, specifically concerning yard equipments.

5. Among the various possibilities considered for example by Gamsky (ref.6), a revetment consisting of geotextile and bitumen to be carried out "in situ" had been chosen, so that the bituminous primer was a first sealing treatment for microcracks occurred in the bituminous concrete layers spread ten years before.

6. Such a revetment was tested in summer 1981 on a bank surface (1,000 m^2) exposed to South. This experimental application, kept under observation for two years, involving two complete cycles of max thermal ranges, showed quite positive answers.

7. The completion of the revetment. The revetment already tested was carried out in summer 1983 as follows:
- Preparation and cleaning of the bank: the bank cleaning was made by blown compressed air, so that incoherent parts of the surface were removed and the anchorage of the revetment was easily made.
- Primer placing: a bituminous primer at low penetration (40 $\div$ 50) of about 1 kg/m^2 was hot-sprayed (temperature 180°C - 200°C) on the surface so prepared, using an equipment purposely arranged, making it possible to act on the whole development of the bank (see Fig.3).
- Geotextile laying: nonwoven sheets, 5.50 m width, were placed on the still hot bituminous surface (see Fig.4). Even if a horizontal laying would have reduced the total length of joints, a vertical laying of the sheets was preferred so that no cuts or folds occurred, because of mixed line development of banks close to the lock.
- Finishing of the revetment: further bitumen at low penetration (40 $\div$ 50) of about 3 kg/m^2 was sprayed, in order to obtain the completion of the revetment (see Fig.5). At last fine dryed sand of about 1.5 kg/m^2 was spread so that the exceeding bitumen was fixed.

A machine was used in order to improve impregnation quality and to reduce construction times. The equipment consists of a simple pipe provided with nozzles which run through the whole bank length and enable the bitumen, introduced from a thermal tank by a volumetric pump, to be uniformly spread.

Fig.3. The hot-spraying of the bituminous primer on the bank.

Fig.4. The laying of a geotextile sheet on the bituminous primer.

Fig.5. The completion of the flexible armoured revetment by hot-spraying of further bitumen.

- Joint construction: in order to ensure a complete waterway waterproofing the geotextile sheets laid vertically were sealed with bituminous elastomerized membranes (0.30 m width) fire-revived and overlapped to the joint points already bituminized.
- Realization of the revetment anchorages to the bank edge and to the waterway bottom: the geotextile sheets were anchored to the bank edge in a longitudinal furrow and, after impregnation, were covered with gravel; then, a concrete curb was built in the furrow so prepared (Fig.6a).
  As to the junction between the bank revetment and the waterway bottom, a finishing layer, consisting of a bituminous emulsion primer of about 0.8 kg/m^2 and of a bituminous concrete course (25 mm thick), was overlapped to the geotextile already impregnated for a section of about 0.25 m (see Fig.6b).
- Realization of the junctions between revetment and structures: a special problem arose from the junctions between the impervious revetment and the engineering structures.
  In the considered section, this problem appeared downstream at the point where the waterway is connected to the lock walls and in a middle point where it is underpassed by a siphon. Both structures, being founded on piles, were little affected by secondary settlements of the peat bed.
  This fact caused differential settlements between the area on the concrete structures and the remaining area of the waterway bottom, made evident by large cracks of the bituminous concrete revetment at the concrete wall level.
  Joint construction by a bitumen-rubber mix suitably designed

would have enabled the waterproofing to be ensured, facing differential settlements, as shown for a similar case by Cazzuffi-Puccio-Venesia (ref.7).
As in the Milano-Cremona-Po inland waterway we only had to make a restoration intervention, we decided to adopt a different solution, based on geotextile use: sheets at the discontinuity line level were folded, according to the arrangement shown in Fig. 7, and impregnated with bitumen only on the external face.
A new covering was used on the waterway bottom by the same bituminous concrete of the impervious layer; one more covering of the banks was made of the same flexible armoured revetment incorporating geotextile.

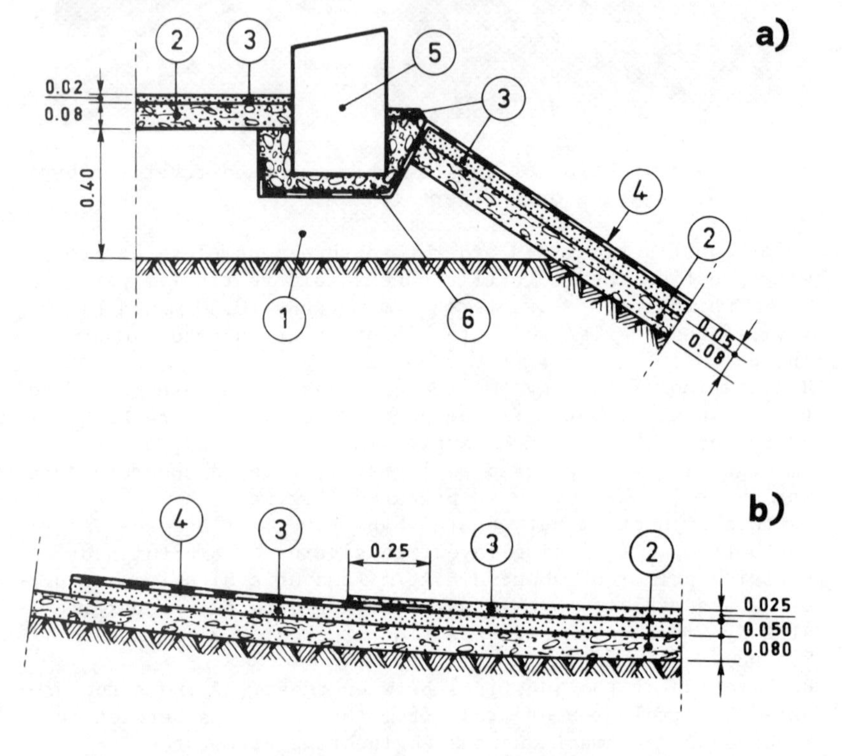

Fig.6. Details of the revetment anchorages to the bank edge (above) and to the waterway bottom (below).

1. Compacted gravel base
2. Pervious mixed bituminous gravel
3. Impervious bituminous concrete
4. Flexible armoured revetment incorporating geotextile
5. Concrete curb
6. Granular material

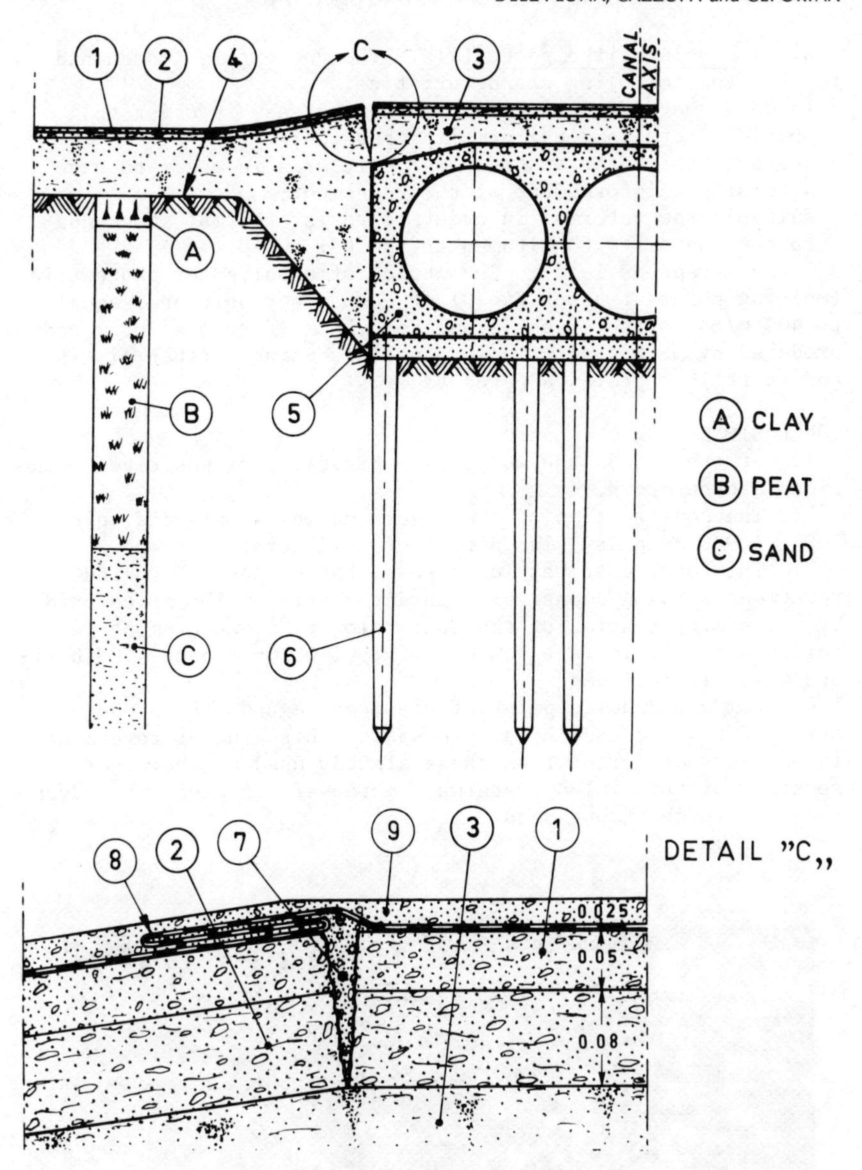

Fig.7. Cross-section of the waterway bottom point affected by differential settlements (above) and a detail of the restoration intervention (below).

1. Impervious bituminous concrete
2. Pervious mixed bituminous gravel
3. Sand
4. Country level
5. Concrete structure
6. Piles (length: 6.00m)
7. Crack filling by bituminous concrete
8. Geotextile impregnated with bitumen
9. Impervious bituminous concrete

8. The role of the geotextile. The geotextile chosen had to show the following characteristics:
- large diameter fibers for porosity increase in order to obtain a total and uniform impregnation;
- suitable thickness and mass per unit area in order to have a tensile reinforcement of the revetment;
- suitable raw material in order to resist thermal shock due to the contact with bitumen at 180°C ÷ 200°C.

All these reasons lead to the choice of a polyester geotextile (melting point at about 240°C) of a mass per unit area equal to 400 g/m^2, of thickness equal to 3.5 mm (for σ = 2kPa) and produced by mechanical needle-punching, using a fiber mix (6 and 15 tex) in prefixed percentages.

CONCLUSIONS

1. At the end of the works (see Fig.8), some positive conclusions can be drawn.

2. The construction of the revetment was simple and quick (700 - 1,000 m^2 a day with a team of 4 workers).

3. The total cost was lower than that of any other kind of revetment showing comparable characteristics. The price paid by the Administration of the Consorzio of Milano-Cremona-Po inland waterway in fact was of It.L. 6,500/m^2 (about 4 US Dollars/m^2), all included.

4. From a hydraulic point of view, as regards the propagation of waves caused by ship passage, this kind of revetment is not very different from those already used on the other sections of this inland waterway: moreover, it gives the advantage of a higher roughness.

Fig.8. A general view of the yard at the end of the works (in particular the mixed line development of the banks close to the Acquanegra lock may be seen).

5. From a geotechnical point of view, the flexible armoured revetment incorporating geotextile will be able to face the effects due to future secondary settlements, keeping waterproofing and continuity.

6. As regards maintenance problems, the former experience makes clear that, whatever restoration occurs, it will be limited and easy to do.

REFERENCES

1. BLAUW H.G. and WERHEY H.J. Design of inland navigation fairways. ASCE Journal of Waterway, Port, Coastal and Ocean Engineering, 1983, 1, February, 18-30.
2. VAN ASBECK W.F. Bitumen in hydraulic engineering. Vol. 2. Elsevier, Amsterdam, 1964, 70-96.
3. DELLA LUNA G. and AGOSTINI R. New methods of countering erosion generated by vessels in canals in Italy and other countries. Proceedings of the XXV Congress Inland & Maritime Waterways & Ports, Edinburgh, Vol. 1, 1981, May, 71-87.
4. MEARDI G. Relazioni geotecniche zona torbosa a monte della biconca di Acquanegra. Consulting report, 1970, June.
5. JAKOBSON B. The design of embankments on soft clays. Geotechnique, 1948, 2, December, 80-90.
6. GAMSKI K. Parts of waterproofing membranes. General report. Proceedings Colloque sur l'Etanchéité des Ouvrages Hydrauliques, Paris, Vol. 2, 1983, February, 61-73.
7. CAZZUFFI D.A., PUCCIO M., VENESIA S. Model study of a joint of bitumen-rubber mix for hydraulic works. Proceedings Colloque sur l'Etanchéité des Ouvrages Hydrauliques, Paris, Vol. 1, 1983, February, 53-57.

19 ProFix mattresses–an alternative erosion control system

W. H. TUTUARIMA, MSc(Civ.Eng.), Zinkcon International BV, The Netherlands, and W. van WIJK, MSc(Civ.Eng.), Amoco Fabrics, The Netherlands

SYNOPSIS. The ProFix mattress is a new and flexible type of revetment system, developed from experience gained at the Delta-plan project in the Netherlands. Based on tests, design criteria are given for the mattress. Moreover filter criteria are given for the geotextiles used to construct the mattress. The first large project executed with this system is in Nigeria. In total 1.1 million m^2 embankment is planned to be protected with the ProFix mattress. The project is described briefly and also the experiences until now.

INTRODUCTION

1. The application of mattresses in bed- and bank protection works has grown tremendously in the Netherlands during the last decades. The execution of the Dutch Delta-plan has been one of the major fields where comprehensive application took place. Extensive application also took place in bank protection works of navigation canals and other waterways. This led to far-reaching improvements and mechanization in the manufacturing of classical mattresses on the one hand, and the development of new types of mattresses on the other. The sandfilled mattress is one of the latter types. The mattress is formed by two layers of polypropylene filter cloth, sewn together at predetermined intervals, forming tubes which will be filled on site with e.g. local available sand. The mesh openings of the geotextile are carefully matched to the grading of the used sand to ensure sand tightness. Since the mattress is often filled with sand, it has a good water permeability and excellent filter properties. The required strength of the filter used for the mattress depends on the slope, the exposed loads, method of construction, the thickness and weight of the fill material. Moreover the upper cloth is stabilized against ultra violet radiation. It is provided with a felt layer to promote and develop vegetation giving extra protection against u.v.-radiation. The cloths allow plant roots to penetrate into the subsoil thus providing extra stability to the construction. The sewn seams of the ProFix mattress are shielded from mechanical damage due to the strong curving of the cloths.

The distance between the seams depends on the required height of the mattress.

CHARACTERISTICS OF THE PROFIX MATTRESS

2. The requirements of the fabrics of which the ProFix mattress is constructed depend largely on the design and the site conditions. For proper functioning of the ProFix mattress it has to be designed by taking into account the following properties:
- filter requirements
- permeability
- tensile strength
- durability

Filter function

3. The top and bottom fabric of the mattress have to act as a filter in order to prevent loss of fill material. The wanted maximum opening size of a fabric can be calculated with the formulas which are developed through the work of Teindl (ref.1) and the Franzius Institute (ref.2). This work has led to the following recommendations.

For cohesive soils:

$$090 \leq 10 \times D50$$
and
$$090 < D90$$

For uniform non-cohesive soils (U < 5):

$$090 < 2,5 \times D50$$
and
$$090 \leq D90$$

For well graded non-cohesive soils ($U \geq 5$):

$$090 < 10 \times D50$$
and
$$090 < D90$$

For soils having little or no cohesion and more than 50% by weight of silt Calhoun (ref.3) has recommended:

$$090 \leq 100\ \mu m$$

090 = opening size of fabric

U = coefficient of uniformity = $\frac{D60}{D10}$

Investigations in Holland (ref.4) have shown that for the correlation between 090 and D90 we may take

$$090 < (1 \text{ à } 2)\ D90$$

Therefore the above mentioned correlation is very conservative and gives a possibility to be more flexible in the design of the fabric.

Permeability requirement

4. The permeability of the soil may be estimated from the empericism K soil = $[D10\ mm]^2 \times 10^{-2}$ m/s, or the formula of Kozény:

$$K = \frac{1}{\alpha} \cdot \frac{\varepsilon^3}{(1-\varepsilon)^2} \frac{g.D}{\gamma}$$

in which:
α = dimension less coefficient, $\alpha \approx 500$ (-)
g = porosity for sand $\varepsilon \approx 0,4$ (-)
D = characteristic grain size, D = D20 (m)
γ = kinematic viscosity, $\gamma = 1.3 \times 10^{-6}$ m^2/s at 10 Centigrades

When the permeability of the fabric is in the same order of magnitude as the fill material or subsoil, then there will be no problems with the occurance of excessive overpressures under the mattress.

Durability

5. It is a well-known fact that plastics are all sensitive to degradation when they are exposed to e.g. u.v.-radiation, temperature, water and oxygen. Since the ProFix mattress is constructed from polypropylene fabrics this is a matter of concern. Also an important factor is the chemical resistance. Polypropylene is unaffected by soil chemicals, acids and alkalies over a pH range of 3 to 12. The life time expectancy we will divide into two categories, i.e.
1) u.v.-life time
2) thermo-oxidative life time

6. u.v.-life time. Aging of polymers is caused by the u.v.-section of the light rays. The rate of aging is determined not only by the intensity of the radiation but also by temperature and humidity. The intensity of the radiation is expressed as an annual irradiated energy on the surface of the earth. The unit for this is kLy (kiloLangley). 1 kLy is 1 $kcal/cm^2$ irradiated energy. In Fig.1. we see the annual energy distribution in kLy on the earth. In Northern and Middle Europe there is an annual energy incident of 60-80 kLy (i.e. 60-80 $kcal/cm^2$/year). In Nigeria e.g. the energy incident is approximately 140 kLy. This means that it is very important to know in what area of the world the material will be used. Polypropylene can be stabilized against u.v.-degradation to the required degree. Since outdoor exposure tests would take too much time, weathering devices have been developed to provide accelerate weathering tests. Several types of test equipment are available, Amoco Fabrics use a Xenotest 1200 for their accelerated weathering tests and the tests are carried out according to the specifications of ASTM G 26-70. Although it is very difficult to extrapolate the test results to outdoor exposure one can say that approximately 10-20 hrs in Xenotest 1200 corresponds with 1 kLy outdoor exposure. The degradation is generally noticed as a change in colour and deterioration of properties such as surface cracking and reduction of tensile strength. The results of the artificial weathering tests are normally expressed in time necessary to reduce the tensile strength

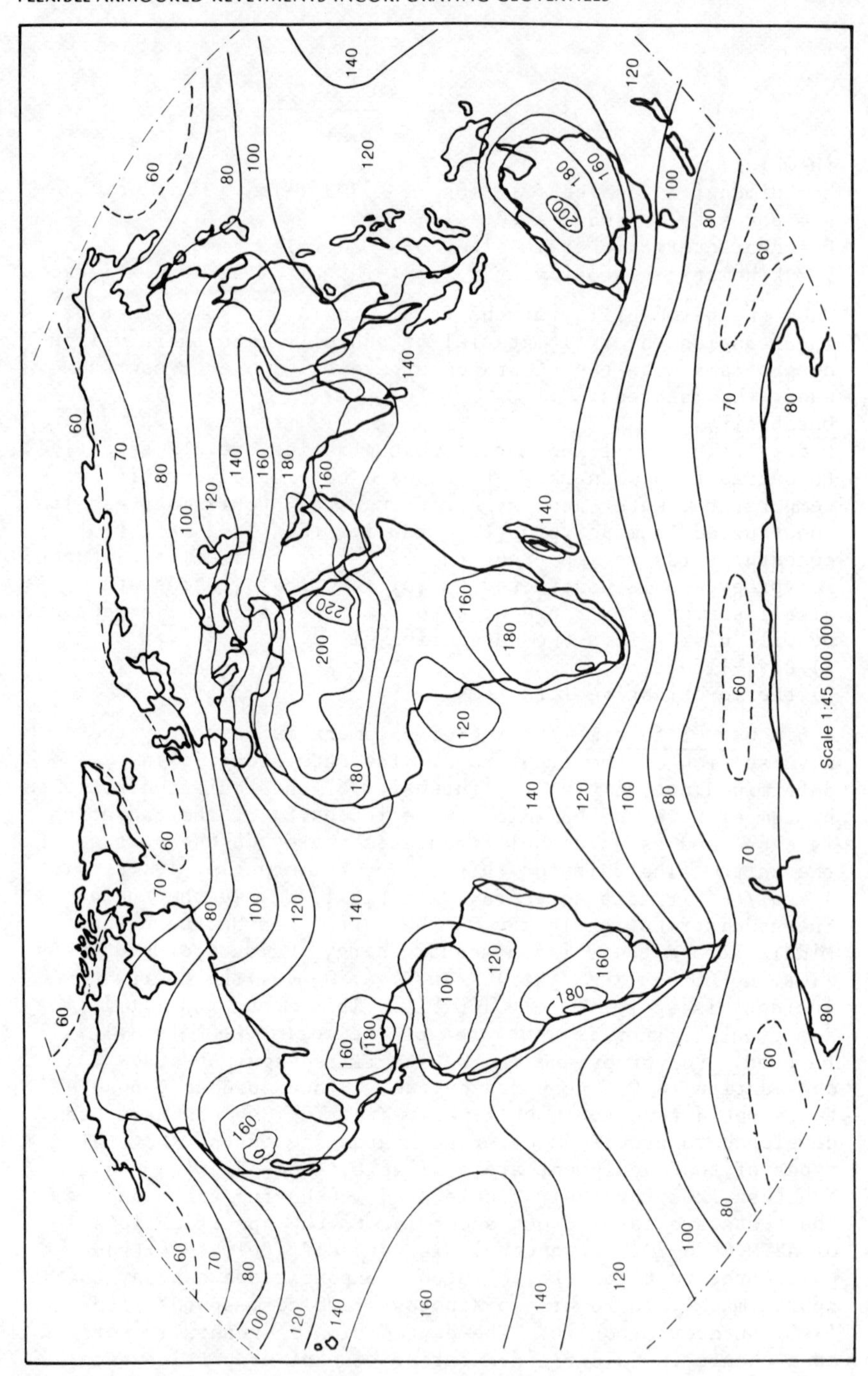

Fig.1. Generalized isolines of global radiation in kLy ($kcal/cm^2/year$)- After M.G. Landsberg

by 50% or by 10%. Since 10-20 hrs in the Xenotest 1200 corresponds to 1 kLy outdoor exposure it is possible to get an idea of the life time expectancy in a certain area. The ProFix mattress has now been used for approximately two years in Nigeria. Since a good part of the mattress is exposed permanently, the top fabric has been given an extra protection against u.v.-radiation. This extra protection is a highly u.v.-stabilized nonwoven fleece needled onto the top fabric. The function of this felt layer is also to enhance the start of vegetation. In the felt layer sand dust will be very easily trapped, this will give again extra u.v.-protection. After two years exposure in Nigeria until now the fabrics of the mattresses are still performing satisfactory.

7. Thermo-oxidative life time. For projects that will have to last long periods it is necessary to be able to give a life time expectancy of the material. Especially when a life time expectancy is wanted of over a hundred years it is necessary to have an accelerated test to determine this period. For a life time of this length one must be sure that the material resists oxidation degradation sufficiently. The normal polypropylene fabrics may well have a life time expectancy of at least 60-100 years, it is however not yet possible to guarantee this because these materials have only existed 20-25 years. Tests have shown that after burial polypropylene fabrics can lose 10-24% of their original strength after 10-14 years (ref.4). To enable us to give a reliable life time expectancy for a period of 200 years, TNO has developed a special test (ref.5). The test is carried out in an oven on 150°C temperature. The oven life test consists of two parts, one test is carried out on tapes as they are taken out of a sample. These tapes have to resist the thermo-oxidative degradation at 150°C temperature for more than 12 days. Another test is carried out on tapes after they have been extracted for 7 days in boiling seawater. After this treatment the tapes have to resist at least 7 days in the oven at 150°C. When the tapes meet these requirements then the life time expectancy is put at 200 years. In order to meet the stringent demands special antioxidant systems have been developed to fulfil these requirements. This so-called heatstabilized fabric is now the basis for the ProFix mattress.

STABILITY TO FLOW AND WAVES

Stability in flow conditions

8. When the shear stress τ_b, exerted by the flow, exceeds the frictional resistance force F of the mattress at the bed, the stability becomes critical. The force F can be formulated as:

$$F = f.N \qquad (1)$$

in which:
f = frictional factor ($0.3 < f < 0.8$)
N = submerged weight - average lift force due to turbulent flow - uplift due to ground water pressure.

From pressure measurements in a flume, Einstein and El-Sami (1949) concluded that the average lift force Δp due to turbulent flow can be related to a reference velocity U_r at some distance above the bed. Taking into account the relationship between the flow velocity and the shear stress τ_b, it can be concluded that a reasonable estimation of the lift force due to the flow can be expressed as:

$$\Delta p = 5\ \tau_b\ (N/m^2) \qquad (2)$$

The uplift of the mattress P due to the ground water pressure can be estimated from the expression:

$$P = \rho w.g.d.\frac{Ks}{Km}\ .\ i_s\ (N/m^2) \qquad (3)$$

in which:
ρw = density of water (kg/m^3)
g = gravitational acceleration (m/s^2)
d = average height of the mattress (m)
Ks = coefficient of permeability of the subsoil (m/s)
Km = coefficient of permeability of the mattress (m/s)
i_s = gradient of the ground water flow (-)

Considering uniform flow and taking the Chézy coefficient as the friction parameter for flow the equations (1), (2) and (3) together will result in an expression for the critical average flow velocity U_{cr} for the stability of the mattress:

$$U_{cr} = 0.28\ C \sqrt{\Delta' d} \qquad (4)$$

in which:
C = Chézy coefficient = $18 \log {}^{12R}/ks$ ($m^{\frac{1}{2}}/s$)
R = hydraulic radius (m)
ks = roughness length (m)
$\Delta' = \Delta - \frac{Ks}{Km}\ .\ i_s$ (-)
Δ = relative density of the submerged mattress (-)
d = average height of the mattress (m)

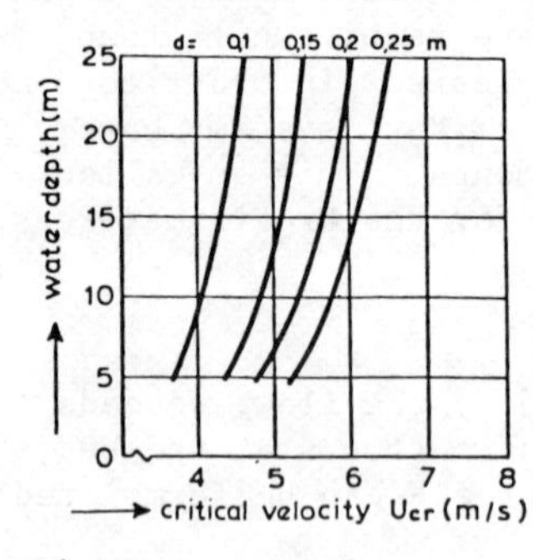

Fig.2.a. $\varepsilon = 0$

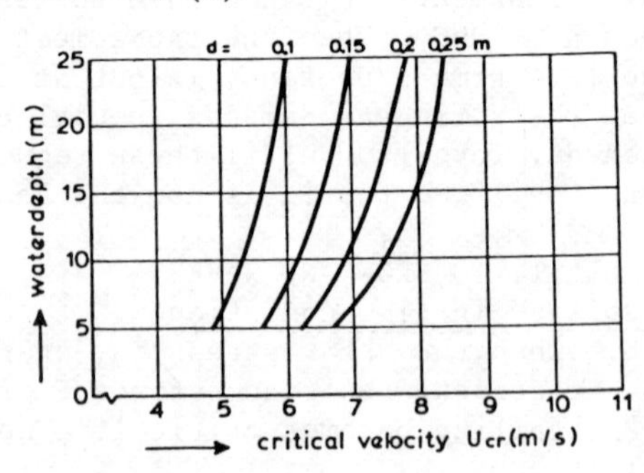

Fig.2.b. $\varepsilon = 1$

Fig.2. Critical flow velocity as function of waterdepth and mattress height for $i_s = 0$.

The design relation (4), in which has been incorporated a safety factor 2,5 for the frictional factor f = 3, is shown graphically in Fig.2. for mattresses filled with sand (ρs = 2,650 kg/m^3) in two conditions: (a) pores filled with air and (b) all pores filled with water.

Stability overlapping edge

9. At an overlapping mattress the stability of the edge is determined by the balance between the submerged weight of the mattress and the negative pressure caused by flow separation at the edge.

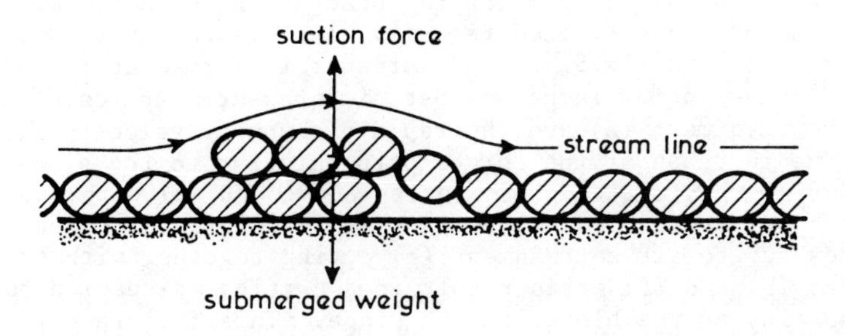

Fig.3. Forces upon an overlapping edge

The stability of the overlapping edge can be expressed in terms of a critical velocity:

$$U_{cr} = a. \sqrt{\Delta d.g.} \quad (m/s) \qquad (5)$$

in which:

a = factor depending on the shape of the edge and the flow conditions ($1.4 < a < 2$)

a = 2 for favourable flow conditions when the edge is directly laying on the underlaying mattress

The relation (5) is shown on the next figure for a = 2 and pores filled with air and water:

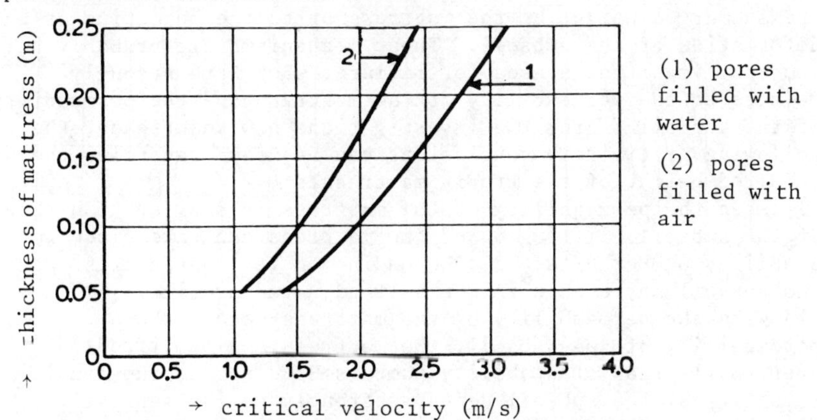

Fig.4. Critical flow velocity for the overlapping edge

Stability to wave action

General

10. Wave action causes various hydrodynamic loads on a slope. A breaking wave (plunging or collapsing) results in impact forces by the wave tongue hitting the structure. Following the impact, high current velocities due to the up and down rush may cause considerable hydraulic loads. Below the breaking wave sudden changes in the velocity field may occur and a rapid increase of the pressure inside the structure can develop. The investigations into the stability of the ProFix mattress under wave action are not completed, however, as wave forces acting on the mattress are not quite known but considered to be almost similar to those acting on a block type revetment, investigated recently, a review of those wave forces is shown in Fig.5. When a breaker is formed at the lowest run-down point large hydrostatic pressures are developed below the revetment (b) and the rapidly changing velocity field causes the increase of the forces perpendicular to the slope (d). During the impact uplift forces above and below the point of impact (e) are caused by the mass of falling water. Low pressures due to air entrainment (g) acting together with the pressures (b) and (f) often result in a critical situation for the stability of the blocks on the slope. This failure mechanism is shown in Fig.5.b. Maximum uplift pressures can be expected at the point of maximum down rush which may occur down to a level of two times the wave height below the still water level.

Stability criterion

11. The stability criterion for revetments is often described by a so-called stability number $H/\Delta d$, in which:
H = design wave height (m)
Δ = relative density of the submerged mattress (-)
d = average height of the mattress (m) (or the stone or block size of a revetment)

For sandfilled mattresses the critical $H/\Delta d$ value can be related either to the condition when uplift pressure exceeds the submerged weight of the mattress or to the beginning of the deformation of the subsoil. These mechanisms are probably the two most important causes for failure. Both are strongly related to the permeability of the mattress and the permeability of the subsoil. From the investigations now undertaken, the following preliminary conclusions can be drawn regarding the required height of the ProFix mattresses:
(1) When the permeability of the mattress is smaller than that of the subsoil, lifting up of the mattress can take place for stability number $H/\Delta d \geqq 2$. Depending on its characteristics the subsoil may change into a self-adjusted profile.
(2) When the permeability of the mattress equals the permeability of the subsoil, the mattress can be lifted up over small areas at stability numbers $H/\Delta d$ = 3-4, the local stability of the subsoil depends strongly on its sensibility for the ground water flow.

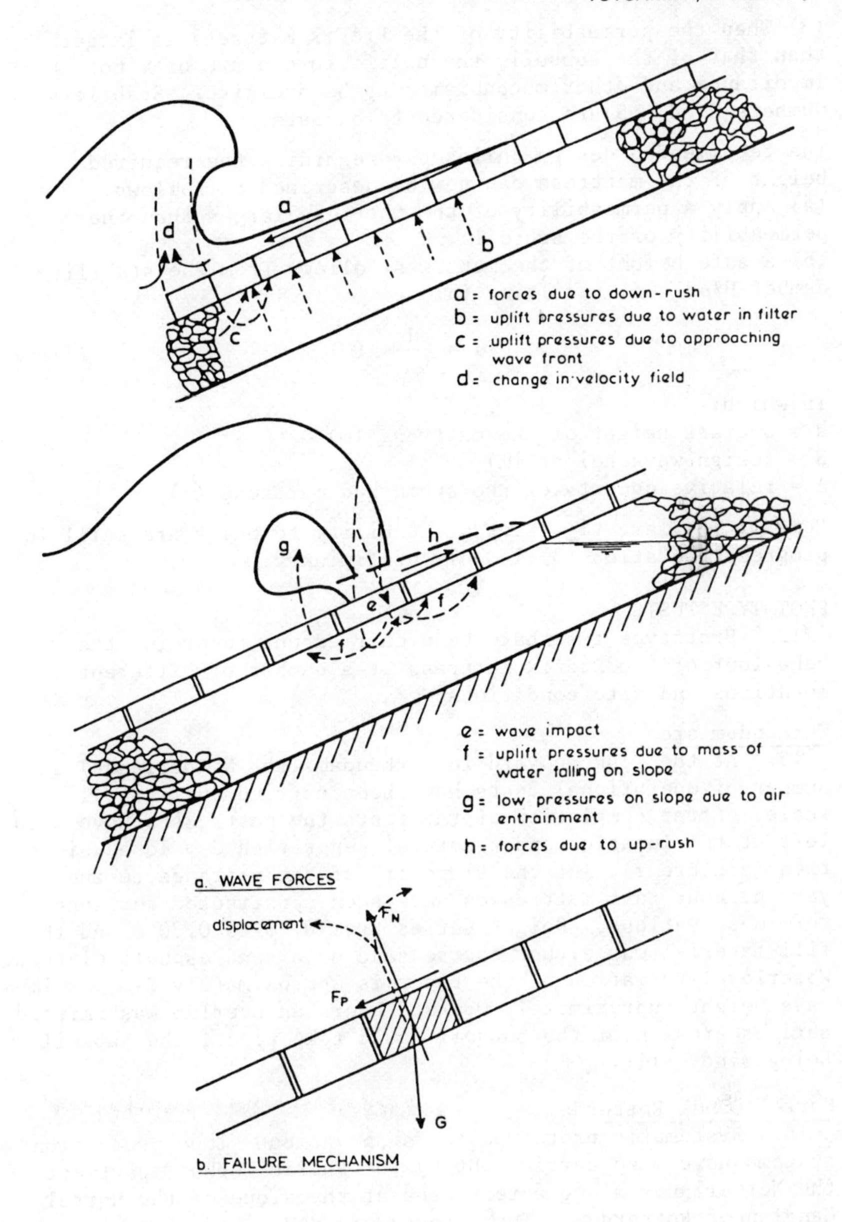

Fig.5. Hydraulic loads on placed block revetments (ref.7)

(3) When the permeability of the ProFix mattress is larger than that of the subsoil, the uplift forces are of minor importance and other mechanisms may be decisive. Stability numbers $H/\Delta d = 5$ are considered to be safe.

The recommended design philosophy regarding the required height of the mattress can now be described as follows:
(a) Apply a permeability of the mattress larger than the permeability of the subsoil.
(b) A safe height of the mattress follows from the stability number $H/\Delta d = 5$, resulting in:

$$d = \frac{H}{5\Delta} \quad (m) \qquad (5)$$

in which:
d = average height of the mattress (m)
H = design wave height (m)
Δ = relative density of the submerged mattress (-)

Remark: As investigations of stability to waves are still in progress, equation (5) is only preliminary.

PROTOTYPE TESTS

12. Prototype tests have been carried out regarding the behaviour of the ProFix mattress at a number of different locations and site conditions.

Werkendam area

13. At the Zinkcon yard in Werkendam (the Netherlands) a number of operational tests have been carried out at full scale. After completion of the tests the mattresses have been left at the location and a natural vegetation developed since then (picture 1). At the slope 1:3 of the entrance to the yard harbour test mattresses have been constructed for long term observations. Height varies from 0.13 to 0.30 m and the fill material was either coarse sand or a sand asphalt mixture. Waterlevel variation of the river is approximately 2 m, maximum wave height approximately 0.5 m. Where an overlap was omitted serious erosion of the subsoil could take place, the subsoil being sandy silt.

Hartel Canal Rotterdam

14. Systematic prototype tests on various slope protection systems have been carried out by the Public Works Department of the Netherlands along a test area at the slope of the Hartel Canal near Rotterdam. These tests are described by a paper of Mr. K.W. Palarczyk (Prototype test of slope protection systems). The mattress was constructed at a slope 1:4 from M.S.L. - 0.4 m up to M.S.L. + 2.0 m total width 20 m. Tidal waterlevels varied between M.S.L. - 0.7 m up to M.S.L. + 1.3 m. The existing subsoil was a layer of gravel 30-80 mm, layer thickness average 0.3 m. The ProFix mattress had an average height of 0.20 m. The mattress was filled with medium to coarse sand contained by tubes formed by a flat base filter cloth and a curved top cloth stitched together at regular intervals of 0.40 m.

The top cloth was provided with a needled polypropylene felt layer to give extra protection to u.v.-radiation.
Measurements have been carried out regarding the water movement along the revetments and in the subsoil including the water pressures as a result of passing ships. Ship waves to a maximum of approximately 1 m have been observed and current velocities up to approximately 1 m/s on top of the slope. The results of the measurements were still not available while preparing this paper. The mattress appeared to maintain its stability under the test conditions. It was of interest to experience the high importance of a sufficient degree of sand density inside the tubes, especially under dynamic load conditions. Due to breakdowns during filling this degree was not obtained immediately and as a result of waves and currents migration of grains could take place downwards the slope, causing densifying the lower part and affecting the filling degree of the upper part. The upper part had to be partly refilled afterwards. Up to now, after 3 months, the mattress appears to perform satisfactory and a natural vegetation has started to develop.

Pict. 1 Test mattress

Pict. 2 ProFix mattress Hartel Canal, Rotterdam

BANK PROTECTION WORKS IN NIGERIA

General

15. On behalf of the Government of Rivers State, Nigeria, a contract has been awarded to Zinkcon International B.V., Papendrecht, the Netherlands, regarding the construction of erosion and flood protection works at villages located in the lower Niger delta. The total length of banks envisaged was 18.5 km requiring approximately 1.1 million m^2 of bank protection. Most of the villages are located at the outerbend of the river where an enduring erosion is highly stimulated by the run-off from 3 m annual rainfall. The inevitable creep of the river into the banks forced the villages for many years to replace the few brick and many timber structures as the old ones fall in. Since there is no rock in the delta available, the ProFix mattress became attractive because bank protection could rely mainly on local materials and labour.

Site conditions

16. The steep eroded banks are mainly clay with stratifications of clay-sand mixtures. The bed material is coarse sand (300-1000 μm) with irregular silt content up to 10%. In July the water rises between 6 and 8 m and the current increases to a maximum of about 2.5 m/s. In November the waterlevel drops and ground water leaks out of the banks, increasing its instability.

Method of construction

17. It was decided to reslope the existing banks by a

refill of sand from the river to a slope of 1:3 and a height of 1 m above the highest known local waterlevel in order to protect the villages also against floodings. The sand was dredged at the opposite side of the river and discharged to the fill area by a floating pipeline. Hydraulic excavators and dozers were used on the fill. The bank protection was extended to a part of the riverbed as to secure the stability of the new slope if erosion at the toe would take place. Required lengths of mattresses vary from 60 to 90 m. Stitching of the filter cloth to create a mattress was carried out in the Netherlands, sewing from the 5 m wide rolls to make 20 m pieces was done at the yard in Nigeria. The accurately folded pieces were spread out at the top of the slopes and pulled in stages out into the river as they are filled. The filling was done pneumatically by blowing dry sand through rubber hoses into flap covered openings every 5 or 10 m along each tube of the mattress. A specially designed beam was clamped into each section of 20 m width and hauled out by cables running to a barge spudded to the riverbed. Also the beam was operated pneumatically to pinch the fabric during towing and allowing for easy release later on when the mattress has reached its final position.

Situation after 2½ years

18. The overall situation of the mattresses after 2½ years is quite satisfactory. The stability to waves and currents has proven not to be endangered, even during the combined action of rather high discharges and outcoming ground water during rapid fall of the riverlevel. Vandalism has not appeared to be a problem up to now, vegetation has rapidly developed and there are no signs that the behaviour of the material in these tropical conditions will cause any concern.

Pict. 3 ProFix bank protection Forcados River, Sagbama, Nigeria

REFERENCES

1. TEINDL H. Filterkriterien von Geotextilien. Dr.Ing. Thesis Innsbruck University, 1979.
2. HEERTEN R.G. Geotextiles in coastal engineering. Matériaux et Constructions, RILEM No. 82, 1982.
3. CALHOUN C.C. Development of design criteria and acceptance specifications for plastic filter cloths. Army Engineer Waterways Experiment Station, Vicksburg, Mississippi, Report No. AD-745085, 1972.
4. Kunststoffilters in Kust- en Oeverwerken. Nederlandse Vereniging Kust- en Oeverwerken, 1982 (in Dutch).
5. Plastic filters in hydraulic structures. Requirements and test methods. Ontwerpnorm NEN, Sept. 1982 (in Dutch).
6. DELFT HYDRAULIC LABORATORY/ZINKCON B.V. Stability of sandfilled mattresses, R. 1698.
7. DELFT HYDRAULIC LABORATORY/ZINKCON B.V. Stability of ProFix sandfilled mattresses under wave action, R. 1903.

T4 Some recent developments in the field of flexible armoured revetments in the Benelux

J. NOMES, Geotextile Division, NV UCO, Belgium and T. J. LUPTON, Geotextile Projects Ltd, Hitchin, Herts

SYNOPSIS. It is the intention of the authors to highlight some typical new developments in armoured revetments carried out in Belgium and the Netherlands. Four different cases are discussed, 2 are based on the use of geotextiles as flexible containers for specific materials. The 2 other cases are based on precast concrete block systems combining flexibility and interlocking features. Each of the developments are discussed in the context of practical construction experience.

THE 'OVOLO MAT' - AN INTEGRATED EMBANKMENT PROTECTION SYSTEM ON THE RIVER SAMBRE IN BELGIUM

Introduction

The river Sambre is an important route for commercial traffic between northern France and the industrial regions of southern Belgium. It was for this reason that Charleroi and more specifically the town of Chatelet was the planned site for a new roro berth. The principal economy of this area was based on its coal mining attributes and not suprisingly much of the available building land consists of mine waste. It was therefore calculated that substantial forms of revetment would be required where the mouth of the dock was cut from the existing embankments.
The design authority proposed cast in place concrete to a thickness of 600 mm based on a specially formulated plasticised mix. Under normal circumstances this quality of concrete can be pumped directly onto submerged sloping embankments without separation occuring. However the practical aspects of preparing the embankment from the loose mining shale in the presence of a continual washing action of passing vessels began to delay the works.
It also became apparent from divers reports that some difficulty would arise in placing conventional shutters due to debris on the river bed.

Approach to design

A new set of design criteria had therefore evolved :

1. A form of protection was required which could armour the prepared slope as soon as it was constructed.
2. The revetment should be homogeneous and of substantial thickness, incorporating relief to hydrostatic pressures at specefied locations.
3. The shutters should be adaptable in shape and form to accomodate the broad tolerances required for the embankment construction. Their function should also not be restricted by the obstructions on the river bed.

Textile based flexible shutters have been used extensively in projects of this nature however the designers were faced with a number of old unsolved problems. The first of these was to provide an envelope which was strong enough to contain a 14 metre column of liquid concrete on a 45 degree slope. Past practice was based on the use of small aggregate mixes having an unconventionaly high water cement ratio. The textile envelope being permiable allowed the expulsion of this free mixing water to occur without loss of important solids. The rapid compaction which results, immediately reduces the hydraulic pressure acting on the textile and therefore diminishes the risk of bursting. The specifiers of the case in question however demanded the use of plasticised (40 mm aggregate) mix design as originaly called for. This compounded the second design problem which had been experienced in other projects. When two layers of fabric are expanded by a liquid mass between the layers on a sloping plane, extensive ballooning occurs at the lowest level where pressure and weight are greatest. Elongation in the textile and sliding of the top layer result in a bulge at the slope base, so large, as to pull the whole mass away from the slope.

Selected design

The solution to both problems was found in a unique modular method of construction. The adjacent sketch (fig. 1) illustrates the envelope with its two layers joined at regular intervals by circular columns of fabric. This large area of contact minimises the stresses where the loads are highest. Helically wound steel bars are positioned in the columns to prevent

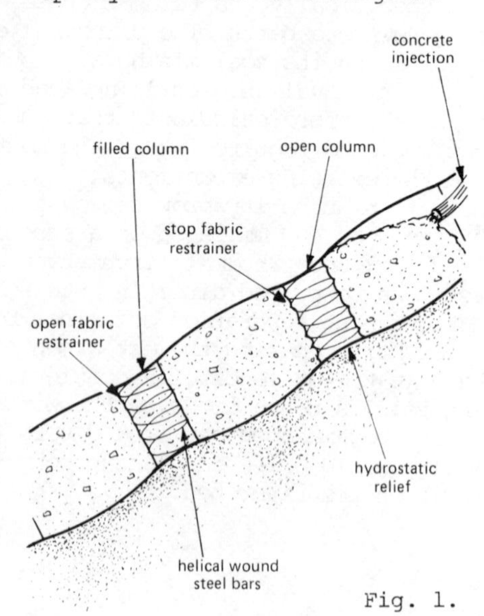

Fig. 1.

sliding and to maintain elongation within defined areas. The textile used was a polypropylene twisted tape construction having a tensile strength of 4400 N/5 cm. The sewn seams in the fabrication equaled the fabric strength and further reinforced its performance by providing an element of stiffness.

Performance

The matresses were delivered to site in prefabricated sheets approx. 50 sq.m. in area. They were tailored to correspond to the planned bank profile on each side of the dock mouth. After each section of the embankment was graded the sheet was lifted into its designated position by crane. Filling of each section followed in sequence bringing the level of concrete to its highest level in one cóntinuous pour using a mobile concrete pump. The total surface area of embankment was 2300 sq.m. and this was completed in 12 days. Some site modification was necessary in those areas where the planned bank profile could not be achieved. This was carried out by a simple hand sewing operation as the section was tailored to the new profile.

Placing of the sheet

A FLEXIBLE AND TOTALLY PERMIABLE CONSTRUCTION COMBINING A WOVEN GEOTEXTILE WITH EXPANDED CLAY GRANULES - " ARGEX ".

Description

In this system a dual layer container has been manufactured from a high strength geotextile and filled with lightweight granular material for the purpose of providing a substantially thick revetment with a controlled drainage capacity. The container is constructed as a matress comprising a series of tubes and laid in the direction of the slope.
Filling can be achieved by air injecting the granules.

Advantage and potential uses

One of the major advantages in the use of this technique is the versatility of its shape and form as well as the materials used. Altough only lightweight materials are used the nett result of the composition is a homogeneous revetment of considerable mass. Eg. unit sizes of 100 sq.mtrs. can easily be placed thereby creating a total mass of more than 15 tonnes.
As the argex can withstand excessive compressive loads the systems designed profile will be maintained throughout the life of the structure.
The open area within the form is approximately 45 % providing a high performance wave energy dissipator combined with lateral drainage capacity. The filter function is controlled in the selection of the geotextiles which form the container. The opportunity exists to use geotextiles with different properties as top and bottom layers to provide added scope to the design. In situations of extreme ultra-violet attack or abrasive conditions, the construction can withstand the load imposed by a protective layer of rip rap.

A case history. The inclusion of the system in the breakwater construction at Zeebrugge harbour in Belgium.

At Zeebrugge on the Belgium coast a major harbour is being constructed into the open sea. The design features two main breakwaters about 3 km. in length which form the protection for the entrance channel to the port installation. A large area behind the breakwater has been reclaimed using hydraulically placed sand fill. The core of the breakwater is formed from 2/300 kg. stones and protected on the seaward side by precast blocks of 25,000 kg. The inner face is protected by a granular fill construction composed of rocks (1000/3000 kg.), stones (200/300 kg.) seagravel and a lateral drain. (see fig. 1).
A full investigation has been carried out to determine the currents and pore water pressures in the section of the breakwater. These effects are critical on the interface between the different granular layers.

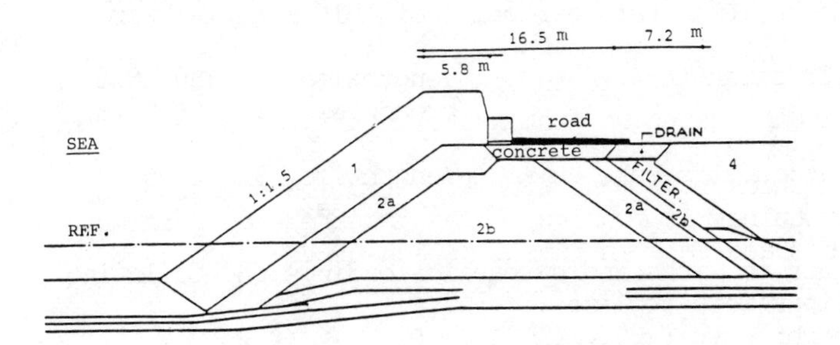

Fig. 2. Cross section of the breakwater
1 :blocks 25,000 kg
2a :200/300 kg
2b :2/300 kg stones
4 :sand fill

The internal currents in the porous dam in combination with the phenomenon of locked air causes increased water levels within. The resulting high water pressures and gradients were calculated for different flow lines. This information was used to ascertain the total stability of the filter structure for both mechanical and hydraulic failure. It became apparent from these studies that the proper functioning of the lateral drain was of the utmost importance to prevent migration of the hydraulic sand fill into the core of the dam.
The system described above was adopted to provide a separating layer between these vital constituents whilst ensuring dynamic equilibrium.
It was envisaged that settlement in combination with severe ground loads would continue to act on the system. A field trial was therefore set up to ascertain the most suitable geotextile to be used as the container.

Field trial: general view

On the results of this trial a woven polyester fabric was chosen.
The mechanical characteristics of this material were :

Tensile strength	: lengthwise	:	180 kN/m
	broadwise	:	80 kN/m
Elongation at break	: lengthwise	:	20 %
	broadwise	:	15 %

In order to withstand hydraulic loadings the following equations were applied :

Sandtightness : $O_{90\ geotex} / D_{95\ sand} < 2$

Permeability : $0,1 < K_{geotex} / K_{sand} < 100$

K : coefficient of permeability.

The selected geotextile has a sandtightness bases O_{90} = 100 microns, a water flow rate Q = 3 l/m2/sec at 10 mm head, $K = 2,7 \cdot 10^{-5}$ m/sec
The matress was fabricated in lengths of 8 m.
Having a thickness of 250 mm and laid to a sloping embankment. Anchoring flaps were incorporated in the factory assembled units which were prepared in a width of 15 m.
The final assembly was carried out by sewing on site.
The argex was fed into the tubes through flexible pipes threaded between the dual layers which were withdrawn as the level was brought up.
The method of moving the argex was carried out using air pressure directly from the containing vehicle. A production rate of 200 sq.m. per hour was achieved .

End of unit

Conclusions

It can be seen from the case study outlined that the system would have equal benefits in more conventional revetment applications.

Containing vehicle

The adjacent sketch fig. 2 illustrates the principal adopted for the field trial detailed above. Although expanded clay appears to offer the most benefits in the studies so far other lightweight materials could be considered with equal cost effectivness.

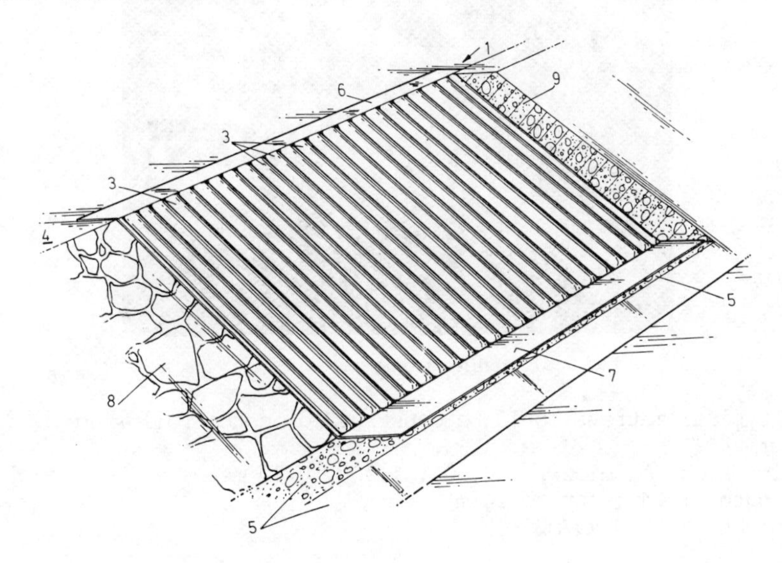

Fig. 3.

1 : Filter construction
3 : Tubes
4,5: Crest and toe of embankment
6,7: Anchoring flaps
8 : Embankment
9 : Rip-rap

ARTICULATING BLOCK REVETMENT INCORPORATING A GEOTEXTILE

Description

A number of successful products exists in this range representing one of the recent advancements in revetment design incorporating a geotextile. The principal feature of the design is the linking and interlocking of proprietory shaped cellular concrete blocks. When in place the blocks act like rip rap conforming to irregular ground conditions with the added advantage of continuity over the entire slope area. The system highlighted in this paper has the unique method of hinged steel bars for linking the blocks together. In other systems this is achieved by various cable lacing techniques. The key advantage in using a hinge type knuckle joint is the ease with which a perfect right angle can be accommodated . The profile and physical dimensions of the block units can be selected to meet specific site conditions. This also applies to the geotextile which can be matched to a specific filter function.

Detail of hinge joint

A typical matress could be constructed on the following format :
Dimensions of blocks : 0,50 . 1,10 . 0,10 m
Length : 4 m (up to 20 m)
Weight : 220 kg/m2

Installation

The mats are placed with the use of an adjustable clamp attached to a spreader bar on a crane or dragline. The matress assembly is firstly placed onto the geotextile and then the clamp enables both to be lifted into position on the slope. Anchoring is achieved by creating horizontal plinths at crest and toe of the embankment or by driven anchors.

Applications

The special hinge technique makes the system ideal when neat, clean profiles are required. Only moderate wave attack can be resisted by the lighter blocks which are more suited to drainage channels and ditches. These blocks have an open area of around 3,5 percent and a very low friction factor.

Case history

Hazewinkel in Belgium has an important recreation lake providing water sport facilities to a large population. In order to maintain a natural appearance to its shores revetments were excluded from the original construction. Due to regular wind wave attack serious erosion had occured in one location. The local authority were anxious that any protection works should not disturb residents or the functioning of the lake. Selection of the funda-mat was made since the water depth of about 2 m. could easily be accommodated from the foreshore. The inbuilt simple anchoring ability assured rapid and economic completion of the works. The adjacent photograph highlights the products pleasing appearance.

Hazewinkel, Belgium

IMPROVED TECHNIQUES IN THE ATTACHMENT OF CONCRETE BLOCKS TO HIGH STRENGTH GEOTEXTILES

Introduction

Current methods of fabricating concrete blocks with geotextile support can be divided into two main categories. The first being those which are cast directly onto the fabric. In this method the anchoring devices are driven through the geotextile on predetermined centres before the concrete is poured. The second group rely on adhesives to affix precast blocks . The alternative method described below was designed to combine the controlled mechanical advantages of the first group with the versatility of the second. A company in the Netherlands devised the anchoring technique which makes use of synthetic nails driven through the geotextile directly into holes preformed in the block. The laying up of the geotextile and the fixing method is fully automated providing an economic and accurate result.

Construction details

1. Concrete blocks (two current designs are discussed)
   a) the 8 recessed sides of this rectangular block assist in reducing hydrostatic pressure and serve to break up wave action and run-off. The sides are tapered to give flexibility to the mat. The openings between the blocks are filled after placement with a coarse material such as gravel or sand, to avoid direct wave action on the fabric. This assures that flapping and subsequent failure of the filter function does not occur.

   Technical data : weight : 170 kg/m2
   dimensions of blocks : 33.30.9 cm
   open area : about 10 %
   concrete : resistance : > 50 N/mm2
   waterabsorption : < 10 %
   nails : 3 nails/block

   b) a cellular block design provides the open areas necessary to promote the growth of natural vegetation. Openings are filled with suitable soil after placement.

   Technical data : weight : 165 kg/m2
   dimensions of blocks : 40.40.9 cm
   open area : about 40 %
   nails : 4 nails/block

2. Synthetic nails
   The illustration shows the non corrosive polyamide nail with its frictional ridges. Extensive tests have been carried out to qualify the anchoring ability.
   Loads applied to the geotextile in the axis of the nail yielded results in the order of 2000 N per nail.

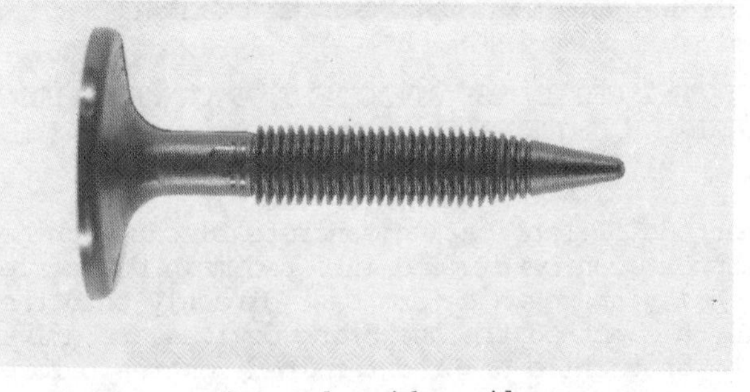

Non corrosive polyamide nail

3. Geotextile
   The normal criteria for selection apply with the added importance of minimum elongation through the loads imposed during the installation. Test results identified a heavy woven polypropylene geotextile as the most suitable.

   Technical data : weight : 500 g/m2
   tensile strenght : 75 kN/m
   elongation at break : 15-20 %
   trapezoidal tear : > 1,40 kN

Applications
Type (a) is suitable for most embankment protection schemes on lakes, rivers and drainage ditches with the exception of those subject to very severe wave attack (coastal works). Applications under water are equally facilitated. The completed surfaces above water are immediately available for foot and vehicular traffic.

Type (b): placing on embankment

Type (b) is most suited to the upper parts of the embankment where environmental aspects are of paramount importance. It is not suitable for regular wave attack but is resistant to surface erosion and affords adequate protection against lateral currents.

GENERAL CONCLUSIONS

The success of each of the topics included in this paper clearly indicate the suitability of modern geotextile products to scour and erosion problems.
Their economic benefits suggest that further research should

be instigated to enable fundamental design concepts to be derived with the geotextile as a qualified component. Such an approach being more cost effective than merely adding the geotextile as a solution to a problem that need not have occured.

ACKNOWLEDGMENTS

Our gratitude is expressed to the following companies for their permission to include technical details relating to their products.

FLUVIO LABOR and SICALEX	BELGIUM
V.O.R.	BELGIUM
ROOK-KRIMPEN	THE NETHERLANDS

T5 Revetment construction at Port of Belawan, Indonesia

E. LOEWY, A. C. BURDALL and A. G. PRENTICE, Sir William Halcrow and Partners

SYNOPSIS. This paper describes the revetments used to protect a fine sand reclamation situated in the estuary of the Belawan river in Indonesia. A substantial length of the revetment is situated under a piled quay where grouted mattresses have been used to protect the 1:2.3 sand slope. Significant post construction settlement of the sand slope is expected and special measures were taken to enable the mattresses to accommodate differential settlement. The installation method in difficult environmental conditions and resulting modifications are presented.

INTRODUCTION

Background

1. Belawan is situated on the north-east coast of Sumatra,(See Figure 1). The present port and township is bounded by the River Belawan to the north and the River Deli to the south. The rivers share a common estuary but the main and deeper channel is formed by the Belawan River. The coastal areas surrounding the estuary are low-lying mangrove swamps intersected by a network of small creeks.

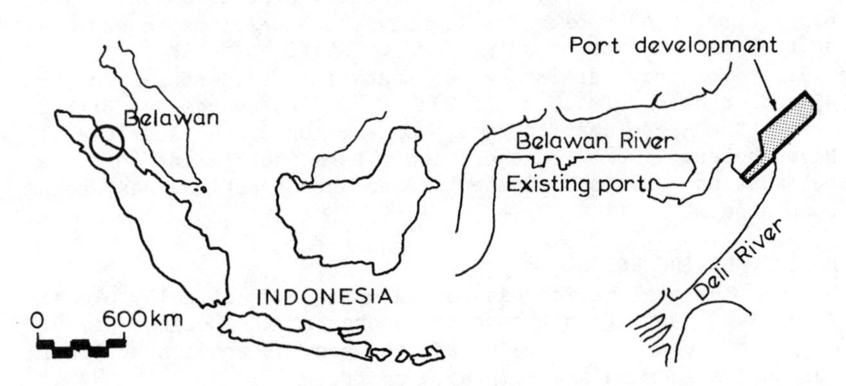

Figure 1. Location map for Belawan

Figure 2. Port layout

2. The port of Belawan is the third largest port in Indonesia and currently handles around 5 million tonnes of cargo annually. To accommodate increasing trade the facilities for handling dry cargo and container traffic are being extended. The first phase of the extension will provide 850m of additional deepwater quay with associated access, cargo storage and handling areas. The layout of the existing port and the planned extension work is shown on Figure 2.

3. Site Description. The reclamation covers some 30ha over an area where the existing sea bed was at a depth of up to 6 m below mean sea level but was largely within the tidal zone. It is located some 15km from the open sea and the dredged approach channel is maintained to a depth of 8m (see Photograph 1).

Photograph 1. Aerial view from the north of the reclaimed area

4. Extensive soils investigations had been carried out to determine the nature of the compressible clays and the volcanic ash and sand, which underlie the reclamation. The clays are highly plastic, lightly over-consolidated and are soft at the surface becoming firm to stiff at depth.

5. Vertical drains were installed to depths of up to 45mand centres of 1.5 m and 2 m of sand surcharge was placed over the reclamation to speed up the settlement. Nevertheless some further long term settlement of the reclamation is anticipated after construction has been completed.

DESIGN REQUIREMENTS

6. The reclamation was formed using hydraulically placed fine sand won from the upper reaches of the Belawan River. Slope protection is required to prevent erosion by tidal current, ship wash and rainfall runoff.

7. The reclamation material is a fine sand with a silt content typically between 10% and 15%. Hydraulic placing achieved a relative density of 40% to 50% and along the edge

of the reclamation to be occupied by the quay, vibroflotation and dynamic compaction was carried out to achieve a minimum relative density of 60%.

8. The spring tidal range is 1.0 m and the maximum tidal currents observed during the various hydraulic investigations were of the order of 1.0 m/sec. Wave action due to winds is usually negligible but for brief periods with waves of up to 0.5 m are experienced. Bow waves from fast moving vessels also reach heights of 0.5 m. Underwater visibility was virtually zero because of the sediment content.

9. Rainfall records indicate six hour maximum rainfall intensities of 14mm/hour with the highest monthly rainfall during the monsoon season towards the end of the year.

10. To accommodate the anticipated post construction settlement the revetment was designed to accept,

- Up to 500mm general settlement
- Up to 200mm more settlement along the rear edge of the quay compared with the toe of the slope (i.e. a differential settlement of the top of the slope relative to the toe)
- a 250mm differential settlement midway between the pile bents relative to the area adjacent to the piles (i.e a 'dishing' of the slope between the piles).

11. The final revetment had to withstand this movement whilst retaining its structural integrity and prevent leaching out of fines from the reclamation material.

ORIGINAL DESIGN

12. During the design phase consideration was given to alternative forms of revetment both for under the quay and elsewhere. For the underquay slopes the selected revetment was specified to be synthetic filter cloth, overlain by 300mm of filter rock and 750mm of armour rock. The rock was required to be angular with 85% of the filter rock being between 5 and 20kg. Armour rock was to be between 50 and 200kg. Filter cloth was specified as Nicholon 66475 and was required to be lapped to adjacent sheets and at piles. The remainder of the reclamation perimeter was to be protected by rock filled gabions on Nicholon 66475.

ALTERNATIVE DESIGN

13. Subsequent to the award of the contract, to overcome the difficulties of obtaining suitable rock, the Contractor proposed the use of fabriform mattresses filled with concrete as an alternative to the filter and armour rock on the underquay slope. No change to the underlying filter fabric was envisaged (see Figure 3).

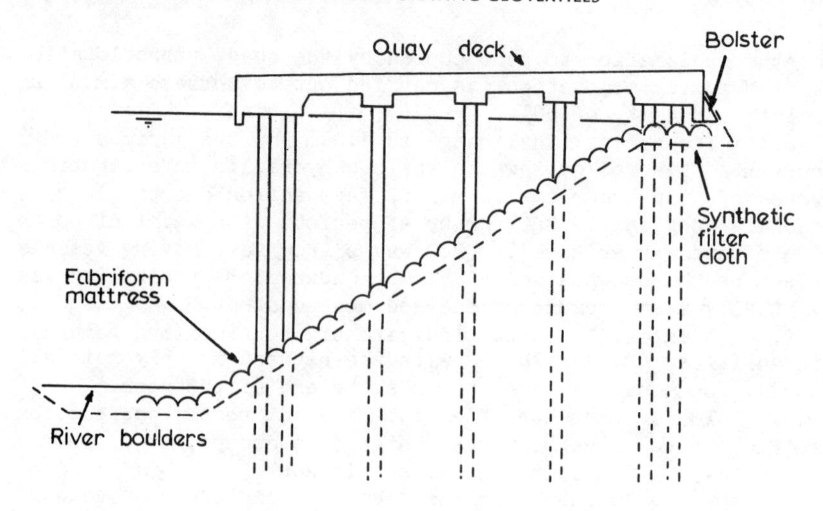

Figure 3. A cross-section of the protected under quay slope using fabriform mattress

14. The fabriform mattresses were to be made up from woven panels of filter fabric connected to the fabric of adjacent mattresses by means of zip fasteners. Construction of the mattress would be such that, on filling, a mechanical joint would be developed between adjacent panels which takes the form of continuous "ball and socket" joints. In addition, a quilt pattern appears on the surface of the panel caused by fastening together the upper and lower layers of the envelope at regular intervals. This, apart from acting as a control on mattress thickness, allows the insertion of a coarser meshed material to create "filter points" at the nodes. The arrangement of the mattresses is illustrated in Figure 4 and on Photograph 2.

15. Filling of the mattresses was to be with a pumpable small aggregate concrete mix, known as micro concrete, with excess water being expelled through the fabriform material. The filling of the mattress results in a reduction in the length and breadth dimensions of the mattress and this reduction has been termed "shrinkage". The proposed 100mm nominal thickness of the mattress gives rise to a consequential 17% shrinkage which had to be allowed for in the fabrication of the mattresses.

16. Various panel and collar arrangements were considered for the underquay works at Belawan to allow the mattresses to be fitted around the piles, and to provide the necessary flexibility to accommodate the anticipated settlement. Initially it was intended that slope protection works would commence ahead of deck construction and advantage would be taken of this by lowering the mattress panels onto the slope, with preformed holes for piles. This arrangement was revised to include a horizontal joint on one side of each pile to

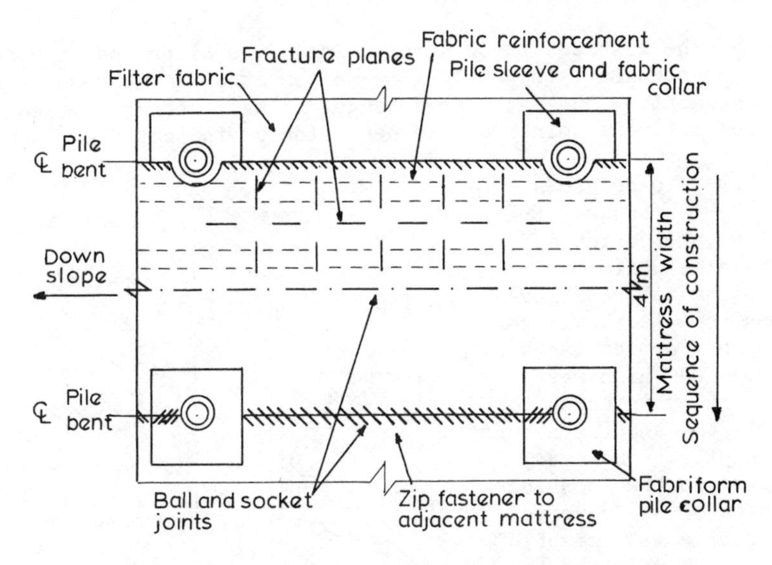

Figure 4. Typical fabriform mattress arrangement

Photograph 2. A length of fabriform mattress during construction

allow the mattress to be unrolled down the slope and unfolded to encompass the piles. After the commencement of construction the mattress arrangement was further revised to place mattress joints on the bent line as indicated on Figure 4.

17. To secure a close fit between the slope protection and the pile while permitting mattress movement due to shrinkage, a fabriform collar was proposed which was initially intended to fill the annulus around the pile. However, after further consideration of the expected relative settlement of mattress relative to quay the structure, a steel sleeve and "top hat" of filter fabric were proposed with a collar of fabriform laid on top of the mattress and tight to the sleeve as indicated on Figure 5 and shown in Photograph 3. As relative settlement occurs the sleeve was expected to slide down the pile and the "top hat" to expand, bellow fashion with a collar for protection.

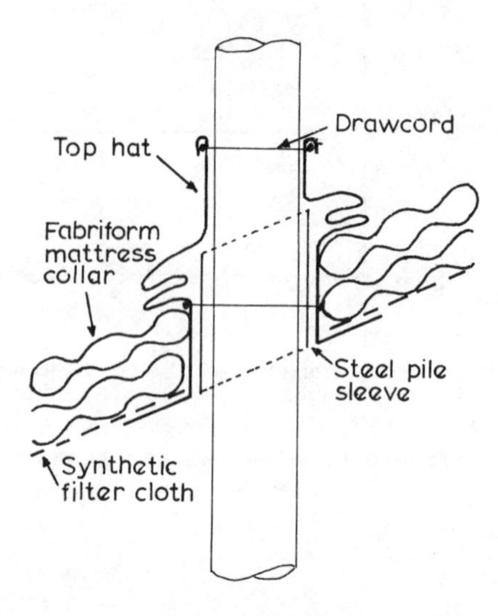

Figure 5. Fabriform mattress arrangement at a pile

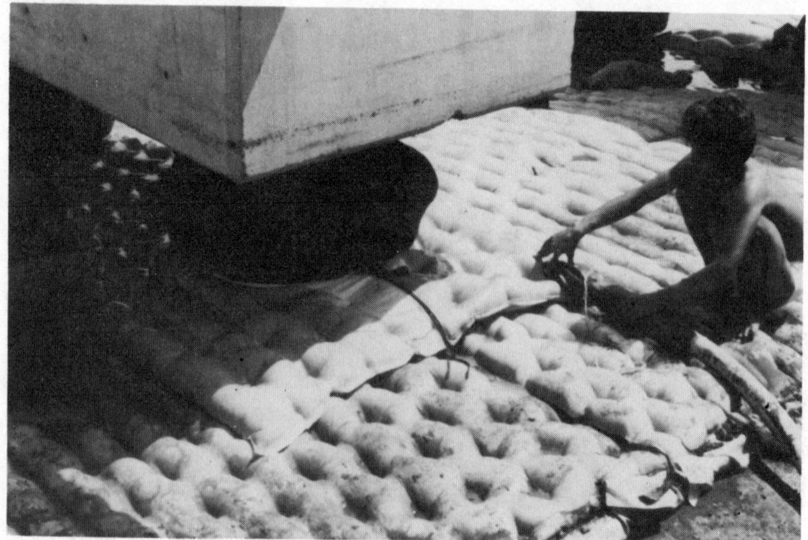

Photograph 3. Fabriform collar installation

18. To give the mattress more flexibility, crack inducers were provided within the panels. Crack inducers comprised lengths of sewn seams joining upper and lower layers of the mattresses giving rise to a section weak in bending and these are indicated on Figure 6.

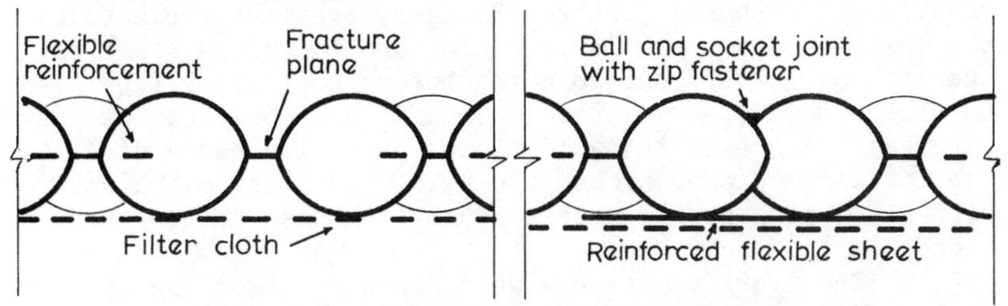

Figure 6. Fabriform mattress details

19. A concrete filled fabric bolster was proposed for installation behind the rear beam to accommodate settlement without loss of material around the beam.

20. The use of a grouted mattress was initially considered for the river bed, in the berthing area where armouring is required to prevent undercutting of the quay slope due to ship wash. However, the original design was retained except that since the area was horizontal, rounded river boulders were substituted for angular rock.

INSTALLATION PROCEDURES

21. The installation of the mattresses, onto the prepared slope, was planned to be carried out after pile driving but before construction of the suspended deck. However, for various reasons deck construction commenced before mattress installation and, as a result, approximately one-third of the underquay slope protection was placed below a framework of precast beams and some 50 m were installed below the completed deck (see Photograph 4).

Photograph 4. Quay construction

22. After excavation of the slope by clamshell grabs sand pumps and, hand trimming was undertaken by divers using water jet for removal of surplus material and gravel filled bags for filling low areas. Care in trimming proved essential to avoid distortions of the mattress with consequent unsatisfactory panel joints. Sleeves, previously placed over the piles were set in position and the slope was finally checked by divers and sounding before the filter fabric was unrolled down the slope. The filter fabric was weighted with bags of gravel to avoid flotation. "Top hats" of filter fabric were placed around the piles and sleeves and secured with steel pins approximately 500mm long pushed into the sand slope.

23. The fabriform panels were drawn out from the shore by divers each new panel being secured to the free end of the previous panel by a zip fastner.

To hold and control the position of the mattresses while they were being filled, steel poles were inserted into sleeves on the mattresses and the poles were held in position by ropes attached to anchors at the top of the slope. As the mattresses shrank in plan during filling, the poles were progressively allowed to move down the slope to allow the mattresses to attain their designed shape.

Micro concrete was pumped into panels through tubular inlets set at intervals down each half panel. Several half panels were filled concurrently, filling commencing at the toe of the slope and progressing up the slope. The leading half panel was not filled until the next panel had been attached.

Subsequently to filling of the main mattresses, overlaying collars of fabriform were placed round the piles and pumped full of micro concrete. The fabric "top hat" was then adjusted to provide slack material which could be taken up as settlement occurred.

The construction sequence is illustrated in photographs 3 to 5.

INSTALLATION DIFFICULTIES

24. Difficulties arose when the first mattresses moved too far down the slope during filling.

25. Initially, the woven filter fabric was laid with the warp along the slope for reasons of economy. As a result of the problems with the first mattresses the filter fabric was rearranged with the warp down the slope as its coefficient of friction was much greater in this direction.

26. Mattress panel joints were initially placed mid-way between the pile bents. An advantage of this arrangement was that the piles would provide additional resistance to the movement of mattresses down the slope. During the early stages of the work a series of trials were carried out to check the sliding resistance between the filter fabric and completed mattress. From the full scale trials it was

established that there would be adequate resistance even with the fabric laid down the slope. Ways of providing increased resistance during the temporary phase when the mattresses were being filled were devised. From then on the filter fabric was laid down the slope. Also the panel points were moved to the pile bent lines so that the shape became a

Photograph 5. Filter fabric and pile sleeves

simple rectangle with semi-circular cut outs for piles on each long face.

27. During trimming of the underquay slope, a minor surface slip occurred affecting a 25 m length of slope. The cause of the failure was not immediately apparent and the progress of work was disrupted while investigations were made. It was concluded that failure occurred as a result of a combination of circumstances the dominant features being wave action from ships and local areas of silt. A second slope failure occurred adjacent to the standing edge of fabriform slope protection. In this case, 12 m of slope were affected but there was no evidence that the failure extended under the grouted mattress. No re-design of the permanent works was considered necessary. Repairs were effected by removing the loose surplus material and making good the hollows with gravel fill fabriform protection was then placed as elsewhere.

PLACING OF ROCK TOE

28. Removal of silt and trimming of the quay trench was carried out by a 200mm diameter submersible pump suspended from a barge. Placing of filter fabric and filter rock was

carried out together. Steel frames 2m x 4m were placed over rectangles of filter fabric laid on the quay deck. Filter rock was then placed in the frames to the required depth and the filter fabric temporarily secured to the frame sides. Frames were placed on the river bed by crane with divers assisting in locating frames against those previously placed. Lapping of filter fabric was achieved by laying out the leading edges. While frames were retained on the leading edge, others were subsequently retrieved as the work face advanced. The frame loaded ready for lowering into position is shown in Photograph 6.

Photograph 6. Frame loaded with river boulders and ready for lifting

29. Rock armouring was placed by net and levelling achieved by re-arranging armour using an orange peel grab and various weights.

PROGRAMME

30. Underquay slope protection works commenced in February 1983 and were programmed for completion in eleven months. Production rates for laying the fabriform mattress and grouting averaged two bays per day. Work on fabriform installation was normally based on an eight hour day, working alternate Sundays. Occasionally, early morning or late

evening tidal work was carried out to take advantage of slack water.

31. Work was normally on only one leading edge although, for a period, up to three leading edges were being worked. The only leading edge allowed to stand for a substantial period of time was at the location where the second minor slip occurred.

32. Quay trench protection works commenced during July 1983 and section 1 (the first 350 m) was effectively completed by October of that year and work was continuing in section 2 at the time of writing this paper and was expected to be completed in early 1984. Placing of frames containing filter fabric and filter rock peaked at about 16 bays per week with work alternating between six and seven days each week. Placing of armour rock and its subsequent trimming to level generally kept pace with the frames.

CONCLUSIONS

33. This use of grouted mattresses and geotextiles is believed to be one of the largest yet carried out for underquay slope protection and it provided valuable experience on some of the construction difficulties that can be met in an estuarine situation. While environmental conditions hampered placing and monitoring, effective installation proved practicable after modifications had been made to the panel arrangement.

34. Placing of the slope protection is not yet completed under the quay but, for the finished areas, no erosion or instability is apparent. The berths have not yet been put into use and subject to the effects of shipwash. Post construction settlement to date has been small.

ACKNOWLEDGEMENTS

35. The authors wish to thank the Belawan Port Administrator for permission to publish this paper. The main contractor for the work was Dharsamrin Indonesia Group with specialist sub-contractor Dowsett Prepakt Limited for underquay slope protection work.

Discussion on Session 5: Construction

Mr I. M. Walker, British Waterways Board

Could Mr Abromeit please explain more fully the function of the roughness layer, and define the criteria used in design.

Dipl. Ing. H.-U. Abromeit

The main function of the roughness layer is to prevent motion of soil to toe of slope on banks steeper than 1:4 if there is endangered subsoil. Opening size of roughness layers therefore has to achieve quick indentation to the subsoil. The roughness coefficient will be ameliorated at the same time.

The opening size of roughness layers has to be tuned to subsoil by filter criteria. Upper and lower limits for opening size ($\sim O_{90}$) of roughness layers have been determined for standard soil types used in BAW for filter tests, paying regard to the present practical experiences. Roughness layers have been used in West Germany since 1974.

Mr B. T. Rathmell, Dowsett Prepakt Ltd

The advantages of the fabriform system can be seen from the wide range of projects carried out throughout the world over the past twenty years. The Belawan installation is especially interesting because of the settlement to be accomodated.

As a specialist contractor we had to take particular care over the conditions of contract under which we carried out the work. Terms of payment were especially important. New and improved systems designed, manufactured and installed by specialist companies will not obtain the international acceptance and use they deserve unless satisfactory contract arrangements can be negotiated.

There has been much interest in the cost of the Fabriform technique. In the UK, the installed cost per square metre is generally between £12 and £15. The cost multiplier for overseas work is generally between 1.25 and 1.50.

Mr D. M. Vick, Consulting Engineer, Houston, Texas

Experience with monolayer flexible revetment mat systems design and construction in the USA has been gained with revetment systems composed of individual concrete cellular units laced together with synthetic fiber tendons running in both the longitudinal and transverse directions. Improvements in these systems over the past 3 years have yielded a 5 fold improvement in the performance of stability of these revetments.

The flexible revetment failure that occurred at Holly Beach, Louisiana, USA, during the construction process in 1983 illustrates how the lack of sound engineering design can (and will) result in disastrous failures in this type of revetment system. In this case, the design engineer with the Louisiana Department of Transportation allowed a poorly draining geotextile to be used beneath the revetment, and a minor weather event with limited wave activity caused the granular embankment slopes to liquify and fail the revetment. Holly Beach is exposed on the open coastline of the Gulf of Mexico.

Dipl.-Ing. W. Mühring, Neubauamt fur den Mittellandkanal

When using geotextile filters as part of a revetment on slopes, it is necessary to take care of the foundation of the filter itself.

It is known that a woven or a thin non-woven geotextile can not act as a sufficient filter if the soil particles can move beneath the fabric by wave attack or other hydraulic forces. These geotexiles are acting only as thin sieves, and are not forming a fibre stabilized boundary layer to the subsoil. The migrtion of soil particles beneath the fabric mainly depends on the fabric structure (thickness, fibre fineness), the grain size distribution of the soil, the inclination of the slope and the current spectrum.

A typical sign (see Fig. 1) of the failure is a bulge beneath the filter between the water level and the bank toe and a depression upslope above the bulge. Similar problems may occur when there is erosion under the filter resulting in small voids or small grooves.

Surface runoff downslope between the filter and the bank must therefore be avoided in any case.

To prevent these failures caused by the migration of the subsoil an additional layer was created as a special foundation layer beneath the non-woven geotextile filter - the roughness layer.

The requirements for the special structure of a roughness layer are:

(a) The dimensions and distribution of the pores in the roughness layer have to guarantee a quick filling up with soil particles in the filter stabilization time to

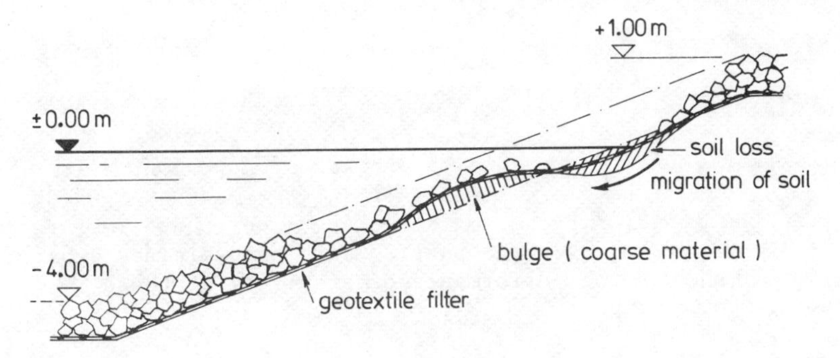

Fig. 1. Erosion under geotextile filter without roughness layer

form a stable boundary layer of soil particles and fibres.

(b) The dimensions and distribution of the pores in the roughness layer have to prevent any soil particle movement in the plane of the fabric after the filter stabilization phase.

To fulfil these requirements the following technical criteria are recommended:

Minimum thickness	10 mm
Effective Opening Size Dw	
BT 3 (silty sand)	0.5 < Dw < 2.0 mm
BT 4 (silt with clay and fine sand)	0.32 < Dw < 1.5 mm

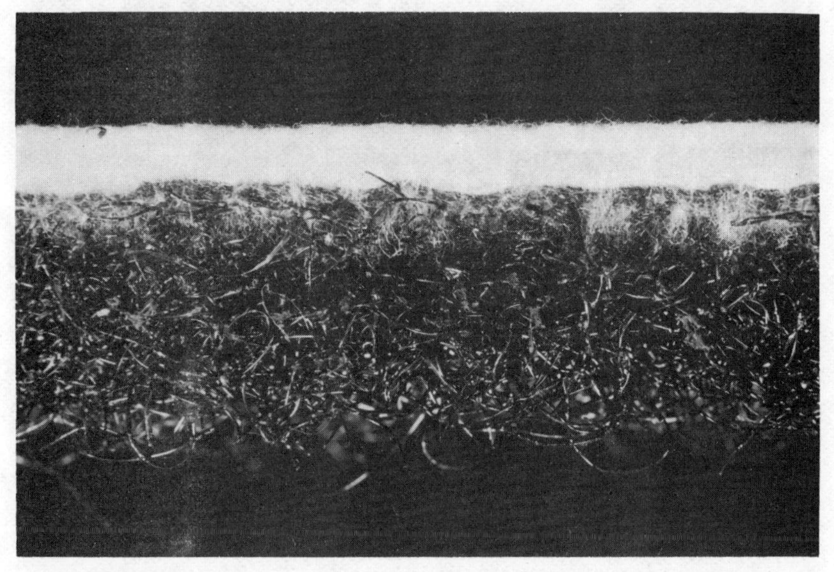

Fig. 2.

Fig. 3.

Fig. 4.

Fig. 2 shows a cross-section of a multi-layer nonwoven geotextile with a roughness layer relating to the given recommendations. There has been very good experience in using this kind of heavy filter mat on revetments on German waterways since 1975.

Dr G. Heerten, Naue-Fasertechnik, Germany

Mr Mühring told us about the advantages of using multi-layer geotextiles with a special roughness layer to prevent the movement of soil particles beneath the filter fabric. The high filtering efficiency of this product is combined with another big advantage: the very high robustness of these heavy needlepunched nonwoven fabrics.

There is a great difference in the robustness of different geotextile types, as shown for example in Figs 3 and 4. These figs show the deformation of a needlepunched nonwoven fabric (Fig. 3) and the damage of a heatbonded nonwoven fabric (Fig. 4) after dropping a stone from a height of about 1 m. Because of the mobility of the fibres in a needlepunched structure only a deformation results, whereas the heatbonded fibres, which are fixed at the crossovers, are cut and the fabric is punctured.

Figures 5 and 6 show the installation of a heavy multi-layer fabric on a construction site. These installation techniques are not in accordance with given recommendations - but it is construction practice. Sometimes we can not imagine what can happen at the site, but we always have to consider that our workmen are not clockmakers and that there is a demand for heavy fabrics of high robustness. There is no necessity for highly sophisticated filtration rules for geotextiles if the fabrics are punctured during installation.

Probably the very good experience with the use of heavy needle-punched multi-layer fabrics on revetments of German waterways after BAW recommendations results not only from the high filtering efficiency but also from the high robustness of these fabrics.

Mr W. van Wijk

I would like to make some further comments concerning techniques of studying the weathering behaviour of polypropylene.

Degrading problems can arise when polymers are exposed to sunlight, rain, temperature and oxygen. The degradation is generally noticed as a change in colour and deterioration of physical properties, for example surface cracking and reduction of tensile strength. It is very important, therefore, to have information about lifetime expectancy of polymers in the open. The speed of the degradation process depends on the type of polymer used and the way it is

Fig. 5.

Fig. 6.

stabilized.

Degradation tests can be carried out in two ways:

(a) outdoor or natural exposure

(b) accelerated or artificial tests.

It is very time consuming to do only outdoor exposure tests. This means that tests have to be carried out which will give quicker results. To be able to do so it is necessary to know the mechanism which causes the ultraviolet (UV) degradation.

The speed of degradation, for instance, depends largely upon the annual irradiated energy on the surface of the earth. This energy is expressed in $kcal/cm^2$ or kLy (kiloLangley). The wavelength of radiation is important. Every type of polymer is sensitive to a different wavelength (Table 1).

The results of an artificial weathering test depends on the testing device, the light source, and the test conditions. UV degradation tests are carried out in a Xenotester. At the moment there are two test methods available.

(a) According to DIN 53 398

Black panel temperature	45°C
Relative humidity during dry period	60 - 80%
Rain cycle	3 min. out of 20 min.

(b) According to ASTM G26-70

Black panel temperature	63°C
Relative humidity during dry period	100 - 25%
Rain cycle	18 min. out of 120 min.

These different conditions will also give different test results (Fig. 7).

From Fig. 7 we see that in the ASTM test the polypropylene fabric is degrading much quicker. This means that this test method gives much quicker results. However, to be able to use these tests it is necessary to translate the results to outdoor exposure. Amoco Fabrics have carried out comparison tests in a Xenotest 1200 to find the correlation between test results and life time expectancy. Fig. 8 shows a comparison between a fabric under test conditions and one under outdoor exposure.

The value used mostly in these tests is the half-life time. This is the time needed for a 50% loss of strength in the fabric. In Fig. 9 we can see that there is a very good relation between the half life time obtained with the Xenotest 1200 and the results of natural weathering tests.

During these natural weathering tests the irradiated energy was also recorded. From this, it was possible to find the correlation between natural and artificial weathering. Table 2 gives the number of hours in different types of Xenotest which correspond to 1 kLy outdoor exposure. With the

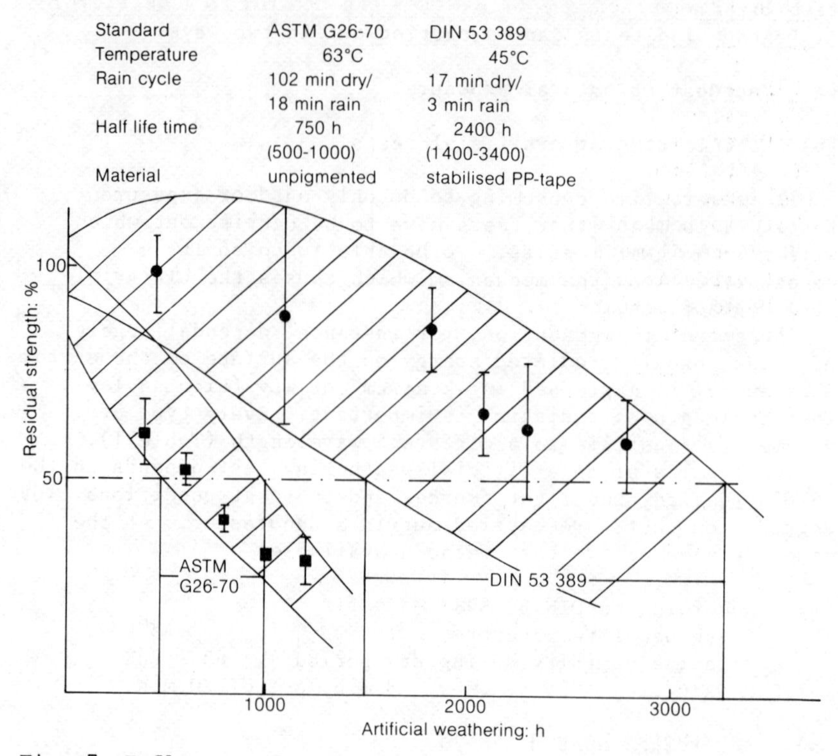

Fig. 7. Influence of test conditions on degradation process

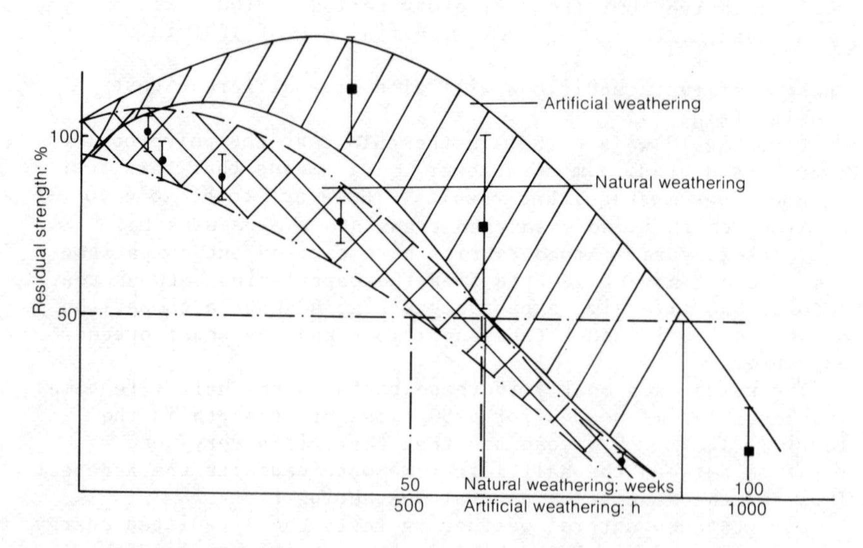

Fig. 8. Comparison of artificial and natural weathering results

Table 1. Wave lengths of maximum photochemical sensitivity

Polymer	Wave length
Polyester	325
Polystyrene	318
Polyethylene	300
Polypropylene	370
Polyvinylchloride	320
VC-VA-copolymers	327/364
Polycarbonate	295/345
Cellulose acetate butyrate	296
Styrene acrylonitrile	290/325

Table 2. Correspondence between tests and outdoor exposure

Type of test	Hours of test corresponding to 1 kLy outdoor exposure	Scale factor
Xenotest 150	25-30	2
Xenotest 450	35-45	3-3.5
Xenotest 1200	15-20	1
Weatherometer	12-18	0.8-1

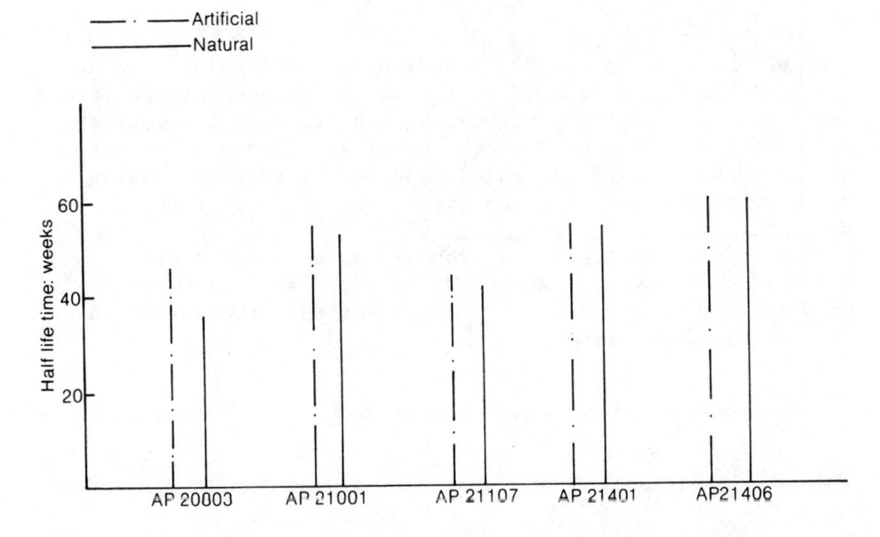

Fig. 9. Artificial weathering compared with natural weathering

help of this information it is possible to determine the life time expectancy of a material in a certain area. Amoco Fabrics have obtained the best results by using a Xenotest 1200. We therefore strongly recommend that this method is used for accelerated weathering tests.

With the map of the annual irradiated energy on the earth given in our paper it is possible to predict the lifetime expectancy of a polypropylene fabric in a given area of the earth. For example, a temporary construction in the UK is executed with a polypropylene fabric which will be exposed permanently.

Annual energy incident in the UK is 70 kLy
Life time construction : 5 years
Total energy incident: 350 kLy.

If we accept 50% loss of strength then we can determine the half-life time in a Xenotest 1200 with Table 2. According to the weathering test ASTM G26-70, the half-life time in Xenotest 1200 is (15 to 20) x 350 = 5000 to 7000 hrs. This means we must find a fabric that has at least this performance in a Xenotest 1200.

The same applies for situations on normal building sites where the geotextile will be exposed over shorter periods. It must be borne in mind that life expectancy can vary enormously even between different types of polypropylene. The material can be stabilized relatively lightly with carbon black (with half-life times in Xenotest 1200 ranging from 1000 hrs to 2500 hrs depending on the type of fabric), or heavily, e.g. in the grass yarns for artificial grass carpets. Tests must therefore be carried out on each individual fabric.

Mr A. C. Burdall

When we wrote our paper construction was proceeding, and no experience could be quoted as to how the embankments performed under the anticipated settlements. In the last few months 30 mm of toe settlement and 100 mm of settlement at the top of the slope has been experienced, and the mattress is moving relative to the quay structure. So far it is performing satisfactorily.

The grouted mattress was proposed as a contractor's alternative. However, due to rock supply and placement problems for this project it is evident that they provided an economical alternative.

Discussion on Session 6: The ongoing role of the working group

Mr M. E. Bramley, Construction Industry Research and Information Association, London

Two aspects of CIRIA's present programme are of relevance to the working group. CIRIA is an independent association of member firms which promotes, finances and manages civil engineering research work. Each piece of research is contracted out to a recognized specialist in that particular field. Membership is mainly, but by no means exclusively, drawn from the UK, and currently comprises about 600 organizations covering government departments, local authorities, public authorities, consultants, contractors, research organizations, universities and manufacturers.

Research must be interpreted broadly, and indeed, much of CIRIA's current work relates to production of the type of information document which the design guidelines to be produced by the PIANC Working Group would comprise. CIRIA has found in recent years that, with the increasing amount of technical literature being published, there has been a corresponding demand for the state-of-the-art document which reviews and appraises all the available information.

The first area of work relates to the maintenance of coastal revetments on the British coast. The word 'revetment' is used here in the traditional sense of protection to a sloping embankment-type defence, as opposed to a heavier type of sea wall incorporating, for example, structural concrete elements. There are about 1100 km of such defences in Britain, under the control of about 50 authorities. In general, these defences are formed along the less-exposed lengths of the coastline. The CIRIA project is aimed at preparing guidelines relating to the maintenance of these existing systems - and will cover the possibility, in certain cases, of complete replacement. The work is supported by the Department of the Environment, and the Ministry of Agriculture, Fisheries and Food.

The first phase of the work will be carried out by Mr L. Summers, in liaison with Hydraulics Research Ltd, who also have government commissions for work in related fields. Many of the types of revetment systems discussed at this conference also have an application in this field of coastal engineering.

The second area of work relates to the reinforcement of steep grassed waterways, such as might be used for emergency spillways from reservoirs, or similar installations where a high, but short duration, discharge occurs. Particular relevance to this project in Britain is given by the forthcoming implementation of new legislation on reservoir safety. It is likely that some of the estimated 1500 earth dams in Britain will require additional spillway capacity to comply with the new law. Such capacity is most readily provided by an additional low-cost emergency spillway acting alongside the existing service spillway. A reinforced grass protection system (ie a greened-over geotextile or open concrete system) provides an environmentally acceptable method of surface treatment for the emergency spillway. Large areas of concrete would be an eyesore. The CIRIA project will provide design guidelines for such waterways, and is supported jointly by UK manufacturers and suppliers (most of whom are exhibiting at this conference), and the Department of the Environment. The research contractor for this work has yet to be appointed.

In both cases, the design guidelines will be drawn up as one part of a state-of-the-art review. A parallel objective of this review will be to identify future areas of research. CIRIA thus accepts that the design guidelines may, in some cases, be interim - pending the results of further research which might be carried out as a second phase to each project. The situation is, I imagine, similar to the proposed PIANC design guidelines.

As regards specific overlapping areas of interest, I should like to mention (a) performance under wave attack, (b) the composite interaction of the armour layer, the geofabric underlay and the subsoil, (c) durability, and (d) the type of integrated flexible revetment system described by Major Wise in Paper 6. The term 'design guidelines' is preferable to 'code of practice', which suggests a rigid framework which would be inappropriate to rapidly developing technologies such as these which we are discussing. A CIRIA member has described design guidelines as a tool-kit - the function and method of use of each tool being fully described, together with illustrations and examples of their use.

I am impressed by the extent of common areas of interest of the PIANC working grouup and CIRIA. In view of the initiative being taken by the British Section of PIANC in organizing technical support for the Working Group, I am particularly keen that CIRIA should explore with PIANC areas in which we can co-operate. This is even more important because funding for the work of our respective organizations is limited. Also, most if not all of the organizations supporting the present initiative of the British Section are in fact members of CIRIA.

Mr I. R. Whittle, Ministry of Agriculture, Fisheries and Food

Common problems exist between the concepts of protection of coastline, sea defences and estuary banks.

This conference has indicated the need to understand design parameters and to understand durability of materials.

Such information is required in order to consider the cost-effectiveness of prospective systems, because central government invests between 60-80% of the cost of projects.

I am pleased that CIRIA has been so ready to come forward to coordinate a full research project leading to correct usage of low-cost protection systems.

Mr K. Pilarczyk, Delta Department, The Netherlands

Research strategy on revements in the Netherlands

Numerous types of revetments have been developed in the past for shore and bank protection of navigation channels against erosion by waves and currents (rip-rap, blocks, asphalt, etc.) The reason for this is the increase of problems in respect of the defence of the shores (i.e. more rigid safety requirements for sea-dikes) and banks of navigation channels (i.e. increase of size and speed of motorvessels), as well as the high cost and shortage of natural materials. The fact that design rules are still limited in quantity has stimulated investigations in the area of rip-rap, artificial blocks and bituminous revetments in relation to geotextiles (1-5).

Problems which arise due to these developments require solutions which can often only be found by in-depth specific studies.

In order to control the future sea-dikes and bank-protection problems, the Dutch Ministry of Transport and Public Works (Rijkswaterstaat) assigned the Delft Hydraulic Laboratory and Delft Soil Mechanics Laboratory to carry out a systematic research into these areas. The research on sea-dike revetments is known as the M1795 research programme and the research on bank protection is known as the M1115 research programme (6, 7).

On basis of the analysis of practical design problems and the gaps in the existing knowledge, the required research programmes have been determined.

The basic programmes have been carried out by means of small-scale models. However, it must be pointed out that a small-scale hydraulic model for navigation purposes still needs a lot of space. For example, in the scope of the M1115 research programme the hydraulic model of an inland navigation fairway 1:25 in scale has been built in a 40 x 89 m shed to observe the induced water motions and their erosive effects on the banks.

Since model research has certain inherent technical restrictions known as scale effects, additional required information has been obtained by means of prototype investigations - i.e. the Delta-Flume and some prototype locations in respect of the sea-dike problems, and the Hartel Canal with test embankments in respect of bank-protection of navigation channels.

The result of the prototype tests, in combination with the model results and the calculation methods (including mathematical models) developed in the framework of the systematic research (M1795) on dike protection and systematic research (M1115) on bank protection extended with knowledge gained from practical experience, will lead to preparation of guidelines for reliable dike and bank protection designs. The aim of the total research is to develop design criteria such that the amount of maintenance and construction costs of new revetments is minimized.

Research aspects and research means

In general, to be able to determine the dimensions of involved protection layers of revetments the following design research aspects have to be taken into consideration.

(a) Research on characteristic/representative loads

- (1) water motion due to wind waves, currents and ship movement
- (2) geotechnical load (gradients due to the water motion)

(b) Research on characteristic/representative strengths

- (1) strength of toplayer
- (2) strength of a sublayer

To solve problems involved in (a) and (b) various research means are available, e.g.:

(a) Evaluation of past experience, i.e. report M1115 -IX.
(b) In situ investigations on existing and/or test-revetments, i.e. prototype measurements in the Hartel Canal (project OEBES).
(c) Calculations and mathematical models (desk studies)
(d) Small-scale physical models
(e) Large-scale models, i.e. Delta Flume.

The physical and mathematical models are suitable for basic research within a wide range of boundary conditions, and for the developing of general design rules, while in situ investigations, evaluation of experience and large-scale (prototype) tests are needed for verifying the final results. In the course of the M1795 and M1115 research programmes all of these methods have been applied.

The general research strategy for both programmes and their

interactions is presented in Fig. 1.

The main difference in approach to the problem between these two investigations lies in the fact that for the sea-dikes the large Delta Flume has been used for prototype tests, while for the bank-protection problems in situ prototype tests have been done. (The large scale models for bank-protection problems are not a reasonable solution). Another point of difference is that the boundary conditions related to banks cannot be reproduced mechanically as it is the case for wind-waves (i.e. using a wave generator) but have to be induced by ship movement. That is the reason that such models need a lot of space.

Apart from the difference in reproducing the hydraulic load, both programmes involve some common aspects regarding stability of the toplayer and the sublayer. The integration of the programmes will take place by means of the mathematical

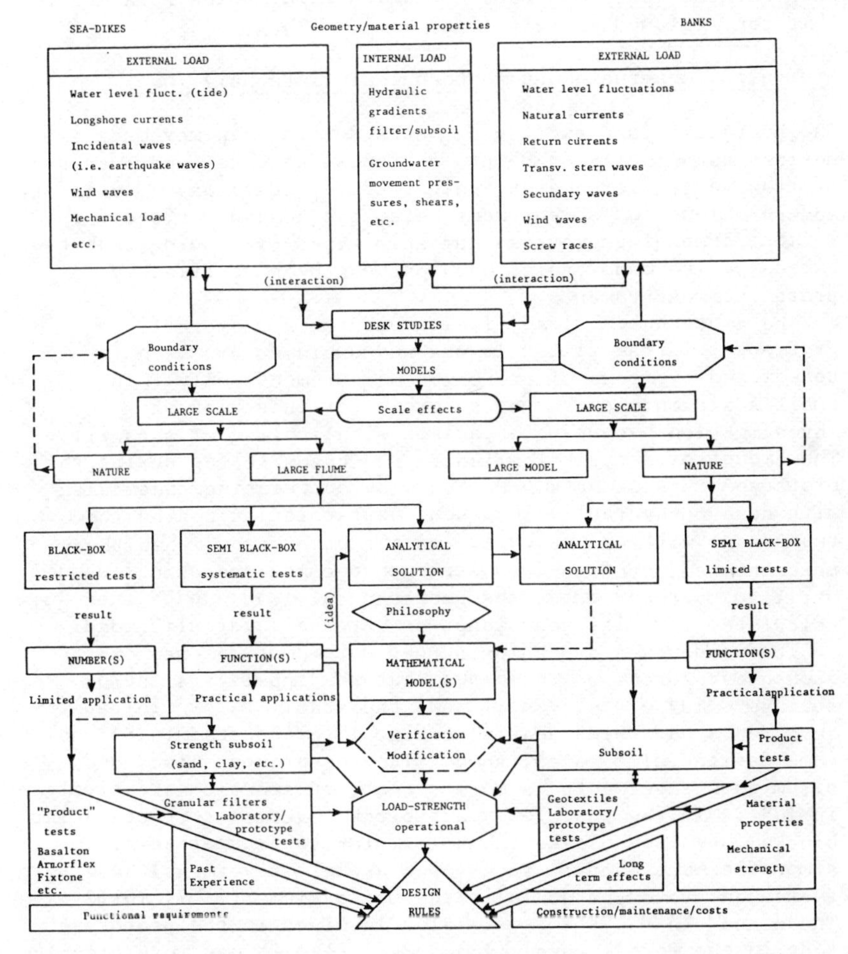

Fig. 1. Research philosophy on load-strength aspects of revetments

model which has to fulfil the requirements of both. This mathematical model might become an important tool in the design of revetments of dikes and banks of navigation channels.

State of research (1984)

In the context of boundary conditions in navigation fairways, by using model and prototype results some experimental relations for speed prediction of the various ships as functions of applied engine power and position of the ship in the fairway have been established.

Simultaneously, the ship-induced water motions in channels have been experimentally described and compared with the existing calculation methods. On this basis the mathematical models have been developed to calculate the ship speed and ship-induced water motions. Detailed information relating to this can be found in refs 8 and 9.

Hydraulic (external) and geotechnical (internal) load

The hydraulic load exerted on banks due to ship-movement, defined as velocity of return currents, heights (and pressure) of transverse stern waves, height of secondary waves, velocity and turbulence of screw races, etc. has been described for various channel geometries and ship types (including push-tow barges). The model relationships have been verified by prototype measurements.

The geotechnical load (i.e. groundfow and hydraulic gradients) in the filter layers and sublayer has been determined by means of prototype measurements only. The small-scale models are not suitable for this kind of investigation because of problems with scaling of subsoil. The amount of data is limited to the cases tested during the prototype measurements. However, by correlating these data with data on hydraulic load some extrapolation behind the test-range will be possible. On the other hand, when the mathematical model on load strength becomes operational and verified by these data, the geotechnical load can be then calculated for different input of external hydraulic load.

The wind-waves and ship-induced water motions can exert such heavy forces on the banks that the top-layers and/or sublayers may erode. Until now, the stability of rip-rap revetments have been nearly completely elaborated (4,8) and some design rules established. The research on block and bituminous revetments is more recent; it started extensively in 1981. In the past few years block revetments (rectangular blocks, basalton blocks (shape simular to natural basalt), armourflex-mats and fixtone (open asphalt revetment) have been tested on prototype scale in the Delta Flume (1, 7, 10-12). These and other systems were also involved in the prototype test in the Hartel Canal (13). Recently, it was more clearly recognized that the groundwater flow and the associated hydraulic gradients can be an important factor for the stability of the top-layer of the subsoil and/or filter, and thus for the embankments. To be able to judge the total stability of embankments information about the strength of filter and subsoil is thus needed. The research on these aspects started in 1984.

Using specially developed test-facilities, the critical hydraulic gradients for filters and subsoils under superimposed total hydraulic (external) load will be measured. In the later phases the mathematical model on load-strength will be extended with these data.

The first idea about critical gradients under dynamic load is already given by the special tests done in the scope of research for the Eastern Scheldt storm barrier (14).

In the research on strength of filters, both granular filters and geotextiles will be involved. The research on the durability of geotextiles has already been done by the Netherlands Coastal Works Association (5).

The results of all these studies and the future economical analysis regarding the execution and maintenance costs will lead to the preparation of guidelines for reliable shore and bank protection designs.

Other activities

After the completion of the short-term measurements in the Hartel Canal it was decided to keep all these prototype test embankments for further studies on long-term behaviour in the next few years. These data, combined with the geotechnical analysis of the data obtained during the short-term prototype measurements, will give more information on he practical applicability of the different protective systems.

In future research attention will be also paid to sheet piling, with its problems concerning pile depth and erosion by overwashing waves, and especially to the natural protective systems (green revetments).

The results of all these studies will bring designers closer to the solution of the typical problem of the choice of protective structure for shore and bank protection works in respect to design load, the abilities of materials, and desired function of construction.

Due to the complexity and diversity of the problems involved it is imperative that all the research is well-coordinated. For this purpose a national working group have been installed recently in the Netherlands.

The same complexity and diversity of the problems involved demands much closer international cooperation in this field. The leading role of PIANC in this process is evident and requires our support.

REFERENCES

1. DEN BOER K., KENTER C.K. and PILARCZYK K.W. Large scale

model tests on place block revetment. Delft Hydraulics Laboratory, Publication No. 288, January 1983 (also Coastal Strutures '83, Washington, March 9-11 1983).

2. DE GRAAUW A., VAN DER MEULEN T., and VAN DER DOES DE BYE, M. Design criteria for granular filter, Delft Hydraulics Laboratory, Publication 287, January 1983.

3. PILARCZYK K.W. and DEN BOER K. `Stability and profile development of coarse materials and their application in coastal engineering. Delft Hydraulics Laboratory, Publication No. 293, January 1983 (also Conference on Port and Coastal Engineering in Developing Countries, Sri Lanka, Colombo, March 20-26, 1983).

4. PILARCZYK K.W. Revetments/filters. In 'Closing Tidal Basins', Delft University Press, 1984.

5. VELDHUIJZEN VAN ZANTEN, R. and THABET R.A.H. Investigation of long-term behaviour of geotextiles in bank protection works. 2nd Int. Conf. on Geotextiles, Las Vegas, USA, 1982.

6. DELFT HYDRAULICS LABORATORY. Systematic research on erosion and bank protection of navigation channels. Reports M1115, 1974 - 1985 (in Dutch).

7. DELFT HYDRAULICS LABORATORY. Systematic research on stability of block revetments under wave attack. Reports M1795, 1982-84 (in Dutch).

8. BLAAUW H.G., DE GROOT M.T., VAN DER KNAAP F.C.M and PILARCZYK K.W.. Design of bank protection of inland navigation fairways. This conference, Paper 4.

9. New aspects of designing bank protection of navigation channels (in Dutch). KIVI-Symposium, Delft, 25 May 1983 (availabel at Delft Hydraulics Laboratory).

10. DELFT HYDRAULICS LABORATORY. Stability of basalton blocks (in Dutch). M1900, 1983.

11. DELFT HYDRAULICS LABORATORY. Stability of armorflex block slope protection mats under wave attack. M1910, 1983.

12. DELFT HYDRAULICS LABORATORY. Fixtone: stability under wave attack. M1942, 1983.

13. PILARCZYK K.W. Prototype tests of slope protection systems. This conference, Paper 13.

14. SPAN H.J.TH et al. A review of relevant hydraulic phenomena and of recent developments in research, design and

construction of protective works. Dutch contribution to XXVth International Navigation Congress, Edinburgh, 1981, Vol I.

Mr R. Filarski, Rijkswaterstaat, The Netherlands

We have 2 working groups working on the same subject, so there is a need for good co-ordination. The first group exists already, working on probabilistic design methods. This group will try to provide us with a new statistical approach, taking into account probabilities and risks. With this approach we hope to be able in a few years to make designs which are cheaper with regard to construction and maintenance costs.

On the other hand, there are some needs which the existing working group doesn't provide for:

(a) to summarize the lessons we have learned from this conference

(b) to detail and investigate further the properties of the new materials which we are using

(c) to establish standard procedures to test the properties of geotextiles.

It would be a benefit for many countries if we had international PIANC recommendations as soon as possible.

I would like to suggest the following approach for the new working group. I would like them to concentrate on the new materials and on the preparation of recommendations for design.

The first task would be to collate the existing knowledge. This conference will provide an excellent basis. Secondly, the existing design recommendations in different countries should be collected, to compare them and to lay the ground for future guidelines. Thirdly, the problems still existing shall be identified.

I would have no objection if the working group did develop guidelines provided that they can be altered according to the results of the probabilistic method, but I doubt if it is possible in such a short time.

If we proceed in this way we will avoid making recommendations without taking into account the progress of the first working group, and yet at the same time we will provide countries lacking design recommendations with a basis for national use.

Finally I think that it will be necessary for the new group and the already existing group to work closely together.